NANO MATERIALS SYNTHESIS AND CHARACTERISATION

NANO MATERIALS SYNTHESIS AND CHARACTERISATION

Editors

V. RAJENDRAN
K. SAMINATHAN
P. PARAMASIVAM
K.E. GECKELER

BLOOMSBURY

LONDON • NEW DELHI • NEW YORK • SYDNEY

First Published, 2013

BLOOMSBURY PUBLISHING INDIA PVT. LTD.
London New Delhi New York Sydney

ISBN: 978-93-82563-36-5

Published by Bloomsbury Publishing India Pvt. Ltd.
VISHRUT Building, DDA Complex, Building No. 3, Pocket C-6 & 7
Vasant Kunj, New Delhi 110 070

Laser Typeset by FORTUNE GRAPHICS
WZ-911/2, Shankarlal Street, Naraina, New Delhi

Printed at Anvi Composers
1 DDA Market, Block A-1B, Paschim Vihar, New Delhi 110 063

डॉ.टी.रामसामी
सचिव
Dr. T. RAMASAMI
SECRETARY

भारत सरकार
विज्ञान और प्रौद्योगिकी मंत्रालय
विज्ञान और प्रौद्योगिकी विभाग
टेक्नोलाजी भवन, नया महरौली मार्ग, नई दिल्ली-110 016

GOVERNMENT OF INDIA
MINISTRY OF SCIENCE & TECHNOLOGY
DEPARTMENT OF SCIENCE & TECHNOLOGY
Technology Bhavan, New Mehrauli Road, New Delhi-110 016

Foreword

I am glad that K.S. Rangasamy College of Technology (KSRCT) is organizing an International Symposium on Macro- Supra- Molecular Architectures and Materials (MAM-12) during November 21-25, 2012. I am also happy to notice the wide variety of topics being covered and the level of participation. I congratulate the Institutions and organizers and wish the Symposium all success.

(T. Ramasami)

Tel. : 011-26510068 011-26511439 • Fax : 0091-11-26863847, 0091-11-26862418 • E-mail : dstsec@nic.in

EDITORS

Dr. V. Rajendran FUSI, FASI, FInstP, is Director, Research & Development, K S R Group of Institutions and Centre for Nano Science and Technology, K.S Rangasamy College of Technology, Tamil Nadu, India. Under his able guidance, 11 scholars have completed and 14 scholars are pursuing their Ph.D. degrees. He has published more than 150 research papers in reputed international and national journals, 50 papers in conference proceedings, 28 refereed books, 4 R&D books and 11 patents. He has won many awards including the Prof. K. Arumugam National Award in 2011, Best faculty award in 2010, Raman-Chandra Sekhar silver medal (2010), Tamil Nadu Scientist Award, NDT Man of Year 2004, DAAD from Germany (2002), INSA, TNSCST Young Scientist, DAE/BRNS Visiting Scientist, Best paper award from MRSI, ASI, ASA and USI and Outstanding Organiser Award for the 7th National Symposium on Ultrasonics, 1996.

Dr. K. Saminathan is Professor and Head, Dept. of Nanoscience and Technology. He has two years experience as research associate (Level-IV), Central Electro Chemical Research Institute (CECRI), INDIA. He did his Post Doctorate from the Arizona State University, Polytechnic Campus, USA. He obtained his Ph.D. in lead-acid battery from Alagappa University, and has to his credit more than 10 research papers in reputed international journals.

Prof. Paramasivam P., Assistant Professor, Centre for Nano Science and Technology (CNST), K.S. Rangasamy College of Technology, Tamil Nadu, India. He has been working in the Centre for Nano Science and Technology, KSRCT for the past four years. He is pursuing his Ph.D. entitled *In-Situ Evaluation of the Mechanical Properties of Nano Structural Materials*, Anna University, Chennai. Presently, he is working in the area of nano materials testing and characterisation.

Dr. K.E. Geckeler is the Founding Chair of the Department of Nanobio Materials and Electronics, World Class University (WCU), Gwangju, South Korea. He is also a Professor in the Department of Materials Science and Engineering and the Department of Medical System Engineering at the Gwangju Institute of Science and Technology (GIST), South Korea and serves as Vice Director of the Gruenberg Center for Magnetic Nanomaterials (GCMN). He received his Ph.D. and M.D. degrees from the University of Tuebingen, Germany (both degrees: "magna cum laude") and spent sabbatical leaves at Harvard University, University of Montana, Clemson University (USA) and at the University of Montpellier (France). He received a series of awards such as the "Fonds der Chemischen Industrie", the "Fritz-Ter-Meer Award", the "Science Price of the President of Korea" and the "Outstanding Research Award of GIST". He has to his credit initiation and organizing the biannual MAM Symposia series. He has published more than 300 research articles and short communications, 12 book chapters, over 10 books, and over 80 patents.

PREFACE

The International Symposium on *Macro- and Supramolecular Architectures and Materials (MAM-12)* with a special emphasis on Nano Systems and Applications is organised by the Centre for Nano Science and Technology (CNST) of the K.S. Rangasamy College of Technology. This conference is jointly organised with the World Class University (WCU), Gwangju Institute of Science and Technology (GIST), South Korea. Having established start-of-art experimental facilities at the CNST, the centre is looking for industrial collaborations and partners for technology transfers in thrust areas of nanotechnology.

More than 525 abstracts were received from India and abroad. Out of the 400 full papers, that were received 260 have been selected and peer reviewed by the expert committee for publication in conference books. All the received papers have been classified under four titles namely – Nano Materials Synthesis and Characterisation, Advanced Nano Materials for Industrial Applications, Nano Biomaterials and Applications of Nano Materials: Electronics, Energy & Environment.

Out of the total 260 full papers accepted for MAM-12, a total of 70 have been identified for the inclusion in the book ***Nano Materials Synthesis and Characterisation***. This book contains various synthesis and characterisation methods of different nanostructured materials such as metal/metal oxides, composites, fibers, etc. We extend our sincere thanks to all the authors for their timely submissions and cooperation.

We are sure that the outcome of MAM-12 will open up new avenues between researchers and industrialists for the development of nano products.

We owe special thanks to members of the technical committee for critical reviews of the contributed papers. Various government funding agencies, private organisations and industries are thanked for their munificent support and sponsorship for the successful conduct of the conference. It is our pleasure to thank the various committee chairs and members for their tireless efforts in making this event a grant success.

The support extended by Mr. Suresh Gopal and Mrs. Jyoti Mehrotra of Bloomsbury Publishing India Pvt. Ltd. in bringing out this book on time is highly appreciated.

V. Rajendran
K. Saminathan
P. Paramasivam
Kurt E. Geckeler

CONTENTS

GROWTH AND CHARACTERIZATION OF NANOSTRUCTURED CdS THIN FILMS BY DIP COATING METHOD

G. VIJAYAPRASATH, M.R. MANIKANDAN AND G. RAVI

Photonic Crystals Lab, School of Physics, Alagappa University, Karaikudi
E-mail: gravicrc@gmail.om; raviganesa@rediffmail.com

ABSTRACT

Nanostructured CdS thin films on glass substrate has been prepared by dip coating technique and the effect of surfactant triethylamine (TEA) on the structural, optical and morphological properties of the CdS thin film has also been investigated. The powder X-ray diffraction (XRD) study reveals that the prepared films belong to cubic system and has prominent peak at plane (111). The crystalline nature of the film increases with the increase of surfactant concentration. Optical property of the crystalline CdS thinfilm has been characterized by UV-Visible study. The uniform deposition of the film over glass substrate has been confirmed by Scanning electron microscopy. The results are discussed in detail.

INTRODUCTION

Over the years, CdS thinfilms have been extensively investigated [1-5]. For a wide and direct band gap (~ 2.42 eV), they have been used in a variety of applications including electronic and optoelectronic devices. CdS thinfilms are also used as n-type window layers in the solar cell thin films. Primary studies are currently performed in order to understand the different characteristics, which lead the CdS thinfilm to be suitable as a window material for solar cells. Researchers have tried different techniques to investigate structural and optical properties, but the identification of the best electronic quality for photovoltaic applications is a challenging due to the thin film nature of the solar cells [6]. Polycrystalline thin films of CdS can be

V. Rajendran, K. Saminathan, P. Paramasivam and K.E. Geckeler (eds.)
Nano Materials Synthesis and Characterisation, pp. 1-4 (2012).

prepared by different techniques such as sputtering, screen printing, physical vapor deposition (PVD), electro deposition, chemical bath deposition (CBD), etc. In conventional solar cells, the CdS window layer was deposited onto the conducting glass substrates, where metallic substrates the CdS layer has to be deposited on the absorber layer in the case of devices [7,8].

The dip-coating method is a step-wise process of chemical deposition of thin films from aqueous precursor solutions. It is a unique method in which thin films of compound semiconductor can be deposited by alternate dipping of a substrate into the aqueous solutions containing ions of each component. This method provides the control of the film thickness at atomic level, which is difficult by any other conventional methods. In this investigation, we report the formation of nanostructured CdS thinfilm on the glass substrate by dip coating technique. The dependence of surfactant triethylamine (TEA) on the structural, optical and morphological properties of the crystalline CdS thinfilm was characterized. As an outcome of this work, the TEA is proven to be very feasible for the preparation of CdS films by dip coating technique.

EXPERIMENTAL

CdS thin films were prepared onto glass substrate by dip coating technique. The deposition of CdS was carried out by a mixture of cadmium acetate (0.025 M) and thiourea (0.05 M) in alkaline medium to yield cadmium and sulphur ions. The various ratio of surfactant TEA (1ml, 2ml, 3ml) was mixed in mother solution in order to characterize their effect in the film formation. For the controlled release of cadmium ion from the source material, ammonia was added into the solution and stirred continuously. In the case of dip coating, etching of the substrate is one of the main criteria, for that the glass substrates were first cleaned in chromic acid solution. The substrates are kept in the solution for 4 hrs. And then washed several times in soap solutions and distilled water, and finally washed thoroughly with de-ionized water. Then it was dipped into the mother solution and hot water for 10 sec to avoid precipitation and also to remove the loosely bounded ions. The substrate was continuously immersed into the mother solution. The procedure was carried out at room temperature.

Successive cycles repeated up to 100 cycles, to get the well adherent and homogeneous CdS thin films in various concentrations of TEA.

RESULTS AND DISCUSSIONS

Structural Analysis

The powder X-ray diffraction pattern of the prepared CdS thinfilm is shown in fig. 1. The XRD pattern reveals that the deposited thin film is mono-crystalline in nature. The strong and sharp diffraction peak indicates the films are well crystalline. The XRD analysis reveals that the obtained film was in cubic phase and (111) plane is prominent phase. The XRD patterns were indexed and the corresponding

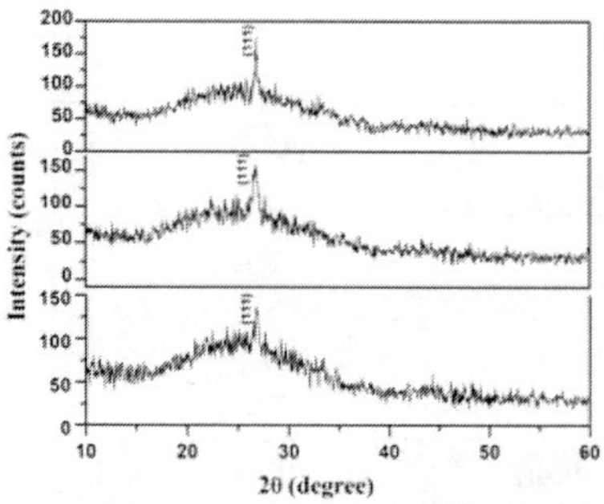

Fig. 1: XRD Pattern of CdS Thin Films

values of inter planar spacings‘d’ were calculated and compared with the standard JCPDS cards (Card No: 80-0019). CdS thin film shows prominent (111) plane**.** The crystalline nature is increased based on the concentration of triethylamine (TEA).

Morphological Analysis

The surface morphology of the thin films was recorded at different magnifications as shown in Fig. 2. Thin film of CdS is homogeneous, well adherent and covers the glass substrate without cracks and pinholes.

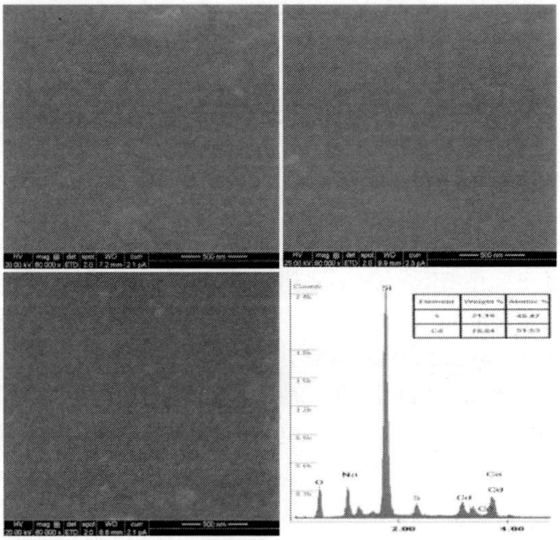

Fig. 2: SEM Micrographs and EDS of CdS Films Deposited using Various Concentration of (1ml, 2ml, 3ml) TEA Surfactant

Compositional Analysis

EDS was recorded in the binding energy region between 0 - 20 keV and is shown in the inset of figure 2. The spectrum peak reveals the presence of Cd and S respectively, which also confirms the stoichiometry of Cd and S in the thin film. The atomic percentage of cadmium and sulphur were found to be 51.53 and 48.47 % respectively,

OPTICAL STUDY

The optical absorption and transmittance spectra of various concentration of TEA added CdS thin film samples were recorded by UV-VIS spectrophotometer in the range of 190-1100 nm. Figure 3 (a) & (b) shows the variation of optical absorbance and transmittance with wavelength. The optical studies show that the films are highly absorptive. The transmittance edge of the films is observed in these spectra at about 482 nm (2.57eV) for the CdS films, which is higher to the literature value [9].

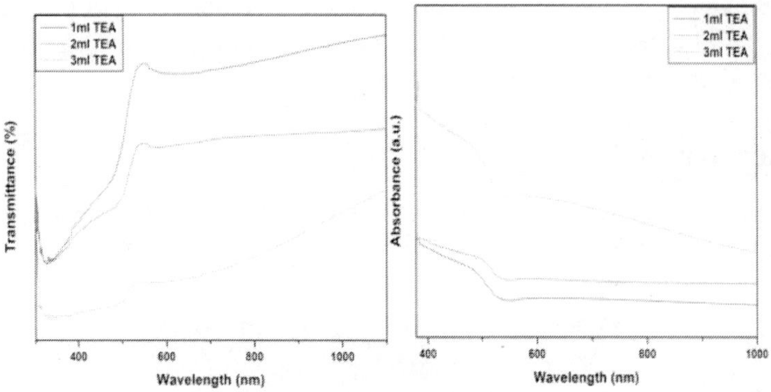

Fig. 3: Optical (a) Transmission and (b) Absorbance Spectra of the CdS Films

CONCLUSION

Thin films of CdS have been deposited successively by using dip coating method. It is an unique method in which thin films of compound semiconductor can be deposited by alternate dipping of a substrate into the aqueous solutions containing ions of each component. Deposited thin films were mono crystalline in nature. Transmittance cut off wavelength for the CdS thinfilm was observed at 482nm and corresponding optical bandgap was 2.57eV. Surface morphology of the thin film has been shown as a cubic like structure and the elements presented in nearly stoichiometric ratio.

REFERENCES

[1] Zehe. J.G., Vazquez Luna, *Solar Energy Materials and Solar* Cells,68,217, 2001.

[2] Oladeji I.O., Chow. L., Liu. J.R., Chu. W.K., Bustamante. A.N.P., et al. *Thin Solid Films*, 359,154, 2000.

[3] Castillo S.J., Mendoza-Galvan, A., Ramirez-Bon, R., Espinoza-Beltran, F.J., Sotelo-Lerma M., et al, *Thin Solid Films*, 373, 10, 2000.

[4] Savadogo O., *Solar Energy Materials and Solar Cells*, 52, 361, 1998.

[5] Oliva. A.I., Castro-Rodriguez R., Ceh O., Bartolo-Perez P., Cabal-lero-Briones F., et al., *Applied Science*, 148, 42, 1999.

[6] Chaure, N.B., Bordas S., Samantilleke A.P., Chaure S.N., Haigh J., et al., *Thin Solid Films*, 437, 10, 2003.

[7] Mathe X., Thompson G.W., Singh V.P., McClure J.C., Velumani S., et al., *Sol. Energy Mater. Sol.* Cells, 76, 293, 2003.

[8] Mathew X., *Semi. Cond. Sci. Technol.*, 18, 1, 2003.

[9] Quiebras J.N.X., Puente G.C., Hernandez J.A., Rodriguez G.S., Readigos A.A.C., *Sol. Energy Mater. Sol. Cells 82*, 263, 2004.

STRUCTURAL AND SPECTROSCOPIC STUDY OF Mn_3O_4 FILMS PREPARED BY DOUBLE DIP METHOD

T. Shrividhya, G. Ravi and T. Mahalingam[1]

Photonic crystals lab, School of Physics, Alagappa University, Karaikudi
[1]Department of Physics, Karunya University, Coimbatore
E-mail: gravicrc@gmail.om; raviganesa@rediffmail.com

ABSTRACT

Manganese oxide is most important transition metal oxide that finds large range of applications in electrochemical materials, high density magnetic storage medium, catalyst, ion-exchange, molecular adsorption and supercapacitors. Hausmannite Mn_3O_4 thin films were prepared by simple and low cost double dip method. The structural and spectroscopic properties of the deposited films were characterized by XRD, FTIR, and micro Raman spectroscopic techniques. The X-ray diffractogram revealed that deposited films are polycrystalline in nature and have tetragonal structure. The results of these studies will be discussed in detail.

INTRODUCTION

Hausmannite Mn_3O_4 has received great deal of attention in recent past due to its extensive applications in electrochemical, electro optical and magneto optical industries. Trimanganese tetraoxide has been used in high density magnetic storage media, catalysts, ion exchange, starting material for ferrites, solar energy conversion, batteries and dilute magnetic semiconductor [1-4]. The tetragonal Mn_3O_4 exhibits spinel structure with tetragonally coordinated Mn^{2+} ion on A site and orthohedrally coordinated Mn^{3+} ion on B site due to Jahn–Teller distortion [5]. Mn_3O_4 exists in two forms: low-temperature tetragonal and high-temperature cubic with a transition occurring at about 1170°C [6].

V. Rajendran, K. Saminathan, P. Paramasivam and K.E. Geckeler (eds.)
Nano Materials Synthesis and Characterisation, pp. 5-8 (2012).

In the present work, Mn_3O_4 films were prepared by double dip technique on glass substrates. This method is employed as it is simple, cost effective and also supports large area deposition. The films were annealed to 200 °C. The structural analysis was performed using the PANalytical Xpert Pro X-ray diffractometer. The micro Raman spectroscopy is a conventional tool that can be used to scan the sample at micro lateral level and distinguish different Raman active phases, thus providing information on phase distribution across the sample. The FTIR spectroscopy is a complementary method. Hence the prepared films were characterized using both this spectroscopic techniques to identify the phase formation.

EXPERIMENTAL DETAILS

Mn_3O_4 thin films were deposited on the glass substrate by double dip technique. The sodium manganate bath solution containing 1:2 ratio of manganese sulphate and sodium hydroxide solution maintained at temperature of 80°C. Well cleaned glass substrate was dipped in the bath solution (step1) for known standardized time and then in hot water (step 2) for same time for hydrogenation. The process of solution dip and hot water dip was repeated for known number of times. The part of Mn_3O_4 is formed as strongly adherent film on glass substrate. The films were annealed to 200 °C in a hot air oven.

RESULTS AND DISCUSSION

X-ray Diffraction Analysis (XRD)

Fig. 1(A & B) shows the XRD pattern of as deposited and annealed Mn_3O_4 films respectively. The diffractograms recorded are in good agreement with the JCPDS card no # 89-4837. The XRD patterns reveal polycrystalline nature of the films with tetragonal structure and space group I41. For both the films the predominant peak orientation is along (211) crystallographic plane indicating perpendicular alignment of c-axis of the grain.

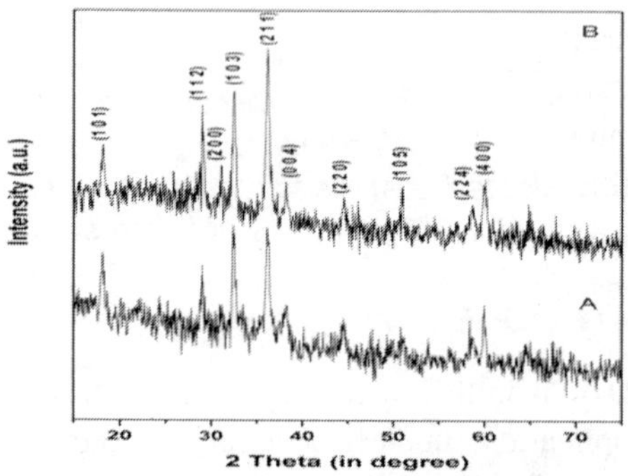

Fig. 1: (A) XRD Diffractograms of as Deposited and (B) Annealed Film of Mn_3O_4

The peak intensities are increased for annealed film compared to that of as deposited film. The micro structural parameters are calculated and tabulated in Table 1.

From the table, it is evident that the crystallite size of annealed film is larger than that of as deposited film which confirms that the crystalline nature increased with heat treatment. The decreased dislocation density and micro strain of annealed films shows that the defects in growth of the film have been reduced due to annealing and hence the quality of film is improved. Further annealing may increase the crystalline nature and reduce defects in the films.

The lattice parameters were calculated and are in good agreement with the standard values. The observed values were a = b = 5.64 Å and c = 9.45 Å. The cell volume calculated from peak fitting the pattern was 312 Å^3 which is nearer to the standard value (V=314 Å^3). This slight deviation may be due to the strain induced during the formation of thin films.

Table 1: Micro Structural Parameters of as Deposited and Annealed Films

S. No	*Sample*	*Crystallite size (nm)* $D = \frac{0.9\lambda}{\beta \cos\theta}$	*Micro strain (lines^{-2} m^{-4})* $\epsilon = \beta \, Cos\, \theta/4$	*Dislocation density (10^{15} lines/m^2)* $\delta = 1/D^2$
1	As deposited	20.90	0.0016	2.3
2	annealed	22.00	0.0015	2.0

Fourier Transform Infrared (FTIR) Analysis

The IR spectra of as deposited film and annealed film are as shown in fig.2. The spectra were recorded using Thermo nicolet 380 spectrophotometer in the range 400 to 4000 cm^{-1}. Two absorption peaks are obtained in the region between 400 cm^{-1} to 700 cm^{-1}. The bands 400 cm^{-1} and 500 cm^{-1} can be assigned to v (Mn -O) vibrations of Mn_3O_4 phase [7-8]. The peaks observed at 1614 cm^{-1} and 3408 cm^{-1} can be assigned to the O-H vibration of H_2O. Further annealing may reduce the moisture content in the film. The peak broadening is observed for annealed film compared to as deposited film is due to the annealing effect. No other organic bonds are observed which indicates the absence of any organic impurities.

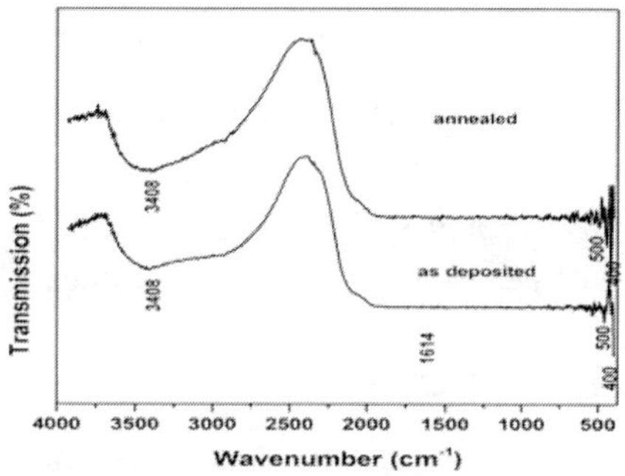

Fig. 2: IR Spectra of as Deposited and Annealed Mn_3O_4 thin Films

Micro Raman Spectroscopic Analysis

The Raman spectra were collected using micro Raman (Princeton Acton SP 2500) instrument with 0.5 focal length triple grating monochromator at room temperature. The Raman scattering measurements were performed using the 514.5 nm excitation line from Ar+ laser. The Mn_3O_4 exhibits two phonon peaks, a T2g symmetry mode at 375 cm^{-1} and a A1g symmetry mode at 660 cm^{-1} [9-11]. This matches with already reported values. The Raman peak shows broadened peaks with decreased intensity in annealed sample compared to that of as deposited sample. This may be due to increase in the crystallite size which can be explained using the phonon confinement effect [12-14].

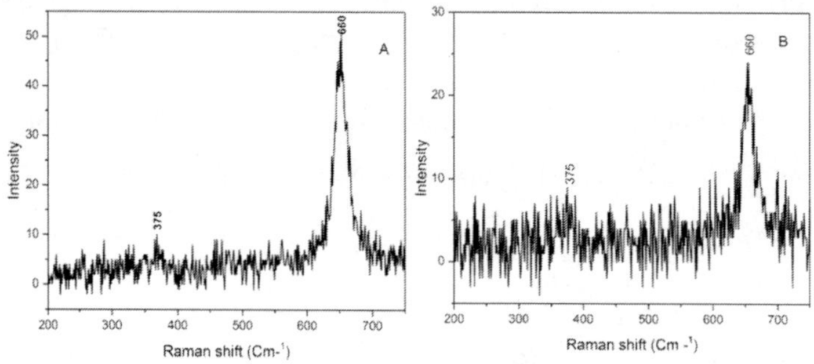

Fig. 3: Raman Spectra of (A) as Deposited and (B) Annealed Mn_3O_4 Films

CONCLUSIONS

Thin films of Mn_3O_4 were successfully deposited using dip coating method. The films obtained were polycrystalline in nature. The X-ray diffractogram of annealed film shows improved structural quality compared with as deposited film. The frequency corresponding to the vibration mode of Mn-O was observed from FTIR spectrum. The Raman signal in tetragonal Hausmannite consists of very sharp peak at 660 cm^{-1} and smaller peak at 370 cm^{-1} which are in good agreement with the literatures values.

REFERENCES

[1] Armstrong A.R., Bruce P.G., *Nature* 381, 499, 1996.

[2] Shen Y.F, Zerger R.P, Deguzman R.N, Suib S.I, Mccurdy L, Potter D.I, et al., *Science*, 260, 511, 1993.

[3] De Vries A.H., Hozoi L., Broer R., *Phys. Rev.* B, 66, 035108, 2002.

[4] Yamashita Y., Mukai K., Yoshinobu J., Lippmaa M., Kinoshita T., et al., *Surf. Sci*, 514, 54, 2002.

[5] Tromel G., Fix W., Koch K. and Schaberg F, Erzmetalle, 29, 234, 1976.

[6] McMurdie H.F., Sullivan B.M., Mauer, *J. Res. Nat. Bur. Stand*, 45, 35, 1950.

[7] Ishii M., Nakahira M., Yamanaka T., Infrared absorption spectra and cation distributions in $(Mn, Fe)_3O_4$. *Solid State Commun*, 11, 209-212, 1972.

[8] Du J, Gao Y.Q., Chai L.L., Zou G.F., Li Y., et al., *Nanotechnology*, 17, 4923-4928, 2006.

[9] Lutz H.D, Muller B., and Steiner H.J., *J. Solid State Chem*, 90, 54, 1991.

[10] Malavasi L, Galinetto P., Mozzati M., Azzoni C., and Flor G., *Phys. Chem. Phys*, 4, 3876, 2002.

[11] Julien C.M and Massot M., *J. Phys.: Condens. Matter*, 15, 315, 2003.

[12] Richter H., Wang Z.P, and Ley L., *Solid State Communication*, 39, 625, 1981.

[13] Campbell I.H. and Fauchet P.M., Solid *State Communication*, 58, 739, 1986.

[14] Jian Z, Cunyi X., Bihui H., Changsui W., Yi X. and Yitai Q., *Journal of Raman Spectroscopy*, 27, 921, 1996.

ACTIVATED CARBON-Fe_2O_3 NANOCOMPOSITES – SYNTHESIS AND MAGNETIC STUDIES

V. Ranjithkumar, S. Sangeetha and S. Vairam

Department of Chemistry, Government College of Technology, Coimbatore, Tamil Nadu
E-mail: vamshen@yahoo.com

ABSTRACT

Activated carbon-iron oxide (AC-Fe_2O_3) nanocomposites have been prepared by in situ thermal decomposition of metal carboxylates into activated carbon matrix using iron lactate as precursor for metal oxide. The morphology and size of the iron oxide particles inserted into activated carbon matrix have been investigated by IR, PXRD, SEM and TEM. The magnetic nature of composites has been investigated by VSM.

INTRODUCTION

Carbon metal oxide nanocomposites are carbon matrix containing dispersed metal oxide in the nanometer range. So far amorphous metal oxide has been synthesised by thermal decomposition of some easily decomposed metal compounds [1-2]. Iron oxide is a well-known magnetic oxide and is widely used as a catalyst, pignut and gas sensitive materials etc [3]. Synthesis of iron oxide in nano size has long been of scientific and technical attentiveness for their unexpected physical and chemical properties compared with bulk materials [4-5]. In this approach, the use of magnetic particles in the different field of sewage coagulation [6], to remove radio nuclides from milk [7], adsorption of organic dyes [8] and for oil spill remediation [9]. The removal of atrazine [10] and organic molecules [11] from an aqueous medium using AC-iron oxide nanocomposites had been reported. In this paper, we report a method for the preparation of AC-Fe_2O_3 by the suspension of activated carbon in the iron (II) lactate aqueous solution followed by the thermal decomposition.

V. Rajendran, K. Saminathan, P. Paramasivam and K.E. Geckeler (eds.)
Nano Materials Synthesis and Characterisation, pp. 9-12 (2012).

EXPERIMENTAL WORK

Preparation of AC-Fe_2O_3 Nanocomposite

Activated carbon/iron oxide composites were prepared from a suspension of commercial activated carbon (AC) (10 g) in aqueous solution of iron (II) lactate (4.5 g) with constant stirring for 12 hours at ambient temperature. The iron(II) lactate entrapped ACs was subjected to control heating at 300 °C for 10 min in the muffle furnace to produce iron oxide in pores of ACs.

RESULTS AND DISCUSSION

Figure 1 shows the IR spectra of Acs (a), iron(II) lactate entrapped ACs (b) and AC-Fe_2O_3 (c). Comparison of Fig 1a and 1b, $\nu_{C\text{-}C}$ at 1110 cm^{-1} in 1a and also $\nu_{C\text{-}C}$ at 1108, ${}^{a}\nu_{C=O}$ at 1551 and ${}^{s}\nu_{C=O}$ at 1351 cm^{-1} in 1b confirms the attachment of the iron (II) lactate in ACs before thermal treatment at 300°C [12]. Fig 1c evidences the intense band at 1115 cm^{-1} can be attributed to stretching of C-C as well as sharp peaks at 622 and 487 cm^{-1} due to stretching of Fe-O along with the absence of carboxylate stretching in final product confirms all the iron(II) lactate completely decomposed into Fe_2O_3 with no loss of carbon content.

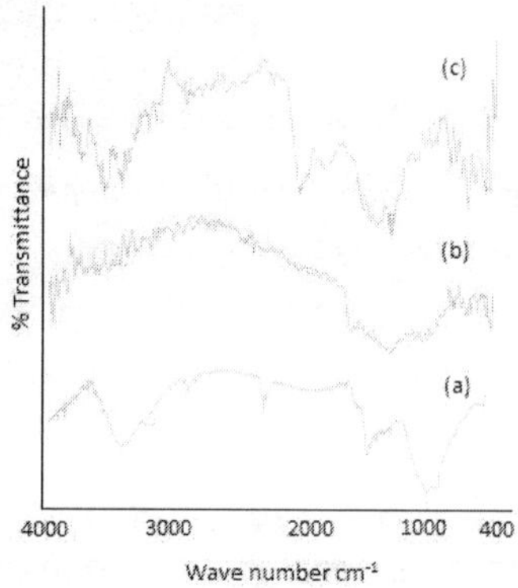

Fig. 1: IR Spectrum of (a) ACs and (b) Iron (II) Lactate Entrapped ACs (c) AC-Fe_2O_3

Figure 2 shows the XRD patterns of the activated carbon (AC) and AC-Fe_2O_3. The position of characteristic peaks in the Fig. 2b show good agreement with JCPDS (89-8104) data of the Fe_2O_3. The broad peaks in the XRD pattern indicate that the crystalline structures of the Fe_2O_3 particles are very small. Diffraction peaks assigned to 2θ = 5.997 and 26.427 can also be clearly seen in the spectrum of Fe_2O_3-AC, indicating that the ACs was not destroyed during experiment.

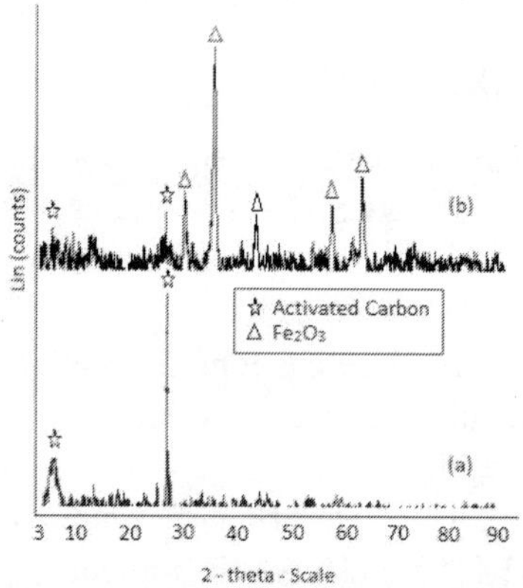

Fig. 2: XRD Pattern of (a) ACs and (b) AC-Fe_2O_3

Figure 3 shows the morphology of the samples was investigated by SEM and TEM (a to f). Although the XRD result showed the presence of crystalline iron oxides of AC-Fe_2O_3, Fe_2O_3 were not easily observed from SEM analysis (Fig. 3b). This may be due to most of the Fe_2O_3 nanoparticles being inserted inside the pores of ACs, as clearly evidenced by the TEM image shown in Fig. 3d to 3f (light shade region is AC and dark spherical shape region is

Fe_2O_3 nanoparticles). The Fe_2O_3 nanoparticle size of the composite is estimated to be about in the range of 12 – 26 nm, which is close to the result obtained by means of XRD.

Fig. 3: SEM Images of ACs (a), AC-Fe_2 (b), EDS of AC-Fe_2O_3(c)

Figure 4a shows the room-temperature magnetization curves of AC-Fe_2O_3. From the graph, M_s of AC-Fe_2O_3 observed at 5.57 and 6.67 emu/g on application of magnetic field of 3000 and 14,500 G respectively, indicate the formation of single domain from multi domain character, supporting the ferromagnetic behaviour of the composite AC-Fe_2O_3 at room temperature. Fig 4b shows the magnetic carbon in the aqueous medium (left) and it was attracted by a permanent magnet. After removing the permanent magnet, the carbon is not easily redispersed in the solution because of the ferromagnetic properties of the AC-Fe_2O_3.

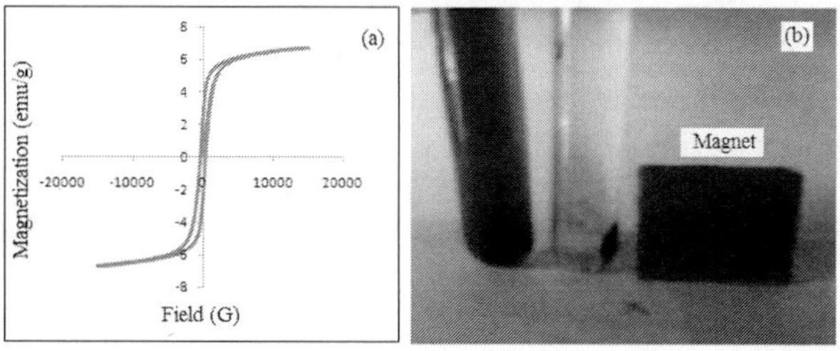

Fig. 4: (a) Hysteresis Loop at room Temperature for AC-Fe_2O_3, and (b) Interaction of the AC-Fe_2O_3 with a Permanent Magnet

CONCLUSION

In this work, magnetic carbon nanocomposites (AC-Fe_2O_3) was successfully synthesized through a simple route, by thermal decomposition of iron (II) lactate entrapped ACs at 300°C. To the best of our knowledge, this is the first report in the synthesis of magnetic carbon shells by using activated carbon and metal carboxylates. Formation of Fe_2O_3 in activated carbon was confirmed by PXRD analysis. The results of TEM analysis of AC-Fe_2O_3 indicating the size of Fe_2O_3 particles in the range 12–26 nm, comply with the calculated values from PXRD patterns. Room temperature M-H curve indicates the ferromagnetism of AC-Fe_2O_3 due to uniform morphology of Fe_2O_3 nanoparticles.

ACKNOWLEDGEMENT

We thank the AICTE, New Delhi (grant-in-aid No. 8023/BOR/RID/RPS-012/2009-10), for financial support of this work.

REFERENCES

[1] Curry-Hyde, H.E., Musch H. and Baiker, A., *Appl. Catal.*, 65, 211-223, 1990.

[2] Cao, H. and Suib, *S.L., J. Am. Chem. Soc.*, 133, 460, 1994.

[3] Cornell, R.M. and Schwertmann, U., VCH Publishers, New York, 1996.

[4] Morrish, A.H., Haneda, K., *J. Appl. Phys.*, 52, 2497, 1981.

[5] Albuquerque, A.S., Ardisson, J.D. and Macedo, W.A.A., et al., *J. Magn. Mater.*, 226-230, 1397, 2001.

[6] Booker, N.A., Keir, D., Priestley, A. and Rithchie, C.D., et al. *Water Sci Technol.*, 23, 1703-12, 1991.

[7] Sing, K.S., Technology profile, *Ground Water Monitor*, 64, 1994.

[8] Safarik, I., Safarikova, M., Buricova, V., *Collect Czech Chem Commun.*, 60, 1448-56, 1995.

[9] Orbell, J.D., Godhino, L., Bigger, S.W., Nguyen T.M., et al. *J. Chem Educ.*, 74, 1446, 1997.

[10] Castro, C.S., Guerreiro, M.C., Goncalves, M.G., Oliveira, L.C.A. and Anastacio, A.S., *J. Hazard. Mater.*, 164, 609-614, 2009.

[11] Oliveira, L.C.A., Rios, R.V.R.A., Fabris, J.D., Grag, V., Sapag, K., Lago, R.M., *Carbon.*, 40, 2177-2183, 2002.

[12] Nakamoto, K., Infrared and raman spectra of inorganic coordination compounds, 4th ed., New York, 1986.

A STUDY ON THE FORMATION OF NANOCRYSTALLINE STRONTIUM HEXAFERRITE PRODUCED VIA A CITRIC ACID AND GLYCINE AS A CHELATING AGENT

Ankush Thakur, R.R. Singh and P.B. Barman

Department of Physics, Jaypee University of Information Technology, Waknaghat, Solan Himachal Pradesh
E-mail: raginirajsingh@gmail.com

ABSTRACT

Strontium ferrite nanoparticles have been synthesized by sol-gel precursor technique using citric acid and glycine as a chelating agent. The citrate-nitrate and glycine-nitrate gels have been obtained from the aqueous solution containing metal nitrates, citric acid and glycine. The calcined powder has been examined by the means of X-ray diffraction (XRD), field emission scanning electron microscopy (FESEM), fourier transform infrared spectroscopy (Mid-IR) and vibrating sample magnetometer (VSM). The XRD analysis confirms that the M-type ferrite phase having hexagonal structure has been formed at 800°C for 8hr. The diffraction pattern also reveals the formation of intermediate phases such as α-Fe_2O_3 *in both the techniques. The crystallite size has been found to be about 42 nm. Surface morphology referred that homogeneous hexagonal platelet-like structure of ultrafine particles has been obtained. The magnetic measurement shows that the values of saturation magnetization and coercivity of samples is suitable for obtaining desired signal to noise ratio (SNR) which can be used for high density perpendicular recording media.*

V. Rajendran, K. Saminathan, P. Paramasivam and K.E. Geckeler (eds.)
Nano Materials Synthesis and Characterisation, pp. 13-16 (2012).

INTRODUCTION

High coercivity materials with magnetoplumbite structure have been extensively use in modern technology applications due to appropriate properties including perfect mechanical, chemical and thermal stability. The discovery of M-type strontium hexaferrite by Philips in the 1950s have been the subject of interest of many researchers for magnetic recording media, microwave filters, radar absorbing materials, telecommunications equipment, injection moulded pieces, plastoferrite, as well as permanent magnets [1]. In addition to these, the increasing exploitation of electromagnetic interference and environmental pollution encouraged for the development of electromagnetic attenuation materials (EAM). The unique and novel properties of nanomaterials with plate like structure can make these materials useful for the absorption of electromagnetic waves as a dielectric or magnetic filler. This work reports the preparation and comparison of different chelating agents on the synthesis and corresponding effects on structural, morphological and magnetic properties.

EXPERIMENTAL

$SrFe_{12}O_{19}$ sample have been prepared with different chelating agents, *viz.,* citric acid and glycine according to sol-gel technique followed by gel to crystallization procedure [2]. Synthesized strontium ferrite have been characterized by x-ray diffraction, field emission scanning electron microscopy, fourier transform infrared spectroscopy and vibrating sample magnetometer for structural, morphological and magnetic study.

RESULTS AND DISCUSSION

The x-ray diffraction reveals that all the apparent intense, sharp and well resolved diffraction planes are well indexed within the hexagonal M-type ferrite, $SrFe_{12}O_{19}$ (*JCPDS*: 80-1198) synthesized *via* citrate and glycine route and have been shown in Fig.1a, respectively. The diffraction peak position at 33.15° corresponds to the presence of hematite as a secondary phase (α-Fe_2O_3) in both citrate-nitrate and glycine-nitrate techniques (*JCPDS*: 89-8104, Fig. 1b). However, the peak intensity of ferrite phase in sample prepared from glycine route is low in comparison to the intensity of sample synthesized with citrate precursor technique. This may be due to production of high thermal energy during combustion process.

The presence of intermediate phase in ferrite system is related to incomplete crystallization reaction and low solubility of Sr^{2+} ions in magnetoplumbite structure. As evidenced from the XRD spectra it takes place dominantly in case of glycine agent. The crystallite size for $SrFe_{12}O_{19}$ nanoparticles has been calculated for the most intense (114) peak using Scherrer's formula, and found to be in the single domain particle size range for both cases. Calculated particle sizes 42.35 nm and 42.37 nm; lattice parameters c= 23.15 Å and 23.04Å; and cell volume V= 699.17 Å^3 and 689.43 Å^3 for citric (C) and glycine (G) chelating agents, respectively. It has been observed that the cell volume of $SrFe_{12}O_{19}$ (G) is lower than $SrFe_{12}O_{19}$

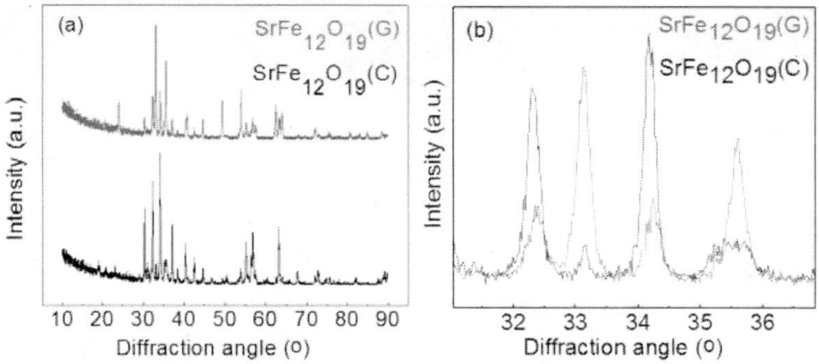

Fig. 1: XRD Patterns of $SrFe_{12}O_{19}$ Prepared from (a) Citric Acid and Glycine and (b) Enlarged View

(C). This slight decrease of cell volume is attributed to reduction in the lattice parameters. FESEM micrographs of calcined samples have been shown in Fig. 2 (a,b). It is evident from the micrograph of $SrFe_{12}O_{19}$ (G) that the particles exhibit porous and agglomerated microstructure. This may be attributed to release of large amount of gases during combustion or decomposition process. The estimated crystallite size has been found to be about 100 nm. The average crystallite size of $SrF_{12}O_{19}$ nanoparticles measured from FESEM is larger than the crystallite size calculated from x-ray peak broadening technique. This is attributed that the FESEM gives the size of secondary particles rather than size of primary particles.

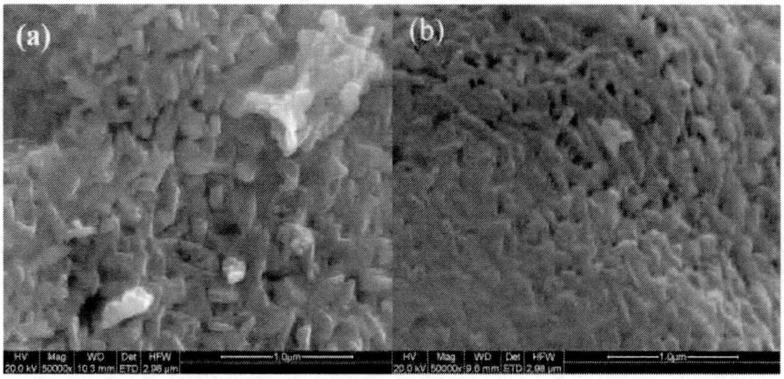

Fig. 2: FESEM-micrographs of $SrFe_{12}O_{19}$ Prepared from (a) Citric Acid and (b) Glycine

Mid-IR analysis of citrate-nitrate and glycine-nitrate system confirms the formation of M-type hexaferrite. The magnetic properties of calcined powders have been measured from hysteresis loops (Figs. 3a,b). The presence of α-Fe_2O_3 is responsible for low value of M_S, as discussed in structural studies. The coercivity of prepared nanoparticles can be explained on the basis of magnetic domain size of particles and the presence of γ-Fe_2O_3.

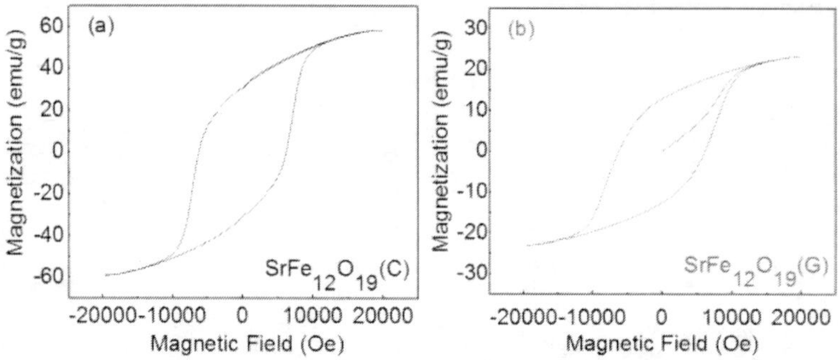

Fig. 3: Hysteresis Loops of $SrFe_{12}O_{19}$ Prepared from (a) Citric Acid and (b) Glycine

H_C is directly proportional to energy barrier (E_A) for single domain nanoparticles [2,3], E_A is in direct proportion to cell volume under the same magnetization direction and thus, the demagnetizing reverse magnetic field is closely related to crystallite size. The crystallite volume of glycine prepared nanoparticles is found to be lower in comparison to citrate route. This result in the dilution of alignment of magnetic spins along the direction of applied field and lower the energy barrier [4]. Consequently, the coercivity decreased. Moreover, the nanoparticles form in the presence of glycine agent is porous as can be seen in FESEM micrographs and hence affects the extrinsic parameters. The results are in good agreement with the XRD analysis and morphological study.

CONCLUSION

$SrF_{12}O_{19}$ have been successfully synthesized using a citric acid and glycine as a chelating agent. The preparation technique, crystallite size as well as the presence of impurities affects the magnetic properties. It has been found that the magnetic parameters for citrate precursor route higher in range and seems to be economical, convenient and versatile for mass production of ferrite nanoparticles in comparison to the other.

REFERENCES

[1] Jacobo, S.E., Hermea, C. and Bercoff, P.G., Influence of the iron content on the formation process of substituted Co–Nd strontium hexaferrite prepared by the citrate precursor method, *J. Alloys Compd.*, 495, 513-515, 2010.

[2] Thakur, A., Singh, R.R. and Barman, P.B., Crystallization Kinetics of Strontium Hexaferrite: Correlation to Structural, Morphological, Dielectric and Magnetic Properties, *Electron Mater Lett.*, 10.1007/s13391-012-2085-z.

[3] *Ai, L.*, Jiang, J., An investigation on synthesis and magnetic properties of ferromagnetic nanoparticles of nickel ferrite coated with TiO_2., *J. Mater. Sci. Mater. Electron.*, 20, 257-261, 2009.

[4] Kumar, V., Rana, A., Yadav, M.S. and Pant, R.P., Size-induced effect on nano-crystalline $CoFe_2O_4$, *J. Magn. Magn. Mater.*, 320, 1729-1734, 2008.

MICROWAVE SYNTHESIS AND CHARACTERIZATION OF ZnO NANOPARTICLES

B. Nageswara Rao, O. Padmaraj, M. Venkateswarlu[1] and N. Satyanarayana

Department of Physics, Pondicherry University, Pondicherry
[1]R&D, Amaraja batteries, Tirupati, Andhra Pradesh
E-mail: nallanis2011@gmail.com

ABSTRACT

ZnO nanoparticles and nanorods have been successfully synthesized by microwave assisted hydrothermal and solvothermal methods. As synthesized ZnO nanoparticles and nanorods were characterized by X-ray diffraction (XRD), scanning electron microscope (SEM), energy dispersive x-ray analysis (EDX) and surface area analyzer (BET).

INTRODUCTION

ZnO is an important photonic and electronic semiconductor because of its wide band gap 3.37eV and a large exciton binding energy of 60 meV at room temperature. ZnO nanoparticles have been attracting much interest because of its potential applications in solar cells, gas sensors, optoelectronics, etc. [1-3]. The day by day development in material science indicates that ZnO has further application in many fields. Over the past decade, a wide range of synthesis methods have been developed for the synthesis of ZnO nanoparticles, such as sol-gel method [4], solution combustion method [5], precipitation [6], solvo/hydrothermal reactions [7], chemical vapor deposition [8], spray pyrolysis [9], thermal evaporation, etc. [10]. However, these processes have disadvantages, such as long duration of synthesis, low yield, etc. Recently, microwave assisted synthesis approach is extremely appealing as it can shorten the reaction time from several hours to minutes with enormous energy savings and cleanliness [11, 12]. Hence, in the present work, ZnO nanoparticles and nanorods were synthesized using microwave

V. Rajendran, K. Saminathan, P. Paramasivam and K.E. Geckeler (eds.)
Nano Materials Synthesis and Characterisation, pp. 17-20 (2012).

assisted hydrothermal and solvothermal methods. The synthesized ZnO nanoparticles and nanorods were characterized using X-ray diffraction (XRD), scanning electron microscope (SEM), energy dispersive x-ray analysis (EDX) and surface area analyzer (BET).

EXPERIMENTAL

Synthesis of ZnO Nanoparticles

Zinc acetate ($Zn(CH_3COO)_2.2H_2O$, Merck chemicals), Polyethylene glycol (PEG-400,Qualigens), sodium hydroxide pellets (NaOH, s.d fine chemicals), ethanol (s.d fine chemicals) and deionized water were used for the synthesis of ZnO nanoparticles and elongated ZnO nanoparticles.

All the chemicals were used without further purification. In the typical synthesis, 1.054g of $Zn(CH_3COO)_2.2H_2O$ was dissolved in 11.6 ml of polyethylene glycol (PEG)-400 under stirring condition. The above prepared solution was added to 58.4 ml of solvent (water/ethanol). The total solution was stirred for 10 minutes and transferred to a 100 ml teflon liner vessel, which contains 0.364g of solid NaOH. The Teflon liner vessel has an outer high strength sleeve made of advanced composite and it is transparent to microwaves. The vessel was sealed and placed on a turn table for uniform heating in a microwave accelerated reaction system (MARS, CEM corporation, USA) and maintained at 140°C for 60minutes for microwave hydrothermal/solvothermal reaction. After the reaction, the precipitated products were sequently washed with deionized water and acetone to remove remnants by centrifugation. The obtained powders were dried at 60 °C over night.

Powder X-ray diffraction patterns were recorded using X'pert PRO MPD, PANalytical (Philips) X-ray powder diffractometer with Cu Kα radiation of wavelength 1.54 A^0 at a scan rate of 2^0 per minute and 2θ values range from 10^0 to 70^0. The microstructures of ZnO samples were taken using scanning electron microscope (Hitachi, S-3400N model). The surface areas of the synthesized ZnO samples were measured using Nova 3200e surface area analyzer (Quantachrome instruments, USA).

RESULTS AND DISCUSSION

XRD

The XRD patterns of ZnO samples, synthesiszed using microwave assisted hydrothermal and solvothermal methods are shown in Fig. 1 along with standard ZnO JCPDS data (PDF # 891397). From fig.1, the observed XRD peaks were compared with the standard JCPDS data and confirmed the formation of pure crystalline ZnO phase. The crystallite sizes of ZnO samples were calculated using Scherrer's formula

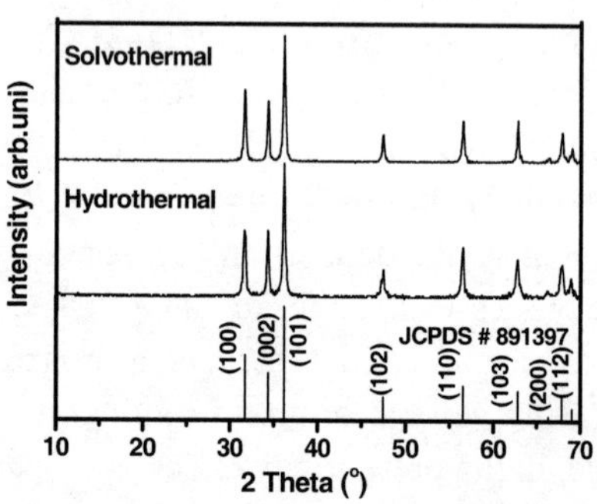

Fig. 1: XRD Patterns of ZnO Samples Synthesized by Microwave Assisted Hydrothermal and Solvo Thermal Methods.

$$D = 0.9\lambda / \beta \cos\theta$$

where, λ is the X-ray wavelength (1.54 A^0) and β is the full-width at half maximum (FWHM) of the diffraction line. The crystallite sizes of ZnO samples synthesiszed using microwave assisted hydrothermal and solvothermal methods are found to be 33.67 nm and 35.34 nm respectively.

SEM-EDX

The SEM micrographs of ZnO samples synthesiszed using microwave assisted hydrothermal and solvothermal methods are shown in Figure. 2a,b. Figure.2a shows the SEM image of hydrothermal synthesized ZnO, which has spherical nanoparticle shape with an average size less than 100 nm. SEM image of solvothermal synthesized ZnO is shown in Fig. 2b, which has nanorod like shape.

Hence, SEM results show that microwave assisted hydrothermal and solvothermal synthesized ZnO has nanoparticle and nanorod morphologies respectively. The EDX spectrum of ZnO nanoparticles is shown in Fig.3c. EDX spectrum confirms the presence of Zn and O elements in the ZnO nanoparticles.

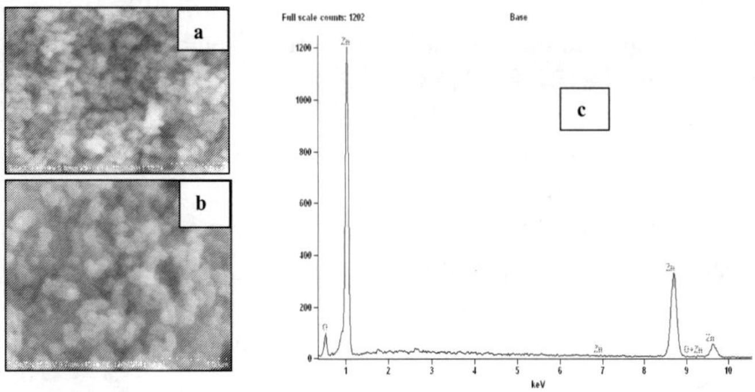

Fig. 2: SEM images of Microwave (a) Hydrothermal, (b) Solvothermal Synthesized ZnO Samples, and (c) EDX Spectrum of Microwave Hydrothermal Synthesized ZnO Sample

BET Surface area Analysis

The surface areas of ZnO nanoparticles and nanorods were measured using BET surface area Analyzer and are found to be 14.810 m^2/g and 10.230 m^2/g respectively.

CONCLUSIONS

ZnO nanoparticles and nanorods were synthesized using a facile microwave assisted hydrothermal and solvothermal methods. The formation of phase pure ZnO nanostructures were confirmed from the observed XRD patterns. The nanoparticle and nanorod morphologies of ZnO were confirmed from SEM results. The existence of Zn and O elements was confirmed from EDX spectrum results.

ACKNOWLEDGEMENTS

NS is grateful to UGC, Govt. of INDIA for providing financial support in the form of research project sanction No.: 39-460/2010 (SR), Dt: 04.01.2011. BNR is thankful to DST, Govt. of INDIA for awarding the INSPIRE fellowship No.: DST/INSPIRE Fellowship/2011/[241], DT: 30-11-2011, for pursuing the Doctoral degree. Authors also acknowledge CIF, Pondicherry university for using SEM-EDX facility.

REFERENCES

[1] Duran, P., Capel, F., Tartaj, J. and Moure, C., A strategic two-stage lowtemparature thermal processings leading to fully dense and fine-grained doped-ZnO,varistors, *Adv. Mater.*,14, 137-141, 2002

[2] Johnsor, J.C., Yan, H., Schaller, R.D., Haber, L.H., Saykally, R.J., and Yang, P., Single Nanowire lasers, J. Phys. Chem. B., 104, 11387-11390, 2001.

[3] Reynolds, D.C., Look, D.C., Jogai, B., Hoelscher, J.E., Sheriff, R.E. Harris, M.T. and Callahan, M.J., Time-resolved photoluminescence lifetime measurements of the ¡5 and ¡6 free excitons in ZnO, J. Appl. Phys., 88, 2152 - 2153, 2000

[4] Seema Rani, Poonam Puri, Shisodia, P.K. and Mehra, R.M., Synthesize of nanocrystalline ZnO powder via sol-gel route for dye sentized solar cell, Solar Energy Materials and Solar Cells, 92(12), 1639-1645, 2008.

[5] Lee, K.R., Park, S. and Lee, K.W., Rapid Ag recovery using photocatalytic ZnO nanopowders prepared by solution combustion method, *J. Mat. Sci. Lett.*, 22, 65-67, 2003.

[6] Demir, M.M., Munoz-Espi, R., Lieberwirth, I. and Wegner, G., Precipitation of Monodisperse ZnO nanocrystals via acid-catalyzed esterification of zinc acetate, *J. Mater. Chem.*, 16, 2940-2947, 2006.

[7] Zhang, J., Sun, L.D., Yin, J.L., Su, H.L., Liao, C.S., and Yan, C.H., Control of ZnO morphology via a simple solution route, *Chem. Mater.*, 14, 4172-4177, 2002.

[8] Shuang, D., Wang, J.B., Zhong, X.L., Yan, and H.L., Raman scattering and Cathodoluminescence Properties of flower-like manganese doped ZnO nanorods, *Materials Science in Semiconductor Processing*,10, 97-102, 2007.

[9] Zhao, X., Zheng, B. and Li, C., Acetate-derived ZnO ultrafine particles synthesized by Spray pyrolysis, *Power Technol.*, 100, 20-23, and 1998.

[10] Dai, Z.R., Pan, Z.W., Wang, Z.L., Novel nanostructures of functional oxides Synthesized by thermal evaporation, *Adv. Funct. Mater.*, 13, 9-24, 2003.

[11] Bilecka, I., Djerdj, I., and Niederberger, M., One-minute synthesis of crystalline binary and ternary metal oxide nanoparticles, *Chem. Commun.*, 886-888, 2008.

[12] Panda, A.B., Glaspell, G., El-Shall, M.S.M., Microwave synthesis of highly Aligned ultra narrow semiconductor rods and wires, *J. Am. Chem. Soc.*, 128, 2790-2791, 2006.

SYNTHESIS AND CHARACTERIZATION OF Ni-Zn FERRITE THIN FILMS PREPARED BY SPRAY PYROLYSIS TECHNIQUE

S.S. Kumbhar, M.A. Mahadik, V.S. Mohite, S.S. Shinde, V.L. Mathe, K.Y. Rajpure and C.H. Bhosale

Electrochemical Materials Laboratory, Department of physics, Shivaji University, Kolhapur
E-mail: bhosale_ch@yahoo.com

ABSTRACT

The Thin films of $NiZnFe_2O_4$ ferrite were prepared by Spray pyrolysis technique. Nickel nitrate, Zinc acetate and ferric nitrate are the starting materials used for these ferrite thin films. These films are deposited at different substrate temperature between 250-450°C. By using X-ray diffraction technique the structural properties were studied which shows amorphous nature. The dielectric constant, loss tangent and ac conductivity with frequency (20Hz-1MHz) were studied for $NiZnFe_2O_4$ thin films.

INTRODUCTION

Thin films of $Ni_{0.1}Zn_{0.9}Fe_2O_4$ having great importance for advanced applications such as microwave control component in GaAs micro monolithic integrated circuit (MMIC) and as insulating layers in the spin–tunneling devices because of their high permeability and high resistivity [1]. Ni-Zn ferrite exhibits favorable properties such as high resistivity, low eddy current losses and high saturation magnetization, which can be used in thin film devices of high frequency [2]. Ferrite has been used commonly as various electronic devices such as inductor and/or electromagnetic wave absorbers in relatively high frequency region up to few hundred of MHz [3]. Nanocrystlline ferrite thin films have been subject of extensive investigation because of their potential application in high density magneto-optic recording

V. Rajendran, K. Saminathan, P. Paramasivam and K.E. Geckeler (eds.)
Nano Materials Synthesis and Characterisation, pp. 21-24 (2012).

devices, color imaging, bioprossesing, magnetic refrigeration and ferrofluds [4]. Ni-Zn ferrite is a type of mixed spinel ferrite in which tetrahedral (A) site are occupied by Ni^{2+} and Fe^{3+} in the cubic spinel lattice . These two antiparallel sublattice from a ferromagnetic structure, in which Ni^{2+} and Fe^{3+} are coupled by superexchange interaction through O^{2-} ions [5]. Therefore many kinds of techniques such as sputtering, sol-gel, CVD and ferrite platting have been attemped for film formation of there ferrite. A spray pyrolysis deposition with a simple apparatus is a versatile technique for producing various thin film materials of a wide range of composition at a relatively low temperature and high deposition rate .The deposited films are characterized by an X-ray Diffraction. To understand the conduction mechanism and dielectric behavior of ferrite the dielectric properties were studied.

EXPERIMENTAL

The Ni $(NO_3)_2.6H_2O$, Zn $(CH_3CO_2)_2$ and Fe $(NO_3)_3.9H_2O$ were used as starting materials for the formation of the Ni-Zn Ferrite thin films. These three metal nitrates are dissolved separately in double distilled water. Final solution was prepared by mixing these three solutions. Ni-Zn ferrite thin films were prepared by spraying the solution on to preheated glass substrates. The solution was automized in air via a pneumatic spray pyrolysis system. The deposited films were characterized using X-ray diffractometer (model PW 1710) using CuKá radiation (k = 1.5406).The diaelectric propreties were carried out by using LCR meter Bridge (model HP 4284 A) in the frequency range 20 Hz-1MHz.

RESULTS AND DISCUSSION

Structural Properties

X-ray diffraction patterns of $NiZnFe_2O_4$ ferrite thin films prepared at a different temperatures ranging from 250-450°C are shown in Fig.1. These XRD patterns of as prepared Ni-Zn ferrite thin films shows amorphous nature. There is no peak obtained from this pattern.

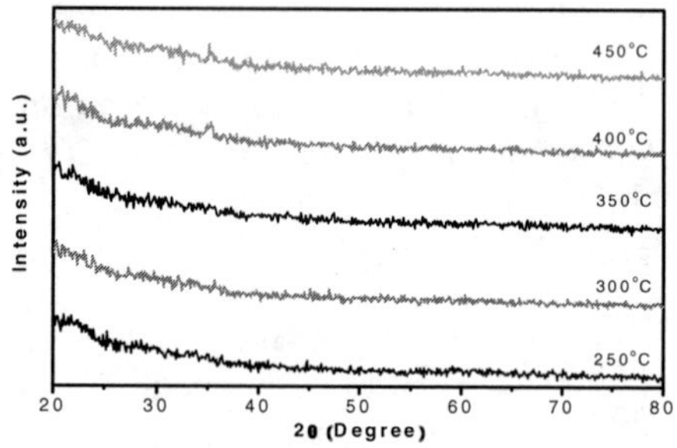

Fig. 1: X-ray Diffraction Patterns of Ni-Zn Ferrite Thin Films

Dielectric Properties

Figure 2 shows variation of dielectric constant of $NiZnFe_2O_4$ samples with frequency which indicates the value of dielectric constant decreases with increase in the frequency, showing the dispersion in the lower frequency range [6]. This dispersion behavior is due to the Maxwell-Wagner type interfacial polarization in accordance with Koop's phenomological theory [7]. In ferrites, electrons are the main charge carriers and the motion of the electrons takes place between Fe^{2+} ions and Fe^{3+} ions present at crystallographic equivalent sites, i.e. octahedral sites. This is known as hopping mechanism of electron transfer. The electrons are through the well conducting grain and the poorly conducting grain boundaries. As the grain boundaries have large resistance, the electrons piles up there and produce large space charge polarization [8].

The Fig. 3 Show the variation of tan σ with frequency of $NiZnFe_2O_4$ thin films prepared at 250-450°C. The dielectric loss arises due to the lag in polarization with the applied alternating electric field and is caused by the impurities and imperfections of thin film [4].

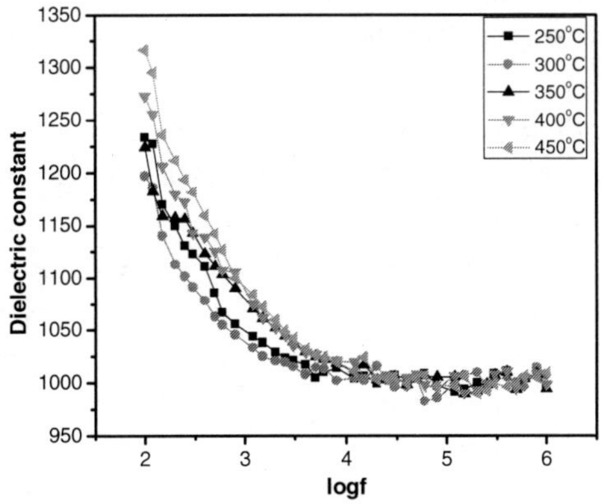

Fig. 2: Variation of Dielectric Constant with Frequency of Ni-Zn Ferrite Thin Films

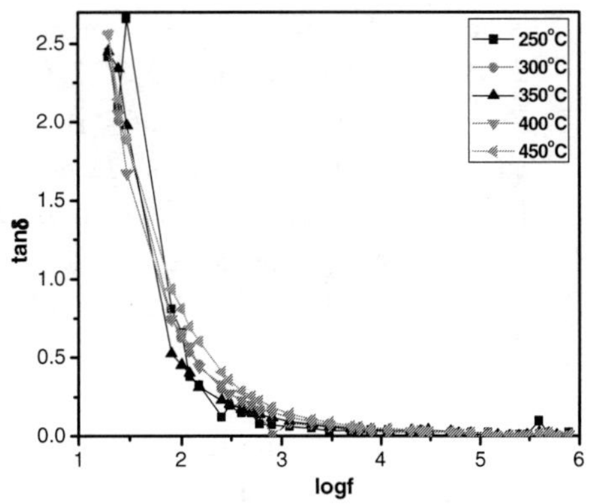

Fig. 3: Variation of Dielectric Loss with Frequency

AC Conductivity

To understand conduction mechanism, AC conductivity was studied. The Fig. 4 shows the variation of log (σ_{ac}) Vs log w^2. In this graph it is seen that the conductivity increases with increase in frequency. It is well known that there are two types of polarons, viz., small polaron and large polaron. In case of small polaron and large polaron model the conductivity increases with increasing the frequency [9]. For the present composite the plot is linear indicating small polaron type of conduction. [10].

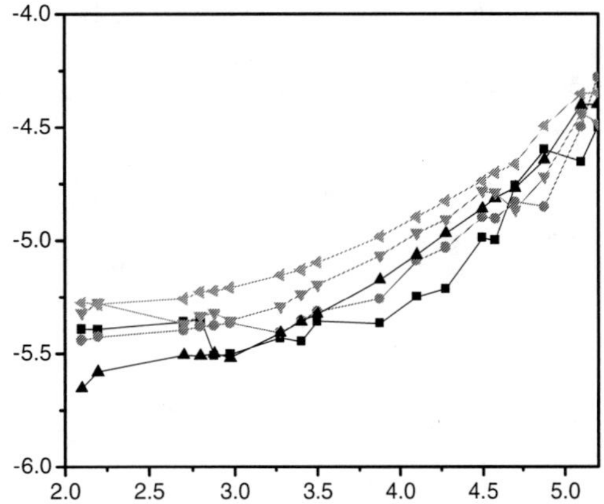

REFERENCES

[1] Lee, J.S., Ha, T.W., Jeong, J.H., Kim, I.W., Yi, S.S., Mgnetic behavior of $Mn_xFe_{3-x}O_4$ thin films grown by ferrite plating method, *Thin Solid Films* 472(2005), 217-221.

[2] Li, L., Peng, L., Li, Y. and Zhu, X., Stuctura and magnetic properties of Co-substituted NiZn ferrite thin films synthesized by sol-gel process, *J. Magn. Magn. Mater.* 324 (2012), 60-62.

[3] Yanagihara, M., Kawano, K., Honda, T., Kyomen, T. and Hanaya, M., Formation of NiZn ferrite nano-crystalline thin films by rf magnetron sputtering with changing the substrate temperature, www. Elsevier.com/locate/tca.

[4] Gupta, N., Verma, A., Kashyap, C. and Dube D.C, Microstructural dielectric and magnetic behavior of spin deposited nanocrystalline nickel-zinc ferrite thin films for microwave, *J. Mang Magn Materials* 308 (2007), 137-142.

[5] Zi, Z., Lei, H., Zhu, X., Wang, B., Zhang, S., Zhu, X., Song, W. and Sun, Y., Influence of annealing temperature on surface morphology and magnetic properties of $Ni_{0.7}Zn_{0.3}Fe_2O_4$ ferrite thin films, *Mate. Sci. Engineering* B 167 (2010) 70-73.

[6] Jadhav, P.A., Shelar, M.B. and Chougule, B.K., Magnetoelectric effect in three phases y ($Ni_{0.5}Cu_{0.2}Zn_{0.3}Fe_2O_4$) + (1-y) (50%$BaTiO_3$+ 50% PZT) ME composite, *J. Alloys Compd.* 479 (2009) 385-389.

[7] Hankare, P.P, Patil, R.P., Sankpal, U.B., Garadkar, K.M., Sasikala, R., Tripathi, A.K., and Mulla, I.S., Magnetic, dielectric and complex impedance spectroscopic studies of nanocrystalline Cr substituted Li-ferrite, *J. Magn. Magn.* Materials 322 (2010) 2629-2633.

[8] Singh, N., Agrawal, A., Sanghi, S. and Singh, P., Synthesis, microstructure, dielectric and magnetic properties of Cu substituted Ni–Li ferrites J. magn. Magn. Mater. 323 (2011) 486-492.

[9] Burange, N.M., Chougule, S.S., Patil , D.R., Devan, R.S., Kolekar, Y.D. and Chougule, B.K., Studies on structural electrical and magnetic properties of y($Ni_{0.5}Zn_{0.3}Co_{0.2}Fe_2O_4$+(1-y)$BaTiO_3$ composites, *J.A. compd.* 479 (2009) 569-573.

[10] Patil, D.R. and Chougule, B.K., Effect of Resistivity on Magnetoelectric effect in (x) $NiFe_2)_4$-(1-x) $Ba_{0.9}Sr_{0.1}TiO_3$ ME composites, *J. Alloys Compd.* 470 (2009), 531-535.

SYNTHESIS AND ELECTROCHEMICAL STUDIES OF TiO_2 NANOPARTICLES

ARGUINE TES RAJ, G. SRIKESH[1] AND V. GUNASEKARAN

Department of Nanoscience and Technology, School of Nanoscience and Nanotechnology, Karunya University, Coimbatore, Tamil Nadu
[1]Department of Chemistry, Karunya University, Coimbatore, Tamil Nadu
E-mail: gunasekaran@karunya.edu

ABSTRACT

Nanocrystalline semiconductor metal oxides especially TiO_2 nano particles have attracted significant interest of materials scientists and physicists due to their special properties and have attained a great importance in several technological applications [1]. In the present work this nanoparticle was synthesized by sol-gel technique. The characterization of nanoparticles was carried out by SEM, UV–Visible spectroscopy, photoluminescence spectroscopy and XRD Electrochemical studies were also done. The synthesis, characterization and applications of TiO_2 nanoparticles are discussed in detail.

INTRODUCTION

Titanium oxide is having great important among the transition metal oxide due to its potential application [1]. Facile synthesis of TiO_2 nanoparticle by Sol- gel process is to be a novel technique and found through sol–gel process, the Physico-chemical and electrochemical properties of TiO_2 can be modified to improve its efficiency [2]. Since this method is a solution process, it has all the advantages over other preparation techniques in terms of purity, homogeneity, felicity and flexibility.

V. Rajendran, K. Saminathan, P. Paramasivam and K.E. Geckeler (eds.)
Nano Materials Synthesis and Characterisation, pp. 25-28 (2012).

EXPERIMENTAL

The reagents (Titanium isopropoxide, ethanol) were obtained from Sigma Aldrich All the chemicals where of analytical grade .These nanoparticle was synthesized by Sol-gel process in which Titanium isopropoxide dissolved in ethanol in the molar ratio 1:4. The solution was sonicated for 4 hrs in order to form sols. To obtain nanoparticles; the gels were dried under 120°C for 12 hrs Also it evaporate water and organic material to the maximum extent. Sintering of dry gel at 450°C for 2hrs were subsequently carried out to obtain desired TiO_2 nanocrystalline material [3].

Physico-chemical characterization of TiO_2 nanoparticle was done by UV-Vis absorption spectra TiO_2 were recorded using a Varian Cary 5E spectrophotometer at room temperature in the range of 200-800nm.The photoluminescence (PL) spectra of TiO_2 were recorded by the Perkin-Elmer lambda 900 spectrophotometer with a Xe lamp as the excitation light source. The XRD pattern of TiO_2 was obtained using a X-ray diffractometer Schimadzu model: XRD 6000 with CuKα radiation in the range of 20-70° (λ = 0.154nm). The surface morphologies of TiO_2 samples were studied using the Hitachi S-4500 Scanning Electron Microscope (SEM). Electro chemical characterization was done by cyclic voltammeter using CH instruments 660.

RESULTS AND DISCUSSION

SEM image of TiO_2 of nanoparticle shows a homogenous structure as seen in Fig. (1). Figure: (2) shows the typical XRD pattern for the TiO_2. The XRD peaks at 25.0099°, 47.8080°, 27.1501°, correspond to the anatase of (115), (024) & (026) which matches with the standard JCPDS 71-0628.

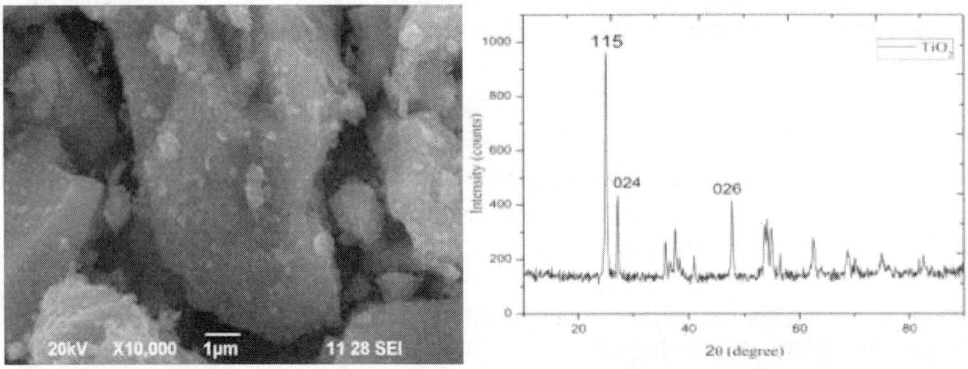

Fig. (1 and 2): Shows SEM Image and XRD Pattern of TiO_2 Nanoparticle

The crystallite size has been calculated by Debye Scherer equation. $D = 0.9\ \lambda/\beta \cos\theta$, Where λ is the X-Ray wavelength, β is the peak width theta is the Bragg's angle. By using above equation TiO_2 crystallite size has been calculated as 43.33 nm [1, 4].

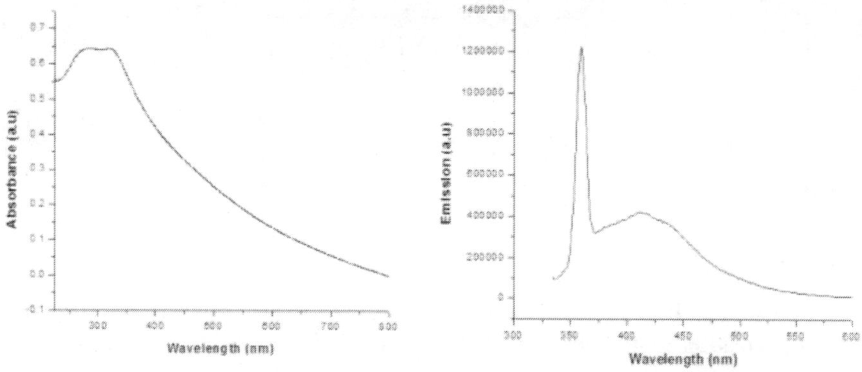

Fig. (3 and 4): UV–Visible s and PL Spectrum for TiO_2 Nanoparticle which Shows Good Absorbance and Emission Properties

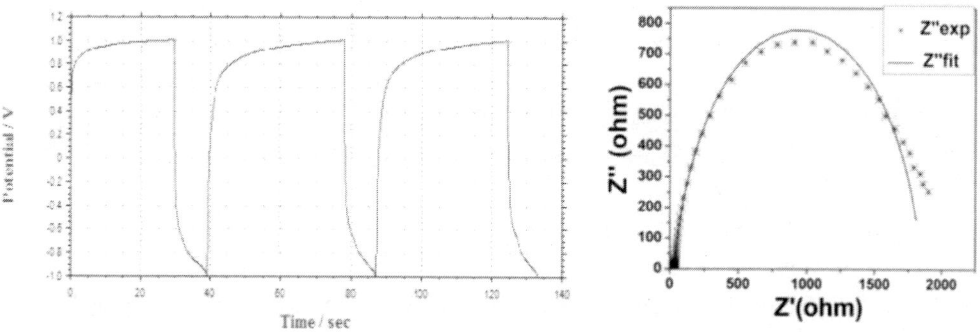

Fig. (5 and 6): Shows Charge/Discharge Impedance Studies of TiO_2 Nanoparticle

The electrochemical performance of TiO_2 samples was further studied by galvanostatic charge/discharge measurements. The charge/discharge curve of the TiO_2 is symmetric & revealing a good capacive behavior and conductivity was stable for every cycles from Fig. (5). Good cycling stability is one of the most important characteristics for high-performance super capacitors TiO_2 electrodes were tested at a scan rate of 100 mVs-1 for 1000 cycles.[5] Impedance spectroscopy helps to resolve the grain and grain boundary effects as well as the electrode kinetic properties. The response of these processes manifests as semi-circles in a Nyquist plot. A typical impedance spectrum of TiO2-gc-imp-1 is shown in Fig. [6]. The plot contains only one semicircular arc which is the characteristic of the behavior of ionically conducting solids. The semi-circle response on the Nyquist plot was fitted with a simple circuit using the Z-View software The equivalent circuit includes a series resistance (Rs) with a Constant Phase Element (CPE) which shown below.

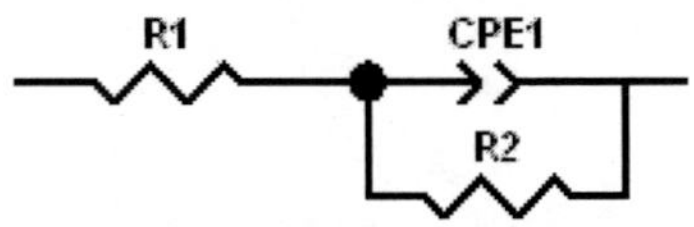

Bulk resistance R_b has been calculated from the intercept of the semicircle at the low-frequency side of the plot with the real axis. The value of bulk resistivity was R2, CPE1–T & CPE1–P was 1812 ohm, 8.3933E-6 F & 1812 ohm respectively.[6]

CONCLUSIONS

Titanium oxide nanoparticles are synthesized by sol-gel method and characterized by various physical characterization techniques. And it is found that this compound is showing good physic-chemical and electro-chemical properties. Hence it is suitable for futuristic capacitance applications.

ACKNOWLEDGEMENTS

The authors thank the Management, Vice-Chancellor, Registrar and Dean (Academic research) of Karunya University, Coimbatore, and DST Nanomission for their kind financial supports and also for their constant encouragement and support.

REFERENCES

[1] A. Ahmad, Gul Hameed Awan and Salman Aziz, Pakistan Engineering Congress, 70th Annual Session Proceedings.

[2] N. Venkatachalam, M. Palanichamy and V. Murugesan, Materials Chemistry and Physics 104 (2007), 454–459.

[3] R. Vijayalakshmi and V. Rajendran, Archives of Applied Science Research, 2012, 4(2): 1183-1190.

[4] Sui-Yang Huang, Ladislav K.Van, Andreas Kay and Michael Gratzel, Nanocrystalline TiO_2 Elecapacity and stability in rechargeable lithium batteriesctrodes exhibiting highstorage, *Active and Passive Elec. Comp.*, 1995, Vol. 18, pp. 23-30.

[5] Ming-Che Yang, Yang-Yao Lee, Bo Xu, Kevin Power, Ying Shirley Meng, and TiO_2 flakes as anode materials for Li-ion-batteries, *Journal of Power Sources,* xxx (2012).

[6] C. R. Gautam, Devendra Kumar and Om Prakash, Dielectric and impedance spectroscopic studies of (Sr1–*x*Pb*x*) TiO_2 glass ceramics with addition of Nb_2O_5, *Bull. Mater. Sci.*, Vol. 34, No. 7, December 2011, pp. 1393–1399. © Indian Academy

SIMPLE CHEMICAL SYNTHESIS AND CHARACTERIZATION OF UNDOPED ZnS

S. ANANDASARAVANAN, O.M. SARAVANA KUMAR AND K. RAJENDRAN[1]

Department of Electronics and Communication Systems, Sri Krishna Arts and Science College, Coimbatore
[1]Government College for Women, Ramanathapuram
E-mail: Sasanand12@yahoo.co.in

ABSTRACT

Zinc sulfide belongs to II – VI group semi conducting compound material with large band gap, widely used as the Cd-free buffer layer for CIGS solar cells. In this paper, we report the simple chemical route for the preparation of ZnS from Zinc acetate and Sodium sulfide as the source materials. White colored ZnS powder was prepared by simple chemical precipitation technique from equal concentration solution of Zinc acetate and Sodium sulfide at room temperature without modifying the pH the mixed solution. Elemental analysis was carried out by EDX spectrum and it confirmed that the prepared material has ZnS. The structure of the prepared ZnS was characterized by X-ray diffraction and scanning electron microscopy and found to have crystal structure with diffraction peaks as (111), (220), and (311). The studies show that crystallite size decreases as the Zn: S molar ratio decreases.

INTRODUCTION

There has been considerable interest in the study of semiconductor nano crystals over last three decades[1] due to their unique structural, electronic, and optical properties which are being applied to diverse fields[2]. There has been growing interest in developing techniques for preparing semi conductor nano particles[3], in which the characteristics are different and attractive from those of bulk materials. It is well known fact that many of the properties of these materials

V. Rajendran, K. Saminathan, P. Paramasivam and K.E. Geckeler (eds.)
Nano Materials Synthesis and Characterisation, pp. 29-32 (2012).

depend largely on the size and shape of the nano materials[4]. Their reduced dimensions enable one to reduce the size of electronic circuitry[5].

Zinc Sulphide belongs to II–VI group compound material, which is an important member of this family, has been extensively investigated[6] with large direct band gap between 3.4 to 3.70 eV[6] depending upon composition, is potentially important material to be used as an anti reflection coating for hetero junction solar cells, for light emitting diode and other opto electronic devices such as blue light emitting diode, electroluminescence devices, nonlinear optical devices, fast optical switches[7] and photo voltaic cells which enable wide application in the field of displays, sensors and lasers. In recent years, nano crystalline ZnS attracted much attention because the properties in nano form differ significantly from those of their bulk counter parts. Therefore much effort has been made to control the size, morphology and crystallinity of ZnS.

Most reported ZnS nano particles are prepared and stabilized by capping materials, the influence of the stabilizer on ZnS nano particles could not be avoided. In this paper, a simple method for preparation of undoped ZnS at room temperature without modifying the pH is reported, which makes it possible to study the real properties of ZnS.

EXPERIMENTAL

The reagents used for these studies are Zn $(CH_3COO)_2$ $2H_2O$ as Zinc source, Na_2S as sulphur source, and deionized water. All the Source materials are of AR grade obtained from Merck and used without any further purification. The undoped ZnS were prepared by simple chemical route in aqueous medium.

Table 1: Ten Samples of Zinc Acetate and Sodium Sulfide and their pH Values

S.No	*Sample ID*	*Molar Concentration of $Zn(CH_3COO)_2 .2H_2O$*	*Molar Concentra -tion of Na_2S*	*pH of Zinc Acetate Solution*	*pH of Sodium Sulfide Solution*	*pH of Mixer of Zinc Acetate and Na_2S*
1	A	1	1	5.49	14	5.4
2	B	0.9	0.9	5.58	14	5.49
3	C	0.8	0.8	5.61	14	5.5
4	D	0.7	0.7	5.73	14	5.51
5	E	0.6	0.6	5.78	14	5.57
6	F	0.5	0.5	5.8	14	5.53
7	G	0.4	0.4	5.85	14	5.57
8	H	0.3	0.3	5.86	14	5.62
9	I	0.2	0.2	5.98	14	5.77
10	J	0.1	0.1	6.2	14	5.8

For the preparation of undoped ZnS, 50 ml of 1 M Zinc acetate and 50 ml of 1 M Sodium sulfide taken in separate beakers and stirred well for 30 minutes. Then Na_2S solution was taken in burette and added drop wise into a beaker containing Zinc Acetate solution with constant stirring for 50 minutes. The precipitates formed in the above process were centrifuged at 6000 rpm for 5 minutes. The centrifugate was decanted. The residue washed with deionised water for several times and finally with ethanol. The precipitates were dried in an oven at 70°C for 24 hours and then crushed to fine powder. The same procedure repeated for the synthesis of other ten samples of undoped ZnS with equal concentration of Zinc acetate and Sodium sulfide. The pH values of precursors and mixer of Zinc acetate and Sodium sulfide are measured and tabulated.

Reaction mechanism

$$Zn(CH_3COO)_2.2H_2O + Na_2S \longrightarrow ZnS + 2CH_3COONa + 2H_2O$$

$$2CH_3COO^{-} + Zn^{2+} + 2Na^{+} + S^{2-} \longrightarrow ZnS + 2CH_3COONa$$

RESULTS AND DISCUSSION

Elemental analysis was carried out by EDX spectrum and it confirmed that the prepared material has ZnS.

The phase purity, crystallographic structure, and size of nano crystallites were determined by powder X–ray diffraction. The product was found to exhibit the characteristics pattern corresponds to face-centered cubic structure, and peaks observed in the XRD patterns match well with those of the cubic ZnS reported. No other impurities such as organic compounds related to recants were detected by XRD analysis indicating the phase purity of the ZnS product. This indicates that preparation condition was sufficient for this material. The diffraction peaks at 2θ values of 29.0, 48.3 and 55.8 correspond to <111>, <220> and <311> plane respectively of cubic ZnS and the lattice constant, a, was calculated. Broadening of the XRD peaks shows the formation of nano crystals of ZnS. The average crystallite size of ZnS nano particles was calculated using the Debye-Scherrer's equation

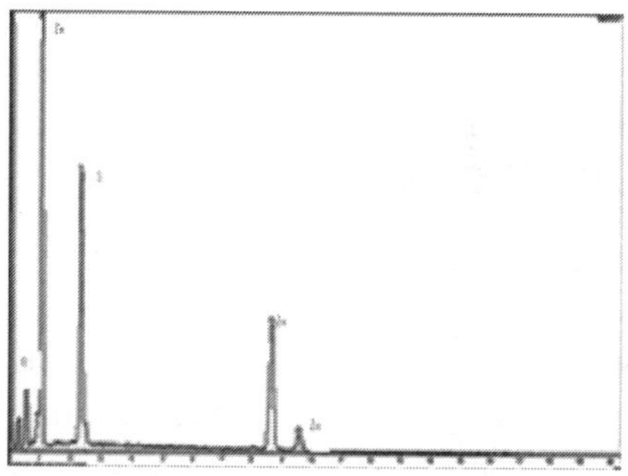

Fig. 1: EDX spectrum of Undoped ZnS

$$D = \alpha\lambda / \beta \cos\theta$$

where D is the mean particle diameter, α is the geometrical factor ($\alpha = 0.94$), λ is the wave length of X-rays used for analysis (CuKα1 radiation and equals to 0.154 nm), and β is the full width at half maxima (FWHM) of peak appearing at the diffraction angle è corresponding to each plane was selected for particle size calculation, and the average particle size was found to be 80-113 nm.

Figure shows the XRD patterns obtained by scanning 2θ in the range 10-90°.

Figure shows the morphology of the prepared sample, it shows that ZnS grains are not uniformly distributed. Also the grain sizes are not same. The average grain size is estimated. It is found to decrease the molar ratio the particle size are also decreased.

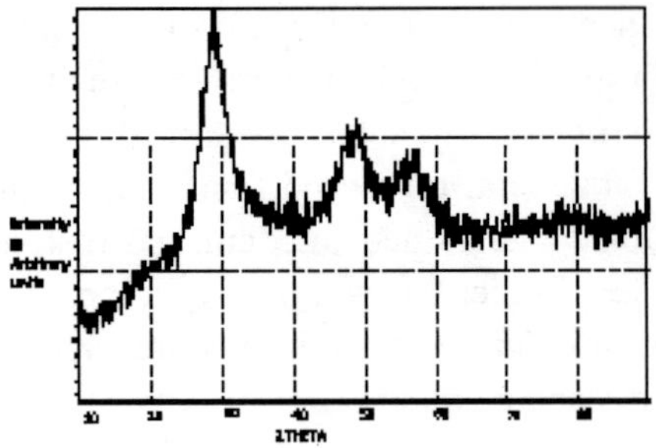

Fig. 2: XRD Pattern of Undoped ZnS

Fig. 3: SEM Images of Undoped ZnS

CONCLUSIONS

In this paper, we report the simple chemical route for the preparation of white coloured ZnS powder sample from equal concentration solution of Zinc acetate and Sodium sulfide as the source materials by simple chemical precipitation technique at room temperature without modifying the pH the mixed solution. Elemental analysis was carried out by EDX spectrum and it confirmed that the prepared material has ZnS. The structure of the prepared ZnS was characterized by X-ray diffraction and scanning electron microscopy and found to have crystal structure with diffraction peaks as (111), (220), and (311).

REFERENCES

[1] U. Baishya & D. Sarkar, *Indian Journal of Pure & Applied Physics*, 49,(2011), 186.

[2] Deng, M.C., Chin, T.S, F.R.J *Applied Physics*,1994, 75,5888.

[3] M.S. Shinde, P.B. Ahirrao, R.S., Patil, *Arch and Appl. Sci. Res.*, 2011, 3(2): 311.

[4] Hu J., Odom J.W. & Lieber C.M., *Acc Chemical Research*, 32 (1999) 435.

[5] V.L. Gayou, B. Salazar-Hernandez, R. Delgado Macuil, G. Zavala, P. Santiago and A.I. Oliva, *Journal of nano research* vol. 9 (2010) 125-132.

[6] Sooklal K., Cullum B.S., Angel S.M. & Murphy C.J., *J. Physical Chemistry*,100(1996) 4551.

[7] J.F. Xu, W. Ji, J.Y. Lin, S.H. Tang, Y.W. Du, *J. Applied physics*, 66 (1998) 639.

STRUCTURAL AND OPTICAL PROPERTIES OF BORON CARBIDE NANOPARTICLES FOR LIGHT WEIGHT ARMOR

Navid Hussain

Centre for Converging Technologies, University of Rajasthan, Jaipur
E-mail: naveedqumar@gmail.com

ABSTRACT

In this paper we have prepared nanoparticles of B_4C using chemical method. A mixture of boron powders, boron oxide powder and carbon powder were placed in the bottom of a graphite crucible. The crucible was then inserted into a high temperature tube furnace. The furnace was vacuumed then filled with flowing argon (50~200 sccm). The furnace temperature was raised quickly to 1000~1150 °C and held at this point for 3~4 hour under argon atmosphere. After the reaction, a grayish product was collected and examined. A vapor-liquid-solid (VLS) mechanism may be involved: The phase and orientation of the sample was characterized by X-ray diffractometry using $CuK_{\hat{a}}$ (λ = 1.5407 Å) radiation. The microstructure was examined by scanning electron microscope. The optical properties were investigated using spectrophotometer. Due to light weight, high strength, hardness it is being used as light weight armor.

INTRODUCTION

Boron carbide is a hard material after diamond and boron nitride and also known as black diamond. It is lightweight (theoretical density: 2.52 g/cm^3) and has a 4-point bending strength of at least 300 MPa [1]. In addition to having a Knoop hardness of 2800 kg/mm^2 [2], using a 100 g load. It has a high melting point (2450°C), high neutron absorption cross-section, and is also resistant to chemical agents. Due to having properties like low specific weight, high hardness and impact resistance it is useful for body and vehicle armor.

V. Rajendran, K. Saminathan, P. Paramasivam and K.E. Geckeler (eds.)
Nano Materials Synthesis and Characterisation, pp. 33-36 (2012).

Boron carbide was first prepared in 1883 but the stoichiometric composition B_4C, known as 'technical boron carbide', was not assigned until 1934 [3].The purity of boron carbide produced by early researchers was less than 75% but Ridgway claimed to have produced crystalline B_4C of 90% purity by carbothermic process in 1933 [4]. Different processes to produce nanoparticles of boron carbide are known. Boron carbide can be synthesized from elemental boron and carbon directly. But this method has become unattractive due to high cost of elements involved in the process. Other synthesis methods include metallotermic and carbothermic processes such as carbon thermal reduction of boron oxide (B_2O_3) over 1000°C, reduction of BCl_3 by CH_4 at a temperature of 1500° *C* with laser, thermal decomposition of a mixture of pure carbon and boron trichloride in an atmosphere of hydrogen, and gaseous reaction between BCl_3 and methane- hydrogen mixture in r. f. argon plasma [5-8]. Recently, boron carbide nanoparticles have been synthesized by co-reduction route using BBr_3 and CCl_4 as the source materials at a low temperature of 480° *C* and new selective synthesis routes to boron carbide and boron nitride based on the use of new polymer systems, poly[2-(vinyal) pentaborane] and poly [B-vinylborozine] as precursors. It was reported that $2(H_2C = CH)\ B_5H_8$ will undergo thermal polymerization to vinly-pentaborane oligomers and that these species can be converted to pure boron carbide with high ceramic yields under mild conditions. Isao et al., reported that boric acid and 2-hydroxy benzyl alcohol (HBA) were used as the B and C sources of B_4C. H_3BO_3 and HBA in DMF solution resulted in a phenolic resin-B_2O_3 hybrid. Among the methods, carbothermal reaction and magnosoisothermal reaction are widely used as main techniques for the production of B_4C powders. The mostly used industrial method to produce B_4C is reduction of B_2O_3 in presence of carbon using Magnesiothermic techniques.

EXPERIMENTAL

Boron powder (Loba Chemie, 200 mesh, 99.00% pure), carbon powder (Fisher Scientific), di-boron tri-oxide powder (B_2O_3, Search Chem) and Cobalt powder (Molychem, 99.50% pure) were used without further changes as purchased. The metal like cobalt (Co) was used during reaction as catalyst to facilitate the formation of boron carbide nanorods. Boron oxide, boron amorphous powder and carbon were used as the precursor agents. Before weighing, carbon powder was baked at 150-200 °C in vacuum. Different ratios of $B/B_2O_3/C$ were mixed for the reaction. Cobalt metal powder was used as the catalyst. Reactions were carried out in a high-temperature tube furnace with a maximum working temperature of 1200°C. Dry powders were mixed by roll milling and put into a covered graphite crucible. The crucible was then placed into the center reaction zone of the tube furnace. The reactor was pre-heated under vacuum at about 200°C for 2-3 hours. Ultrahigh purity argon was then flushed through the furnace several times before filling the reactor with argon at a constant flow of 200 sccm. The furnace was then heated to 400 °C for 30 minutes. After this period, the temperature was rapidly raised to 1000- 1150 °C it took about 30 min and was maintained at these temperatures for 3~4 hours before it was turned off and cooled down to room temperature. After the reaction, the as-formed product inside the crucible was collected and analyzed. The as prepared sample was also characterized that confirmed the existence of crystalline boron carbide.

RESULTS AND DISCUSSION

The phase and orientation of sample were characterized by X-ray diffractometry using CuK_{α} (λ = 1.5407 Å) figure-1. The XRD pattern confirms the formation of B_4C. The crystalline size of the sample was calculated using Scherrer's formula

$$T = 0.91\lambda/\beta \cos \theta$$

Where λ = wavelength of *x*-rays (1.5407 Å), β = Full width at half of the maximum peak (FWHM), θ = Angle of maximum intensity. The calculated crystallized size is about 25 nm.

The surface topography and microstructure were examined by scanning electron microscope (SEM) Figure 2. The average particle size is observed about 36 nm.

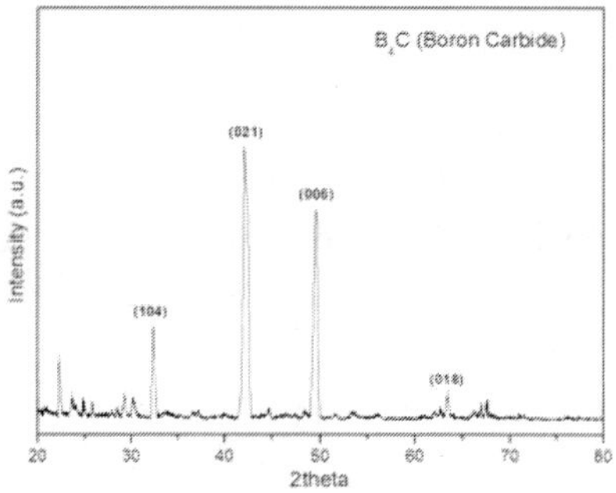

Fig. 1: XRD Data of Sample

Fig. 2: SEM Image of Sample

The optical absorption spectra of Boron Carbide was observed using spectrophotometer. The relation between the absorption coefficient, α, and the incident photon energy, hν, can be written as [9]:

$$\alpha h\nu = A\left(h\nu - E_g\right)^n$$

where α is the absorption coefficient, h is Planck's constant, ν is the photon frequency and E_g is the optical band gap, A is a constant depending on the transition probability and *n* is an index that characterizes the optical absorption process and is theoretically equal to n = 1/2 and 3/2 for direct allowed and direct forbidden transition, respectively, while n = 2 and 3 for indirect allowed and indirect forbidden transition, respectively. To calculate

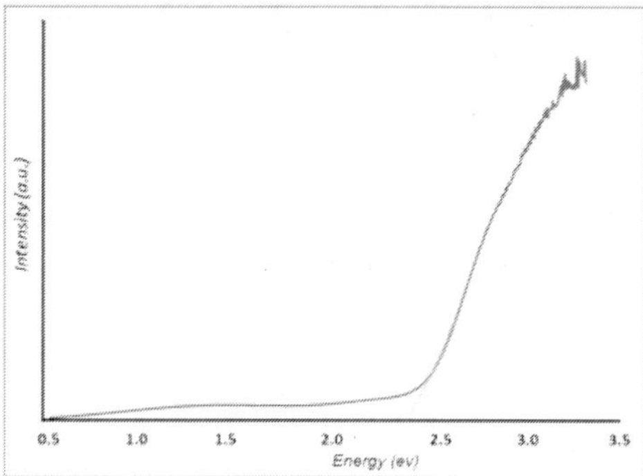

Fig. 3: Shows Energy Band Gap for Boron Carbide Nanoparticles.

the direct band gap of B_2O_3, plot of $(\alpha h\nu)^2$ versus hí is shown in Figure 3. An extrapolation of the linear region of a plot of $(\alpha h\nu)^2$ on the *y*-axis versus photon energy (hν) on the *x*-axis gives the value of the optical band gap E_g. So, the optical band gap of sample (Boron Carbide nanoparticles) was found to be 2.39 ev.

CONCLUSIONS

Different methods are available for fabrication of boron carbide nanomaterials like mechanical milling, carbothermal processes, sol-gel method, classical chemical VD, plasma-enhanced CVD. Carbothermal reduction of boric acid which has been the commercial method for the production of boron carbide was used. Cobalt metal powder was used as the catalyst. A smooth range of nanoparticles of boron carbide were synthesized and characterized using SEM images and XRD data. The optical properties of sample were also looked into. Due to having properties like low specific weight, high hardness it is being used for body and vehicle armor.

REFERENCES

[1] F. Thevenot, “A Review on Boron Carbide,” *Key Engineering Materials*, 56-57 59-88 (1991).

[2] J. F. Shackelford and W. Alexander, CRC Materials Science and Engineering Handbook, CRC Press, Boca Raton, Fl., 1991.

[3] A. Joly, C.R. Hebd, *Seances Acad. Sci*. 97 (1883)

[4] R. R. Ridgway: ‘Boron carbide and the method of making the same’, US patent no. 1897214, 1933.

[5] A. M. Hadian and J. A. Bigdeloo,” The Effect of Time, Temperature and Composition on Boron Carbide Synthesis by Sol-gel Method,” *Journal of Materials Engineering and Performance*, vol. 17[1], pp. 44-49, 2008

[6] S. Mondal and A. K. Banthia, “Low-temperature synthetic route for boron carbide,” *Journal of the European Ceramic Society*, vol. 25, pp. 287-291, 2004.

[7} A. Sinha, T. Mahata and B. P. Sharma, “Carbothermal route for preparation of boron carbide powder from boric acid-citric acid gel precursor,” *Journal of Nuclear Materials*, vol. 301, pp. 165-164, 2002.

[8] K. Yamada, “Boron carbide particles formed from an amorphous boron/ graphite powder mixture using a shockwave technique,” *J Am Ceram Soc., vol.* 79[4], pp. 1113- 1116, 1996.

[9] J. Tauc, Amorphous and Liquid Semiconductors, Plenum Press, New York, 1974.

AN INSIGHT INTO BHARADWAJA'S VYMANIKA ŚASTRA IN THE PREPARATION OF VATADHARANA LOHA

Padmashree Anand, R.S. Kulkarni, E.S. Dwarakadasa[1] and S. Vynateya[2]

CMRTU, RVCE Campus, Mysore Road, Bangalore
[1]IISc., Bangalore
[2]MTD, CPRI, Bangalore
E-mail: padmashreeanand@gmail.com

ABSTRACT

Throughout history, there are records scattered in Sanskrit literature about Indian scientists to be the best materials scientists since they knew how to use the available materials without affecting the balance of materials in nature or the environment. In several instances, proper links regarding sources and references, follow up work, practice and procedures are missing despite the uniqueness and possible utility being entirely new to modern science. Maharshi Bharadwaja's ancient work on 'Vymanika Œastra' is one of the most significant in this regard. A few attempts have been made to decipher and follow details given in this treatise to repeat the experiments and recreate some of the property - specific materials.

The purpose of this research is to re-enact the making of one such material described in the text, reproduce those conditions and check whether there is validity of the claimed properties. An attempt was made to reproduce 'Vatadharana Loha', - a light and strong material as mentioned in the scripture. The methodology adopted, analysis of the microstructures of the samples prepared and certain properties of Vatadharana Loha obtained, are described in this paper.

INTRODUCTION

Vymanika Śastra, a part of the unknown work, 'Yantra Sarvasva'(encyclopedia of machines) authored by Maharshi Bharadwaja consists of nearly 6000 lines in lucid Sanskrit dealing with the construction of an aircraft in about 600 aphorisms embedded in 8 chapters. Most of the

V. Rajendran, K. Saminathan, P. Paramasivam and K.E. Geckeler (eds.)
Nano Materials Synthesis and Characterisation, pp. 37-40 (2012).

scientific information in ancient India has been recorded in Sanskrit language but in the encrypted form. The unique feature in addressing knowledge of ancient Indian origin is in the methodology adopted, distinctly different from the approach of modern scientific methods. The best approach to check the veracity of the claims made in the text is to reproduce at least some of the materials mentioned therein.

EXPERIMENTAL WORK

The property – specific material selected for the study was Vatadharana Loha – a light, soft and strong material. The deciphering of terminologies for the Vatadharana loha was done and an attempt was made to reproduce it whose composition is given in Table 1.

Table 1: Composition of Vatadharana Loha

Sanskrit name	*Modern Chemical Equivalent*	*Proportion*
Simhaasyakam	Iron rust	Equal parts
Shaarana	Juice of jute??	
Surya	Copper	
Varchula	Salt Petre (KNO_3)	
Temperature : 102 kakshyas*		

*Kakshya - appears to be a non-linear scale of temperature.

The coal fired blower type furnace was allowed to attain a temperature of 900°C and A1(~1 kg) SiC crucible was kept inside the furnace. For each trial, when the crucible reached the temperature of the furnace, the charge [Table 2] was placed in the crucible with occasional stirring to obtain homogeneity and the crucible was allowed to attain a temperature of ~1350°C so as to melt the charge. Shaarana probably means juice of jute, and if it is so, the purpose of addition of it probably is to facilitate the reduction of iron oxide to iron. So mother in law's tongue (a highly fibrous plant with the biological name Sansevieria trifasciata) was used in place of juice of jute in Trial # 4 & 5. The molten metal was poured into a metal mold (Trials 1,2 and 4) and into the sand mold (Trial 3 and 5).

Table 2

Vatadharana Loha	*Trial # 1*	*Trial # 2*	*Trial # 3*	*Trial # 4 & 5*
Raw materials	Weight in grams	Weight in grams	Weight in grams	Weight in grams
Iron rust	25	55	75	55
Juice of jute??	Not available	Not available	Not available	110g of Mother in law's tongue
Copper	200	110	150	110
Salt Petre (KNO_3)	200	110	150	110

RESULTS

Chemistry, conductivity and density were checked (Table 3)

Table 3

Trial #	*Chemical composition*	*Electrical Conductivity Sm^{-1}*	*Electrical Conductivity (IACS)*	*Thermal Conductivity W/mK*	*Density g /cm3*
1	Cu 97.39%; Fe 2.61%	1.867×10^7	32.2%	137.22	7.95
2	Cu 97.09%; Fe 2.91%	1.229×10^7	21.2%	90.33	7.81
3	Cu 96.79%; Fe 3.21%	1.036×10^7	17.9%	75.64	7.68
4	Cu 97.15%; Fe 2.85%	1.387×10^7	23.91%	101.26	7.75
5	Cu 97.29%; Fe 2.71%	1.519×10^7	26.2%	110.9	7.87

The samples were prepared for metallographic examination in the standard way and the microstructure was examined (Figs. 1-5).

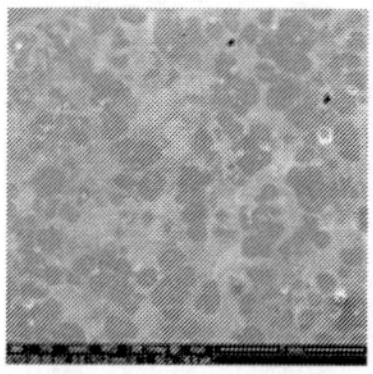

Fig.1 : Trial # 1

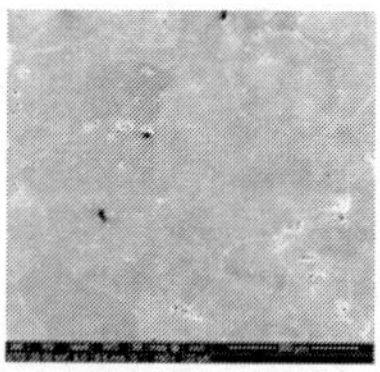

Fig 2 : Trial # 2

Fig 3 : Trial # 3

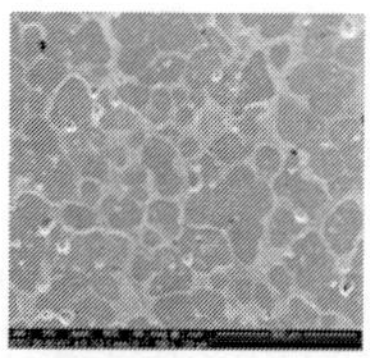

Fig. 4 : Trial # 4

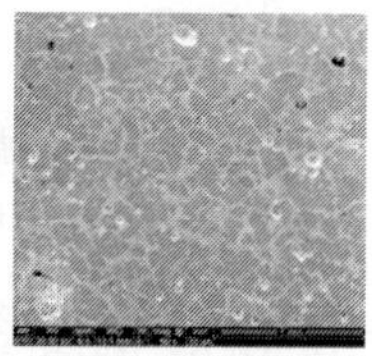

Fig. 5 : Trial # 5

DISCUSSION

Figure 1 indicated iron powder particles were dispersed in the matrix of copper. Fine grains of copper were observed throughout the microstructure. Pores were observed distributed uniformly across the microstructure which may have resulted due to the gas generated by addition of KNO_3. The single phase structure in Fig. 2 shows that iron has dissolved in copper and grains of alpha phase of copper. Many pores have formed flower like group structures. In Fig. 3, the single phase structure indicates that iron has dissolved in copper and there are grains of alpha phase of copper. Pores are more when compared to Trial 1 and 2. It is evident from the microstructure (Fig 4), iron has dissolved in copper and grains of alpha phase of copper are present. Also the pores are evidently more when compared to previous trials. In the microstructure of Trial 5 (Fig. 5), iron powder particles were dispersed in the matrix of copper. The uniform distribution of pores in the structure is much evident and increase in the number of pores is probably due to the addition of Mother in law's tongue in both Trials 4 and 5. The results (Table 3) indicate that with increase in the percentage of iron in the product, electrical conductivity, thermal conductivity and density of the product decreases. There was also a limitation on the temperatures that could be reached as also the rates of heating. With the availability of quickly heating modern furnaces, it is possible that something is amiss.

CONCLUSION

This research is a step towards integration of ancient and modern science and technology. It is expected that the present experiments will lead to a greater clarification and lend credibility about the work reported in ancient literature. This research sheds light on the fact that KNO_3 can be used to produce porosity in copper alloys and thus further research in this regard may culminate in the preparation of copper foam which will be a light material and thus can find lots of applications in aerospace industry.

BIBLIOGRAPHY

[1] Vymanika Shastra, Maharshi Bharadwaja, Translated into English and edited, G. R. Josyer, Coronation Press, Mysore (1973).

[2] Brihad Vimana Sastra, (Ed) Swami Brahmamuni Parivrajaka Gurukul Kandgi, Dayanand Bhavan, New Delhi (1959).

[3] Vimana in Ancient India, D K Kanjilal, Sanskrit Pustak Bhandar, Calcutta (1985).

[4] Naren Sheth, "Was Maharshi Bharadwaja a Solar Scientist?", in 'Science and the Vedas', (Ed.) Ishwarbhai Patel, Maharshi Academy of Vedic Sciences (1984) 36 – 39.

[5] VIMANA Aircraft of Ancient India and Atlantis, David Hatcher Childress, Adventure Unlimited Press (1994) 103-200.

[6] High Technology in Ancient Sanskrit Literature – A Preliminary Report, C.S.R. Prabhu, Samskritavignanavaibhavam, June (1999).

[7] Briefly ……., Hinduism Today, April, 1997.

[8] N.G. Dongre, S.K. Malaviya and P. Ramachandra Rao, The study of various materials described in Amshubodhini of Maharshi Bharadwaja, Final Report, NML, Jamshedpur, (2000).

[9] Amshubodhini, Maharshi Bharadwaja, pp. 69-77.

[10] M.K. Kawadkar, Bharatiya Bouddhika Sampada; May, Aug, Nov (2000); Feb, May, Aug (2001); Feb, Aug (2002).

[11] Damodar Joshi, Rasaratnasamuccaya by Vagbhata, edited with English translation (1991–1992).

[12] N.G. Dongre, S.K. Malaviya and P. Ramachandra Rao, Prakasha Stambhanabhida Loha of Maharshi Bharadwaja, , Journal of History of Sciences, 33 (4), (1998) pp. .273-280.

[13] V K Didolkar and M K Kawadkar, A novel infrared transparent glass material as per 'Amshubodhini' (2002).

[14] Dina Nath Mishra, Wonder that was India, The Pioneer, Feb 06. (2005)

[15] M.P. Rao et al., Vymanika Shaastra Rediscovered – Project Study Report, , AR & DB, New Delhi (2002).

[16] Vimana Ancient flying Machine, *http://www.crystalinks.com*, Ellie Crystal.

[17] Ancient and Lost Civilizations, *http://www.crystalinks.com*, Ellie Crystal.

[18] Proof Advanced Ancient Indian Civilization Existed, *http://www.abovetopsecret.com*

[19[Airplanes in Ancient India, *http://www.SciForums.com*

ELASTIC AND MAGNETIC PROPERTIES OF LANTHANUM – STRONTIU MANGANITES

A.R. BULATOV, KH. G. BOGDANOVA[1], V.A. GOLENISCHEV-KUTUZOV, R.N. KHIZBULLIN AND V. RAJENDRAN[2]

Kazan State Power-Engineering University, ul. Krasnosel'skaya 51, Kazan, 420066 Tatarstan, Russia

[1]*Zavoisky Physical-Technical Institute, Kazan Scientific Center, Russian Academy of Sciences, Sibirskii tr. 10/7, Kazan, 420029 Tatarstan, Russia*

[2]*K S R Kalvi Nagar, Tiruchengode, Namakkal (Dt.), Tamil Nadu, India*

ABSTRACT

The temperature dependences of the transverse and longitudinal hypersound velocities at the frequencies (0.5-0.7) GHz were measured in lanthanum-strontium manganites with the $L_{1-x}Sr_xMnO_3$ (x = 0.125; 0.15; 0.175) composition. Structural phase transitions the position of which is confirmed by electric resistance and magnetic measurements data for the same samples were observed. These results were analyzed within the model of the competing Jahn-Teller distortions and magnetic ordering. Anomalies in the behavior of the longitudinal hypersound velocities were related to the local Jahn-Teller distortions the suppression of which upon magnetic ordering is considered as a possible origin of colossal magnetoresistance.

INTRODUCTION

The study of the influence of the Jahn-Teller (JT) effect on the physical properties of the transition metal compounds was reviewed by Kugel and Khomskii [1] and Meicher [2]. One of the most studied objects having the largest CMR value are slightly doped lanthanum-strontium

---

V. Rajendran, K. Saminathan, P. Paramasivam and K.E. Geckeler (eds.)
Nano Materials Synthesis and Characterisation, pp. 41-46 (2012).

manganites with the general formula $L_{1-x}Sr_xMnO_3$, in which the concentration of the Sr ion varies within $0.1 < x < 0.2$. Such manganites in the above limits of changing x and the temperature range T= 140-400 K experience a series of structural and magnetic phase transitions.

On the basis of neutron and X-ray diffraction data it is considered that in the samples with $x < 0.15$ both local distortion of separate octahedra and cooperative JT distortions, and in the samples with $0.15< x <0.20$ only local JT deformations of the MnO_6 octahedra are present [3, 4].

In addition to methods already used in studying the dynamics of the electron-lattice and spin-lattice interactions, the acoustic research methods, particularly at high frequencies (f = 770 MHz) when the acoustic wave length can be comparable with the size of the structural or magnetic inhomogeneities, are rather attractive. The study of acoustic characteristics (velocities of separate modes depending on the concentration of the dopant centers along with the magnetic and resistive measurements) allows one to get information about the features of structural and magnetic phases, transition between them, mechanisms of electron-lattice and spin-lattice interactions.

EXPERIMENTAL RESULTS

$La_{1-x}Sr_xMnO_3$ (x = 0.125) Sample

As a result of our complex experimental studies, the temperature changes of the elastic modulus $(C_{11}-C_{12})/2$ and C_{44} for transverse and C_{11} for longitudinal acoustic waves were observed in the

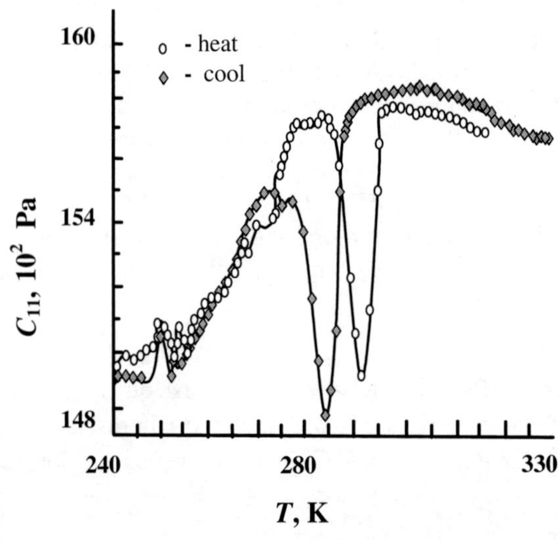

Fig. 1: Temperature Dependence of the Coefficients of Elasticity C_{11} for the Longitudinal Acoustic Waves for $La_{0.875}Sr_{0.125}MnO_3$

$La_{0.875}Sr_{0.125}MnO_3$ sample in the temperature interval of T = 290-270 K and T= 180-150K. Upon the propagation of the transverse acoustic wave the shift elastic modulus $(C_{11}-C_{12})/2$ strongly decreased starting from T= 340K and reached a sharp minimum close to T = 280K. At T < 280 K $(C_{11}-C_{12})$ / 2 increased. As the temperature decreased further, the $(C_{11}-C_{12})/2$ coefficient drastically decreased beginning with the temperature T = 200K. On the contrary, with the sample temperature decrease, starting from T = 340K the elastic modulus C_{11} for the longitudinal acoustic waves increased, and close to T_s = 280K there was a minimum in the C_{11}, below T = 280K the considerable reduction of C_{11} remained (Fig. 1).

According to the neutron diffraction results, the T_{s1} value was attributed to the structural phase transition from the pseudocubic phase O* to the orthorhombic phase O'.

The temperature hysteresis we observed for the elastic moduli close to T_{s1}=280K (Fig. 1) allows us to attribute them to the structural phase transition of the first kind.

$L_{1-x}Sr_xMnO_3$ (*x* = 0.15) Sample

The main result of our studies of the $L_{0.85}Sr_{0.15}MnO_3$ sample was the observation of sharp changes of acoustic, magnetic and electric parameters in the temperature intervals T = 320-340K, T = 230-270K and T = 170-220K. In the first case, there was a sharp decrease in the elastic modulus C_{44} close to the temperature T_{s1} = 320K (Fig.2), which was recorded in the

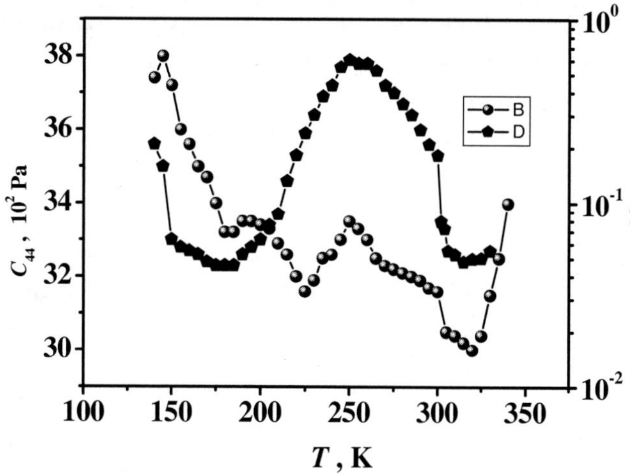

Fig. 2: Temperature Dependence of C_{44} (B) and *ñ* (D) for La $_{0.85}$ Sr $_{0.15}$ MnO.

value of the electric resistance as well (Fig. 2). As the temperature decreased in the interval T = 320-280K, the C_{44} and p values increased. The peak decrease in the C_{44} value was at T = 240K. The increase in the C_{44} value was changed on its sharp decrease close to T = 180K and slight decrease at T = 140K. Similar temperature changes of the C_{11} values were less significant. These sharp changes of the acoustic parameters C_{44} correspond to the temperature changes of the electric resistance p and magnetization M, which agree well with data obtained for the samples with the given composition [5].

$L_{1-x}Sr_xMnO_3$ (x = 0.175) Sample

We also studied the propagation of the transverse and longitudinal ultrasound waves in the $L_{0.875}Sr_{0.175}MnO_3$ sample. As a result, we observed sharp changes of the parameters of the shift elastic modulus C_{44} close to T = 310K, T = 285K and T = 200K. The change of the parameter of the shift elastic modulus C_{44} in the vicinity of T = 285K was of non-hysteresis character and corresponded to the phase transition from the paramagnetic to the ferromagnetic state. This was supported by the magnetic measurements we performed on the same sample [5]. The changes of the C_{44} and C_{11} values close to T = 300 – 310K and 200 – 210K have the hysteresis nature. This allowed us to attribute them to the structural phase transition, probably from the rhombohedral to the orthorhombic state in the paramagnetic and ferromagnetic phases. In the paramagnetic phase T = 310K the incomplete transformation of the rhombohedral structure to the orthorhombic one occurs and the final establishment of the orthorhombic structure occurs at the second structural transition T = 200K already in the ferromagnetic phase.

The hysteresis nature of the changes of elastic modulus C_{44} close to T = 310K can be attributed to the possible phase transition of the first kind, which was not observed earlier.

When the magnetic field of 1T was applied, the phase transition at T = 310K for all elastic modes shifted to the lower temperatures and the magnetic one at T = 285K shifted to higher temperatures.

DISCUSSION OF RESULTS

These experimental data allow us to make some assumptions about the character of the JT distortions in slightly doped lanthanum-strontium manganites and their effect on the transport processes.

First, it follows from the observed features of the acoustic anomalies that upon decrease in the temperature, the MnO_6 octahedra are deformed and the local short-range ordering of the octahedral occurs. These distortions do not change the general symmetry of the crystals and therefore they are very poorly observable by other methods. However, such local JT distortions are responsible for the increase in the electric resistance in the given temperature range. In all ranges of undoped and slightly doped manganites ($0 < x < 0.20$) the distortions of octahedra and their local ordering occur in the paramagnetic phase. Moreover, the temperature of the beginning of the JT distortions as well as their intensity decrease with increase in the Sã ion concentrations with temperature 1050K (at x = 0). There are two scenarios in the character of the JT distortions at the further decrease in the temperature depending on the x value. At *x* d" 0.15, the ordering of the distorted MnO_6 octahedra is cooperative over the whole volume of the crystal. This process is accompanied by the structural phase transition of the first kind. Its temperature is also lowered from 730K ($x = 0$) to 280K ($x = 0.15$) with increase in doping.

For the samples with $x = 0.15$ no cooperative ordering of the distorted MnO_6 octahedra occurs due to the increase in T_s but the local JT distortions are immediately suppressed with increase in the spontaneous magnetization. Such suppression of the JT distortions is also accompanied by the phase transition of the first kind of the "order-disorder" type. For all sample $0.10 \leq x \leq 0.20$ the increase in the JT distortions leads to the increase in the electric resistance and softening of the elastic modulus.

Both types of the JT distortion (local and cooperative) decrease with increase in the magnetic ordering and the mobility of charge carriers. Since the temperature of the ferromagnetic ordering T_s increases with increase in x from 130K ($x = 0.1$) to 240K ($x = 0.15$), for the samples with $x \leq 0.15$ this process occurs simultaneously with the structural phase transition of the first kind.

As to the energy, the suppression of the local JT distortions with increase in the magnetic ordering requires much less expenses than the suppression of the cooperative ordering. This is confirmed by the character of the temperature anomalies in heat capacity. The suppression of the cooperative JT distortions occurs gradually in a large temperature interval. This determines the low CMR value for $x \leq 0.15$. On the contrary, the suppression of the local JT distortions occurs in a narrow temperature interval and the applied magnetic field comparable and even exceeding the spontaneous magnetization value leads to the sharp suppression of the local JT distortions. This process is typical for the samples with $x \geq 0.15$, when the significant reduction of the resistance and decay of the elastic waves occurs somewhat higher that T_c. These processes are enhanced by the applied magnetic field. This suggestion is supported by the similarities of the temperature intervals of anomalies in heat capacity, elastic parameter and CMR of the samples with $x = 0.15$ and $x = 0.175$.

The work was supported by the Russian Foundation for Basic Research (project no. 11-02-00093a).

REFERENCES

[1] Kugel, K.I. and Khomskii D.I. Jahn-Teller effect magnetism transition-metal compounds: *Usp. Fiz.* Nauk. . 36, 621-664. (1982).

[2] Meicher, L. Physical Acoustics, Academic Press, New-York (1976).

[3] Geek, P., Wochner, D., Bruns, B., Buchner, U., Gebhardt, S., Kiele, P., Reutler and Revcoievsch A., Rearrangement of the orbital-ordered state at the metal – insulator transition of $La_{7/8}Sr_{1/8}MnO_3$m *Phys. Rew. B. Condens.* Matter. 69, 104413-104422 (2004).

[4] Uhlenbruck, S. Interplay between Charge Order Magnetism, and Structure in $La_{0.875}Sr_{0.125}MnO_3$, *Phys. Rev. Lett.* 82, 185-188 (1999).

[5] Bogdanova, Kh.G., Bulatov, A.R., Golenishchev-Kutuzov V.A., Elokhina, L.V., Kapralov A.V., Korolev A.V., Neifel´d E.A., Shakirzyanov, M.M. Acoustic and Magnetic Properties of $La_{0.825}Sr_{0.175}MnO_3$ Lanthanum Manganites, *Physics of the Solid State*. Ò.**45**. 284-289 (2003).

HEXAGONAL SYNTHESIS OF LaF_3: Nd^{3+}, Sm^{3+} NANO CRYSTALS AND STUDIES OF NLO PROPERTIES

S.G. GAURKHEDE, M.M. KHANDPEKAR[1], S.P. PATI[2] AND A.T. SINGH[3]

Department of Physics, Bhavans College of ASC, Andheri (W) Mumbai
[1]Material Research Lab, Department of Physics, Birla College, Kalyan
[2] National Institute of Science and Technology,Palur Hills,Behrampur, Odisha
[3]Department of Physics K M Agarwal College of ASC, Kalyan
E-mail: s.gaurkhede@gmail.com

ABSTRACT

Hexagonal shaped LaF_3 nanocrystals (NC) doped by $Nd^{3+}$and Sm^{3+} ions are synthesized using a domestic microwave oven. The powder XRD study confirms that crystalline size of the particle is close about 20 nm (JCPDS standard card (32-0483) of pure hexagonal LaF_3 crystals).The Transmission Electron Microscope (TEM) analysis indicates the size of the primary and secondary particle is in the range between 15nm-20nm.The presence of fundamental groups has been verified by the FTIR spectra. The synthesized nanocrystals are also studied by Non Linear Optical (NLO) properties. It has been found that Second Harmonic Generation (SHG) efficiency of LaF_3: Nd^{3+}, Sm^{3+} containing rare earth elements is less than pure Potassium Dihydroxyl Phosphate (KDP) crystals.

INTRODUCTION

In recent years, the fields of luminescence and display materials have undergone a revival of sorts with the evolution to nano-sized luminescent particles, driven primarily by an ever-

---

V. Rajendran, K. Saminathan, P. Paramasivam and K.E. Geckeler (eds.)
Nano Materials Synthesis and Characterisation, pp. 47-52 (2012).

increasing awareness of the unique physical and optical properties that differ from those of identical bulk materials as the size of the particle reduced to the nanometer region [1]. Studies on the luminescent properties of lanthanide-doped nanoparticles have attracted a great deal of interest since they are considered as potentially useful phosphors in lamps and display devices[2],components in optical telecommunication[3],active materials in lasers[4], new optoelectronics devices[5], up converters [6], magnetic resonance imaging (MRI) [7], biological fluorescent labels [8]. LaF_3 nanocrystals are widely used in lubricants, additives in steel and metal alloy, electrode materials [9] chemical sensors and biosensors [10]. LaF_3 possesses low phonon energy, adequate thermal and environmental stability [11], and hence is as excellent host matrix [12] for investigating luminescence properties. Nanoparticles of LaF_3, doped with lanthanide ions, are studied in the past for their luminescence properties [13].In past several investigations has been carried to study the optical properties of LaF_3:Nd^{3+} [14] for their possible applications in optoelectronics devices. This paper presents a study of LaF_3:Nd^{3+}, Sm^{3+} nanoparticles synthesized in laboratory with a simple microwave radiation method. The study shows that these nanoparticles have hexagonal shape and exhibit luminescence properties.

EXPERIMENTAL

Synthesis of Nanocrystals

LaF_3: Nd^{3+} and Sm^{3+} nanocrystals were synthesized by an aqueous route using microwave oven for heating in low power range. The method is simple and cost effective. Water soluble $LaCl_3$+$NdCl_3$+ $SmCl_3$ (1 unit) and NH_4F (3 units) are mixed to obtain a solution in 1:3 molar proportion [15]. We prepared a 10 ml homogenous mixture (in de ionized water) in a 100 ml beaker using 0.064 mol of $LaCl_3$+$NdCl_3$+$SmCl_3$, allowed a 10 ml solution of 0.576 mol NH_4F to drip into this solution uniformly through a funnel attached with a stopper to facilitate control of dripping. The whole set up can be placed inside a conventional microwave oven during the reaction. The microwave oven was maintained on low power range (in on-off mode set at 30sec) for 30 minutes. The low power range helps to avoid spill off of the solution. A white ultrafine crystalline precipitate identified as doped LaF_3 nanocrystals appears almost instantly at the bottom of the beaker. The precipitate was washed several times with de-ionized water, absolute methanol and acetone, and then dried it in the microwave oven to about 15 minutes. The dried sample was then stored in sealed tubes for further characterization.

LaF_3: Nd^{3+} and Sm^{3+} nanocrystals were also prepared by using methanol in place of deionized water with the method given above.

CHARACTERIZATION

Powder x-ray diffraction (XRD) measurements have been performed using a PANALYTICAL X'PERT PROMPD diffractometer model. Transmission electron microscope (TEM) analysis has been carried out for different magnification by PHILIPS (CM 200). Fundamental groups have been verified by the FTIR spectra using Spectrum one: FT-IR Spectrometer. The fluorescence spectrum was measured on LS 45 luminescence spectrometer (Perkin Elmer Corp).NLO studies for the measurements of SHG efficiency, is obtained through the crystalline powder sample by using Kurtz and Perry technique.

RESULT AND DISCUSSION

The XRD results are shown in Fig.1 which indicates that LaF_3: $Nd^{3+}Sm^{3+}$ nanoparticles are well crystallized, and the patterns are in good agreement with hexagonal structure (Space group: P_3cl (165), Cell=0.7187×0.7187×0.735 nm^3, $\alpha = \alpha = 90°$, $\gamma = 120°$) known for bulk LaF_3 (JCPDS card No.32-0483) [16].The calculated cell parameters $a = b = 0.7126$ nm and $c = 0.7255$ nm for the LaF_3: $Nd^{3+}Sm^{3+}$ nanoparticles, are smaller than those of LaF_3 nanoparticles ($a = b = 0.7187$ nm and $c = 0.735$ nm.

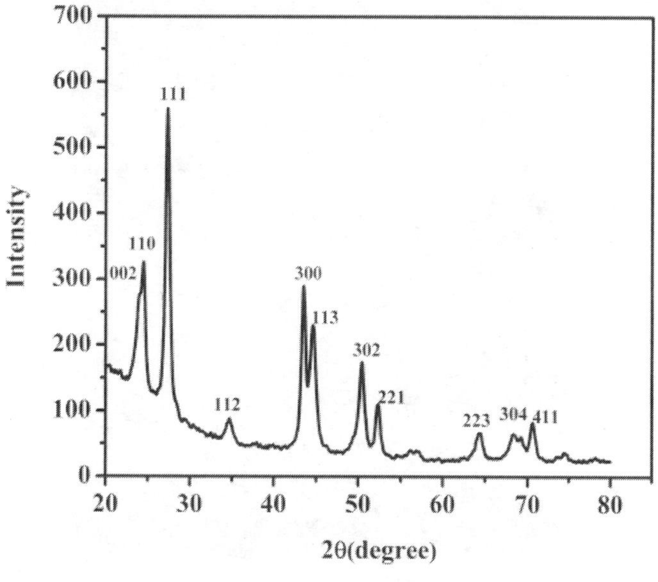

Fig.1: XRD Pattern of LaF_3: Nd^{3+}, Sm^{3+} Nanocrystals

The decrease in the lattice parameters of LaF_3: Nd^{3+}, Sm^{3+} nanoparticles can be attributed to the smaller radius of Nd^{3+} ion (0.99 nm) and Sm^{3+} ion (0.96 nm) as compared with that of La^{3+} ion (0.106 nm) [17]. This indicates that Nd^{3+} ions and Sm^{3+} ions are doped into the LaF_3 lattice and occupied the site of La^{3+} ions, with the formation of a LaF_3: Nd^{3+}, Sm^{3+} solid solution. The broadening of diffraction peaks for LaF_3: Nd^{3+}, Sm^{3+} nanoparticles is also shown by Figure.1, which reveals the nanocrystallites nature of the samples. According to Scherrer equation, $D = 0.90\lambda/\beta \mathrm{cosr}\ \theta$, where D is the average crystallite size, λ is the x-ray wavelength (0.15405 nm); θ and β being the diffraction angle and full width at half maximum of an observed peak, respectively. After subtraction of the equipment broadening, the full width at half maximum (FWHM) of the strongest peak (111) at $2\theta = 27.9°$ helps to calculate the average crystalline size of LaF_3: Nd^{3+}, Sm^{3+} nanoparticles as 15-20 nm.

The transmission electron microscopy (TEM) image in Fig.2 shows that the particles are well separated from each other. The nanocrystals have a hexagonal shape and a particle size of 6–20 nm. When these nanocrystals are incorporated in the polymer matrix, these particles are so small that the Rayleigh scattering can be nearly neglected. The selected area electron diffraction (SAED) pattern in (Fig. 2 inset) shows three strong diffraction rings corresponding to the (002), (111) and (300) reflections, which is in agreement with the hexagonal LaF_3 structure [18], suggesting that the original structure of LaF_3 is retained even after the modification. The results, particle size from TEM measurements in agreement with those obtained from XRD studies.

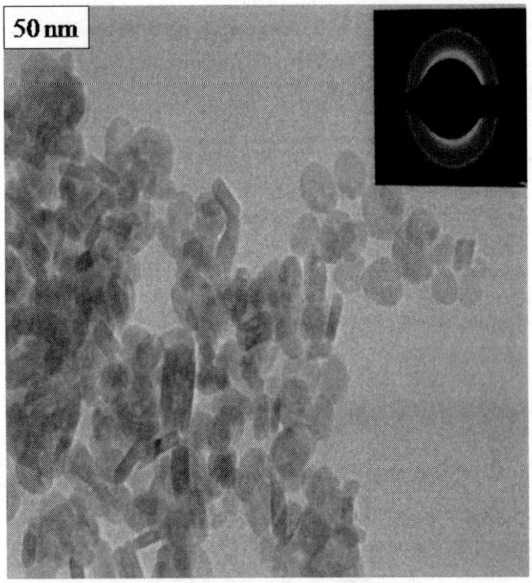

Fig. 2: TEM Image of Nd^{3+}, Sm^{3+} doped LaF_3 Nanocrystals

Figure 3 has shown FTIR spectrum of the LaF_3: Nd^{3+}, Sm^{3+} nanocrystals. The characteristic absorption peaks have been observed in the range of 4000 cm^{-1} to 500 cm^{-1}. The broad absorption band at about 3434 cm^{-1} can be attributed to 5_{as} (O-H) stretching and bending vibrations. The peaks at 2925 cm^{-1} and 2853 cm^{-1} can be attributed to ν_{as}(C-H) group of the long alkyl chain [19]. The characteristic IR peaks located at 1632 cm^{-1} could be assigned to δ (H_2O) bending vibrations from the residual absorbed water and 1439 cm^{-1} can be assigned to the asymmetric (ν_{as}) and symmetric (ν_{as}) bending vibrations of δ (O-H) group from methanol. Other absorptions are obtained due to use of methanol and acetone while preparing the sample.

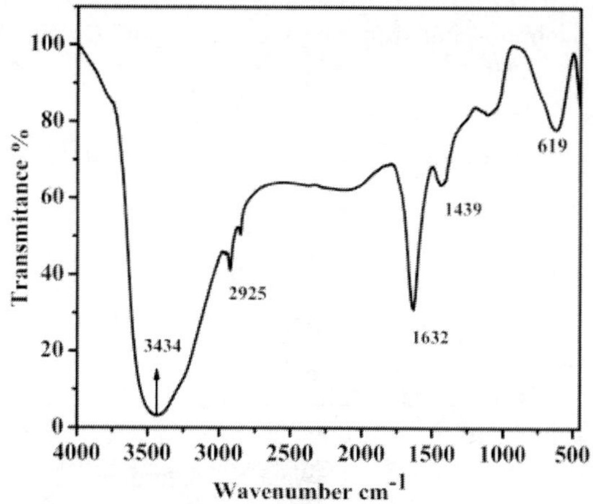

Fig. 3: FTIR Spectrum of the of Nd^{3+}, Sm^{3+} Doped LaF_3 Nanocrystals

It is observed that the measured relative SHG efficiency of LaF_3 doped Nd^{3+}, Sm^{3+} in deionized water with KDP crystal is 0.281 times and that of LaF_3 doped Nd^{3+}, Sm^{3+} in methanol with KDP crystal is 0.513 times by using Kurtz and Perry technique [20].

CONCLUSIONS

Nanocrystals of LaF_3: Nd^{3+}, Sm^{3+} have been rapidly synthesized by chemical route using domestic microwave oven at low power range. Nanocrystals of hexagonal geometry with particle size varying from 15 nm-20 nm have been located by TEM analysis which is in agreement with the average crystalline size obtained from XRD studies. The FTIR spectrum has been used for identification of fundamental groups present in the material. It has been found that relative SHG efficiency of LaF_3:Ln^{3+} (Ln^{3+}: Nd^{3+}, Sm^{3+}) containing rare earth elements are less than pure KDP crystals.

REFERENCES

[1] Tissue B.M., Synthesis and luminescence of lanthanide ions in nanoscale insulating hosts, *Chemistry of Materials*, 10, 2837–2845, 1998.

[2] Song H., Chen B., Sun B., Zhang J., Lu S., Ultraviolet light-induced spectral change in cubic nano-crystalline Y_2O_3:Eu^{3+}, *Chemical Physics Letters*, 372, 3-4, 368–372, 2003.

[3] Nishi M., Tanabe S., Inoue M., Takahashi M., Fujita K., Hirao K., Optical telecommunication- band fluorescence properties of Er^{3+}-doped YAG nanocrystals synthesized by glycothermal method, *Optical Materials*, 27, 4, 655–662, 2005.

[4] Lezhnina M.M., Kaetker H., Kynast U.H., Synthesis and optical characterization of rare earth nanofluorides, *Optical Materials*, 30, 2, 264–272, 2007.

[5] Pan Y.X., Su Q., Xu H.F., Chen T.H., Yang C.L., Synthesis and red luminescence of Pr^{3+}doped $CaTiO_3$ nanophosphor from polymer precursor, *J. Solid State Chem.*,174, 1, 69–73, 2003.

[6] Jia P.Y., Lin J., Yu M., Sol-gel deposition and luminescence properties of $LiYF_4$:Tb^{3+} thin films, *Journal of Luminescence*, 122-123, 1-2, 134–136, 2007.

[7] Evanics F., Diamente P.R., Van Veggel F. C. J. M., Stanisz G.J.,. Prosser R.S., Water-soluble GdF_3 and GdF_3/LaF_3 nanoparticles-physical characterization and NMR relaxation properties, *Chemistry of Materials*, 18, 10, 2499– 2505, 2006.

[8] Wang F., Zhang Y., Fan X., Wang M., One-pot synthesis of chitosan/LaF_3:Eu^{3+} nanocrystals for bio-applications, *Nanotechnology*, 17, 5, 1527–1532, 2006.

[9] Bralic M., Radic N., Brinic S., Generalic E., Fluoride electrode with LaF_3-membrane and simple disjoining solid state internal contact, *Talanta*, 55, 3, 581–586, 2001.

[10] Miura N., Hisamoto J., Yamazoe N., Kuwata S., Salardenne J., Solid-state oxygen sensor using sputtered LaF_3 film, *Sensors and Actuators*, 16, 4, 301–310, 1989.

[11] Kudryavtseva O.V., Garashina L.S., Rivkina K.K., Sobolev B.P., Solubility of LnF_3 in lanthanum fluoride, *Soviet Physics-Crystallography*, 18, 531, 1974.

[12] Zheng H.R., Wang X.J., Dejneka M.J., Yen W.M., Meltzer R.S., Up-converted emission in Pr^{3+} doped fluoride nanocrystals-based oxyfluoride glass ceramics, *Journal of Luminescence*, 108, 14,395–399, 2004.

[13] Dejneka M.J., The luminescence and structure of novel transparent oxyfluoride glass-ceramics, *Journal of Non-Crystalline Solids*, 239, 1–3,149–155, 1998.

[14] Stouwdam J.W., Hebbink G.A., Huskens J., Van Veggel F. C.J.M., Lanthanide-doped nanoparticles with excellent luminescent properties in organic media, *Chemistry of Materials*, 15, 24, 4604–4616, 2003.

[15] Meng J.X., Zhang M.F., Liu Y.L., Hydrothermal preparation and luminescence of LaF_3:Eu^{3+} nanoparticles, *Spectrochimica Acta A*, 66, 1, 81–85, 2007.

[16] Carnall W.T., Goodman G.L., Rajnak K., Rana R.S., A systematic analysis of the spectra of the lanthanides doped into single crystal LaF_3, *The Journal of Chemical Physics*, 90, 7, 3443–3457, 1989.

[17] Liu Y.F., Chen W., Wang S., Joly A.G., Westcott S., Woo B.K., X-ray luminescence of LaF_3:Tb^{3+} and LaF_3:Ce^{3+}, Tb^{3+}water-soluble nanoparticles, *Journal of Applied Physics*, 103, 6, Article ID 063105, 2008.

[18] Wang X., Zhuang J., Peng Q., Li Y., Hydrothermal synthesis of rare-earth fluoride nanocrystals, *Inorganic Chemistry*, 45, 17, 6661–6665, 2006.

[19] Guo H., Zhang T., Qiao Y., Zhao L., Li Z., Ionic liquid based approach to monodisperse luminescent LaF_3:Ce,Tb nanodiskettes: synthesis, structural and photoluminescent properties, *Journal of Nanoscience and Nanotechnology*, 10, 3, 1913–1919, 2010.

[20] Kurtz S.K., Perry T.T., A powder technique for the evaluation of nonlinear optical materials, *Journal of Applied Physics*, 39, 8, 3798–3813, 1968.

SURFACE AREA MEASUREMENT OF CARBON NANOMATERIALS OBTAINED FROM CASTOR OIL

SANDESH JAYBHAYE AND VINOD LOHAKANE

Nanotechnology Research Lab., Department of Chemistry, Birla College, Kalyan
E-mail: jaysandesh@gmail.com

ABSTRACT

Carbon Nanomaterial (CNM's) synthesized from castor oil by direct pyrolysis at 750°C in an inert atmosphere. CNMs was then characterized by (XRD) X-ray diffraction and SEM (Scanning Electron Microscope).The specific surface area measurement of CNMs was done by using adsorption of methylene blue dye (MB).In this method a fixed concentration of aqueous methylene blue solution was allowed to adsorb on fixed quantity of carbon nanomaterial.The amount of adsorbed methylene blue dye was determined by UV spectrophotometer. It was found that the concentration of adsorbed methylene blue corresponds to the surface area of test sample.The surface area of CNM was found to be 32.0 m^2/g. The obtained data was correlated with that of BET surface area measurement and found to be comparable.

INTRODUCTION

A carbon nanotubes discovery in 1991 [1] was a microscopic wonder, due to the porosity with nanometer size and large surface area. The synthesis of carbon nanomaterials from natural precursors receiving the interest because of its unique characteristics of utilization of waste material in different applications via synthesis of different CNM's. from it.Most of the researchers are using petroleum products for the preparation of carbon nano materials;sandesh Jaybhaye et al. [2] at NTRC is able to produce these materials from plant derived precursors. There are various physical and chemical properties of CNM has been studied for different applications like hydrogen storage [3] super capacitor [4] solar cell applications [5] micro

V. Rajendran, K. Saminathan, P. Paramasivam and K.E. Geckeler (eds.)
Nano Materials Synthesis and Characterisation, pp. 53-58 (2012).

wave absorption for lithium ion battery [6] etc. has been carried out.Specific surface area is one of the important property of carbon nanomaterials that can be related to their physical or chemical behavior. The most commonly used methods to evaluate the specific surface area of carbon nanomaterials are based on adsorption of nitrogen [7], water vapour [8], ethylene glycol mono ethyl ether [9] or colour dye [10]. A very simple adsorption of methylene blue hereafter will be called as MB is used in lab to determine the specific surface area of the carbon nanomaterials which is synthesized in laboratory using acastor seed oil. In this method MB is used to adsorb on CNM and is then evaluated by using a UV-Vis spectrophotometer.

EXPERIMENTAL

Preparation of Catalyst

The catalyst is necessary for the synthesis of Carbon nanomaterials. The diameter of the catalyst particals should be in nano range ~ 40 to 60 nm. Particle of such size can be produced by the urea decomposition method[11].

A mixture of Nickel nitrate and urea; 5g and 10geach respectively was taken in a beaker and kept on heater with continuous stirring. A homogeneous liquid mixture was formed then this mixture was kept in muffle furnace in presence of air at 300°C till completely dry.As a result, urea gets decomposed and burned off, nickel wasoxidized to form nickel oxides which wasleft in the form of residue.

Then this residue of Nickel oxide was collected and reduced in Hydrogen atmosphere at 650°C for 2 hours. This gave Nickel catalyst of nano size.

Synthesis of CNM

The CNM was synthesized from castor oil using a Nickel catalyst by chemical vapour deposition method (CVD).In this CVD synthesis method two electric furnaces A and B (vaporizing Furnace A and Pyrolising Furnace B) and one meter long quartz tube kept inside both the furnaces was used. The schematic representation of arrangement of all the apparatus was as represented in fig. 1. Weighed 4 g of castor oil in quartz boat (C) and 200 mg of Nickel catalyst in other quartz boat (D) and kept both the boats in furnace A and B respectively.

The Hydrogen gas was allowed to pass through a quartz tube (Q) for five minutes with constant flow rate so as to remove the oxygen from tube. The temperature of pyrolysis furnace (B) was set to 750°C.Once the furnace B reaches a desired temperature. The furnace A was turned on and the temperature was set to the boiling point of castor oil (approx. 350°C). The heating of furnace A was continued till all oil gets vaporized. The heating of furnace B was continued for 1h. The furnace B was allowed to cool at room temperature and the material was collected from boat D.

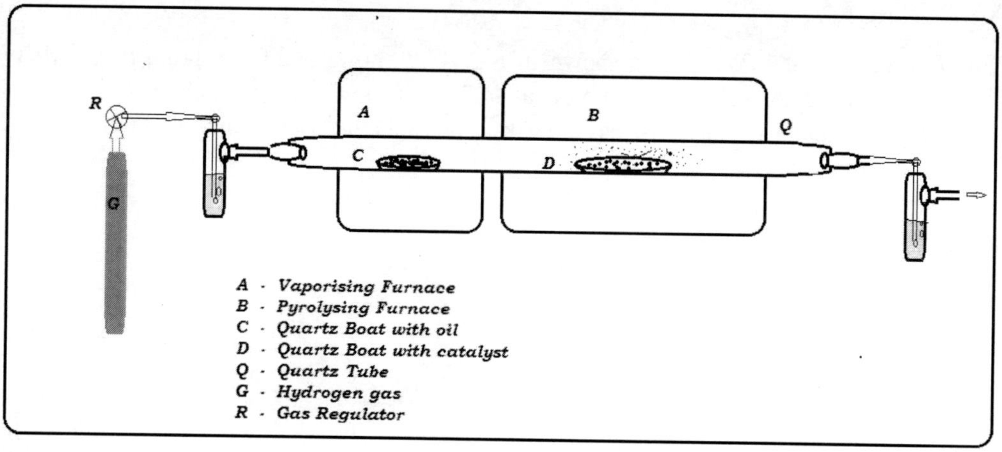

Fig. 1: Schematic Representation of CVD Furnace

Purification of CNM

The obtained CNM was treated with mixture of HNO_3 and HCl so as to dissolve the traces of metal impurities and amorphous carbon. The purified CNM was then dried in furnace at 200°C for 1h.

RESULT AND DISCUSSION

The purified CNM was characterized by using a SEM and powder XRD.

SEM Study

The SEM image shows the CNM obtained from castor oil gives Multi walled Carbon Nanotubes (MWCNT) having diameter 50 nm shown in Fig. 2. It also confirms that purification with acid mixture removes metal particles as well as amorphous carbon.

Fig. 2: SEM Image of CNM Obtained from Castor Oil

XRD Study

X ray diffraction of MWCNTs is shown in Fig. 4. It shows a diffract gram of CNM; two major peaks, one at 26.46° (002) and other at 44.38° (101) which are the characteristics peaks of graphitic carbon [12] The sharpness of peaks represents the crystallinity of graphitic carbon.

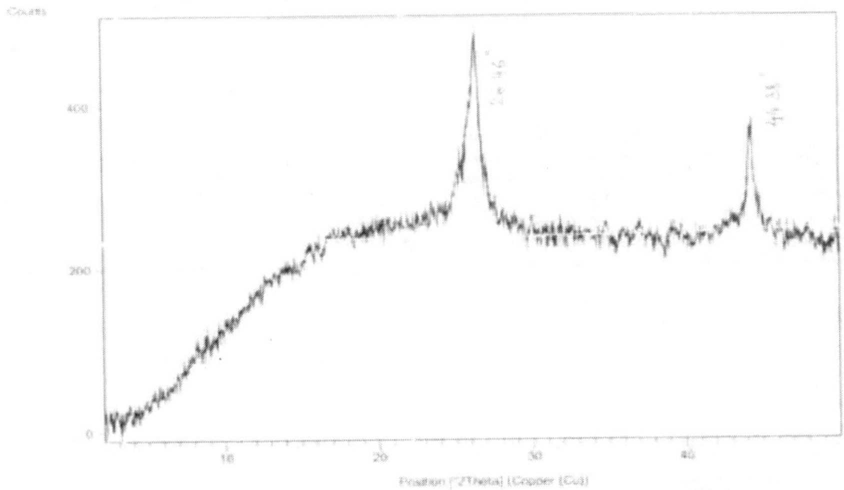

Fig.4: XRD Diffractogram of CNM Obtained from Castor Oil

Surface Area Measurement

CNM prepared by pyrolysis of castor oil is directly used for this experiment. Surface area of this CNM is measured using methylene blue adsorption method. Methyleneblue solution was prepared in distilled water and the concentration of methylene blue solution was analyzed by measuring its absorbance at 662 nm on UV/Vis spectrophotometer. This wavelength corresponds to the maximum absorption peak of the Methylene blue. First of all a calibration curve of O.D. of methylene blue against its standard concentration (1 to 5 ppm) is plotted and is shown in Fig. 5. The regression factor is found to be 0.99

A known weight of MWCNTs (0.050 g) was taken in 50ml conical flask containing 5 ppm solution of methylene blue. The mixture was stirred until all the CNM gets submerged into the solution. This mixture was kept at room temperature and shaken periodically for 10 min and allowed to settle the small particals. After 10 min, the methylene blue uptake onto carbon was calculated from the difference between the methylene blue concentration before and after adsorption onto the carbon. This difference corresponds directly to the active surface area of carbon. The Specific surface area measured by using this technique was found to be 32.0M^2/g

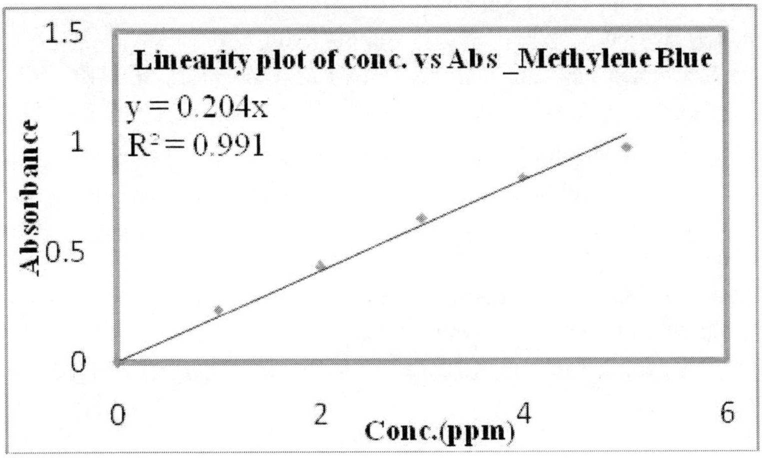

Fig. 5: Calibration Curve of Absorbance Vs Conc.

CONCLUSIONS

Adsorption of methylene blue allows the determination of the specific surface area of Carbon Nanomaterial. This work has shown that the method was simple and requires less elaborate apparatus and time than other methods. Using this technique, we have shown that the CNM obtained from castor oil have different specific surface areas. Consequently, this finding makes it possible to use specific surface area measurements directly as a characteristic in quality control, much like other mechanical properties of CNMs. This method was found to be comparable to that of existing BET surface area measurement technique.

Sample Details	*Surface Area m^2/g*	
	BY BET	*Methylene blue*
MWCNTs	30.69	32.0

REFERENCES

[1] Ijimas, *Nature* (London) 1991; 354 (56).

[2] Vilas Khairnar Sandesh Jaybhaye, Chi-Chang Hu, Rakesh Afre, T. Soga and Maheshwar Sharon, *Carbon Letters*, Vol. 9(3), (2008) 188-194.

[3] C. Dillon, K.M. Jones, T.A. Bekkedahl, C.H. Kiang, D.S. Bethune, M.J. Heben, (1997), *Nature* 386, 377-379.

[4] Vilas Khairnar, Maheshwar Sharon, Sandesh Jaybhaye and Michael Neumann, SRINMC, Vol. 36(2), (2006)171-173.

[5] Sandesh V. Jaybhaye, Maheshwar Sharon, Dattatray E. Kshirsagar Carbon materials for energy application 2005; 171-178.

[6] Sunil Bhardwaj, Maheshwar Sharon, T. Ishihara, Sandesh Jaybhaye, Rakesh Afre, T. Soga and Madhuri Sharon, *Carbon Letters,* Vol. 8(4), 2007, 1-7.

[7] Brunauer S, Emmett P.H. and Teller E.J. *Am.Chem. Soc,*. 60, 309, 1938.

[8] Quirk J.P., *Soil sci.* 8, 423, 1955.

[9] Dechnik I. and Stawinski, *J. Soil Sci.*, 3, 15. 1970.

[10] Maheshwar Sharon, Sandesh Jaybhaye, D. Sathiyamoorthy and Sunil Bhardwaj, Proceedings of the International Conference on Molecules to Materials at Longowal, Punjab, March 3-4, 2006, pp. 50-52.

[11] A.K. Chattarjee, Maheshwar Sharon, Ranjan Bannerjee and Michael Neumann, *Spallart - Electrochimical Acta* 48 2003; 3439-3446.

[12] X.B. Wu, P. Chen, J, Lin and K.L. Tan, *International Journal of Hydrogen Energy* 25, 2000, 261-265.

SYNTHESIS, CHARACTERIZATION OF METAL NANOPARTICALS AND ITS USE IN CNT GROWTH

SANDESH JAYBHAYE AND MAHESHWAR SHARON[1]

Nanotechnology Research Center, Birla College, Kalyan, Maharashtra
[1] NSNRC, SICES College for Arts, Science & Commerce, Ambernath, Maharashtra
E-mail: jaysandesh@gmail.com

ABSTRACT

Metal nano particles have been prepared by a wide variety of techniques. We have synthesized catalyst powders of nano size metal particles by hydro thermally. Nickel (Ni) nitrates or Iron (Fe) nitrate or Cobalt (Co) nitrates with the addition of urea oxidized calcinated and reduced to obtain nano size particles. The effect of proportion of urea and metal nitrate on the crystal structure and morphology of the nano metals have been investigated. Nano sized metal particles with the uniform spherical morphology were obtained by the urea decomposition method, by treating metal nitrate with urea of weight ratio 1:3. These catalysts have been characterized by X-Ray diffraction (XRD) and Scanning electron microscope (SEM). The catalyst was then used to grow carbon nano tubes (CNM) by chemical vapor deposition (CVD) method.

INTRODUCTION

The advent of microscopic techniques with sub nanometer resolution that nanomaterials research has becomes increasingly popular. The growing interest and importance stem from the various unique applications [1–5] that nanoparticles can be used for, such as catalysts [6], electronic [7] and optical components [8]. The properties of nanoparticles gradually change from those of bulk material to molecular with decreasing particle size. The greatest obstacle for the

V. Rajendran, K. Saminathan, P. Paramasivam and K.E. Geckeler (eds.)
Nano Materials Synthesis and Characterisation, pp. 59-64 (2012).

implementation of nano-devices today is the lack of techniques for mass production and assembly of nanoparticles. Neither conventional methods of controlling bulk materials nor common chemical synthetic methods are suitable for nanomaterials. These require a delicate balance between the physical forces, usually defined by their surface properties that determine their behavior. The two main approaches for building nano-devices are the "top down" and "bottom up". In the former, nano-devices are built by miniaturization of conventional techniques, such as lithography. Complete control over the size distribution requires size selection techniques, such as size-selective precipitation [9, 10]. In order to synthesize well-defined mono disperse nanoparticles, control over the nucleation and growth process is essential. There are several different methods for making nanoparticles.

Currently, the emphasis of synthesis has been on the preparation of nano particles, with well-defined size, shape and surface properties. Metal nano particles have been prepared by a wide variety of techniques such as laser ablation, nucleation from vapor, thermal decomposition of organometallic compounds, sonolysis, pulse radiolysis, Electrochemical reduction and chemical reduction of the corresponding metal salts.

In this paper we discussed the preparation of nano size metal catalyst powders by urea decomposition method from inorganic compounds. The effect of proportion of urea and metal nitrate on the crystal structure and morphology of the nanometals has been investigated. These prepared materials have been characterized and used for the preparation of different types of carbon nano materials.

EXPERIMENTAL

Preparation of Nano Metals

Nano metal particles have been synthesized by urea decomposition method. In which different metal nitrates compound have been taken in alumina crucible. It is then mixed with Urea $(NH_2)_2CO)$ in 1:3 by weight proportion and little quantity of water is added to obtain homogeneous slurry. The slurry is then decomposed at 300°C in a muffle furnace for about 3h followed by oxidation at 600°C for 1h. Then it is hydrogenated in horizontal furnace at 850°C for 3h to get nano sized metal particles. The following chemical reaction has been takes place during oxidation and reduction of metal (M = Ni or Fe or Co) nitrates.

Characterization

Nano metal particles obtained by urea decomposition method have been characterized by SEM and XRD. The SEM observation was conducted with FEI quanta 200, and X-RAY was conducted on Philips Panalytical Xpert.

Preparation of Carbon Nano Materials

These nano metal particles were used as a catalyst for the preparation of carbon nano materials from acetylene, as a precursor, by chemical vapour deposition method (CVD). In this, acetylene gas is purged on different catalyst powder (Ni, Fe and Co) at 700°C–800°C in hydrogen atmosphere. Springs like spiral carbon nano fibers (SCNF) were grown on the Ni, carbon nano tubes (CNTs) on Fe particles and Carbon nano fibers (CNFs) on Co metal particles surface. A detailed procedure is described in [11]. These as grown CNMs characterized by SEM and Raman spectra.

RESULT AND DISCUSSION

Although the size of nano particles cannot be exactly controlled during synthesis, it is possible to narrow the size distribution by controlling the reaction conditions and proportions of urea and metal nitrates. The most suitable proportion of metal nitrate and urea was found to be 1:3.

Figures 1(a-c) reveal the SEM images of the particles prepared in this study. Figure 1(a) exhibits the size distribution of nickel particle in the range of 500 nm and as its multiple. The Nickel requires stabilization to avoid agglomeration. Figure 1(b) shows the uniform size distribution of iron particles by about 80 nm. But the grain boundary of the individual particles could be easily traced. The SEM image of cobalt particle Fig. 1(c) shows wider variation in the particle size between 500 nm and 1000 nm. The percentage of the low-scale particles is high. The SEM reveals a uniform particle size of iron particles, short range of distribution of the particles in the case of nickel highlighting the reproducible method of material preparation.

XRD spectrum Figs. 1(d-f) showing the crystal orientation of plane structure of the Ni, Fe and Co nano particles respectively. The most intense peaks are indexed and the data provided confirmation that the methodology that we have optimized gives rise to reproducible products. The Graphite crystallite size was calculated from the XRD line broadening Scherrer relationship, $d = 0.9\ \lambda / \beta \cos \theta$, where d is the diameter in A^0, β is the full width half maxima (FWHM) and λ is the wavelength of X-rays. The crystallite as calculated size was found to be 74, 30 and 55 nm of Ni, Fe and Co respectively.

From SEM and XRD study, it is observed that 1:3 weight ratio mixtures of metal nitrate and urea gave nano sized metal powders with the uniform spherical morphology.

These nano metal particles were used to prepare the CNMs from acetylene by CVD method. The SEM observation of carbon nano materials suggest that over 90% of the product made from acetylene is in the form of either fibers or micro carbon spring or nano tubes or mixture of all. Acetylene pyrolised on Ni at 750°c in hydrogen atmosphere gives large quantity of SCNFs and little CNFs (Fig. 2a). Fe metal particles gave CNTs (Fig. 2b) and over Co gave large quantity of CNFs. The composition of metal nitrate and size of catalyst controls the diameter of carbon nanotubes.

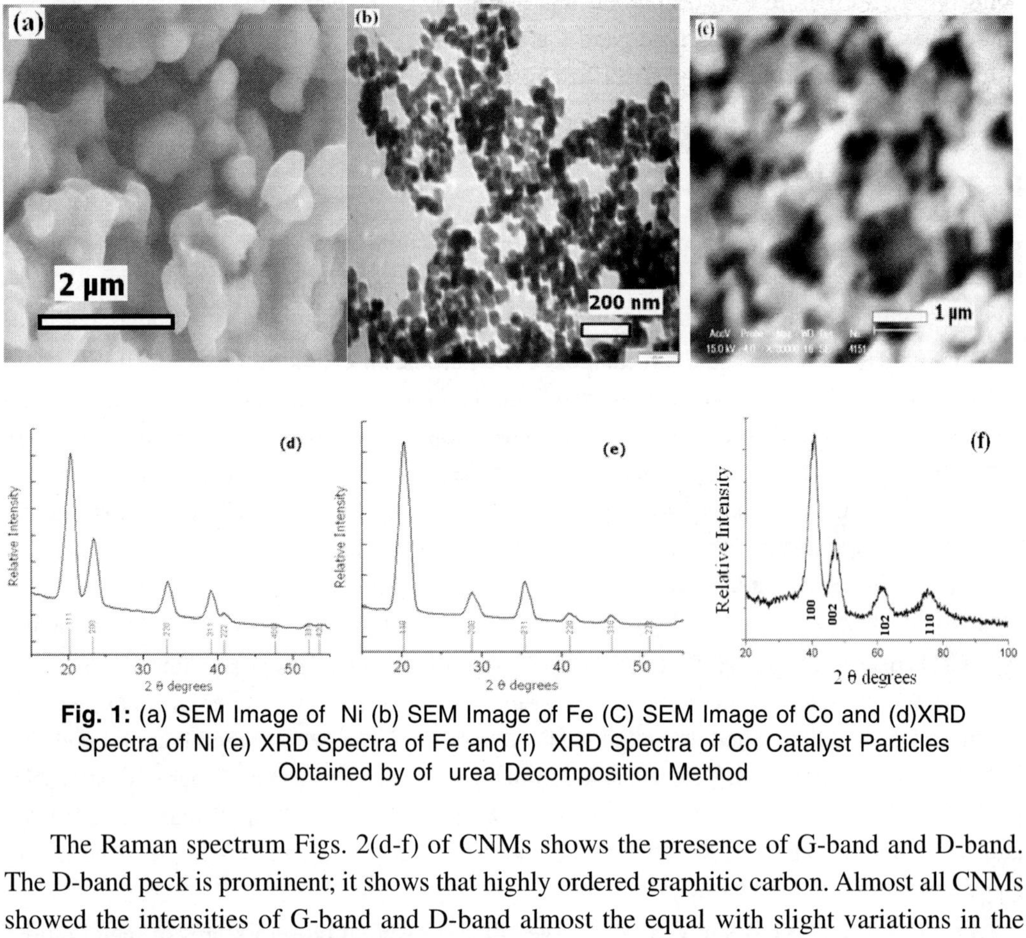

Fig. 1: (a) SEM Image of Ni (b) SEM Image of Fe (C) SEM Image of Co and (d)XRD Spectra of Ni (e) XRD Spectra of Fe and (f) XRD Spectra of Co Catalyst Particles Obtained by of urea Decomposition Method

The Raman spectrum Figs. 2(d-f) of CNMs shows the presence of G-band and D-band. The D-band peck is prominent; it shows that highly ordered graphitic carbon. Almost all CNMs showed the intensities of G-band and D-band almost the equal with slight variations in the range of 0.72 to 0.73. In Fig. 2(d) SCNFs contain more disordered carbon than the CNTs and CNFs in Figs. 2(e) and (f) respectively.

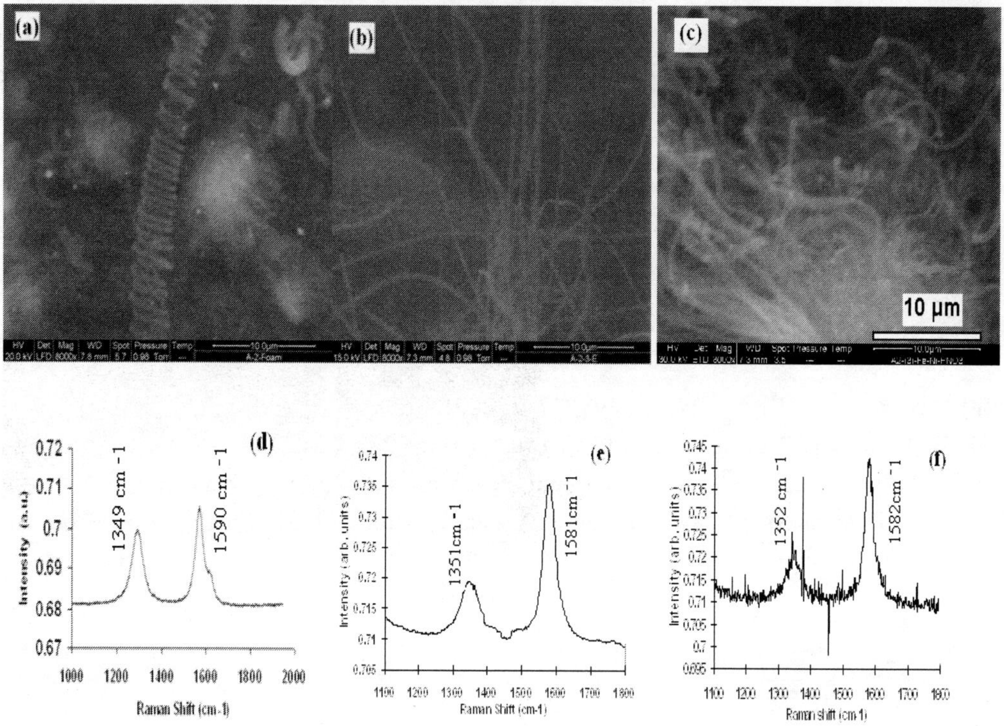

Fig. 2: (a) SEM Image of SCNFs over Ni (b) SEM Image of CNTs over Fe (c) SEM Image of CNFs over Co (d) Raman Spectrum of SCNF (e) Raman Spectrum of CNTs (f) Raman Spectrum of CNFs Obtained from Acetylene at 750°C

CONCLUSION

The experimental results indicate that urea decomposition method is easy and convenient method to get bulk nano metal particles. It was observed that when metal nitrate treated with urea of weight ratio 1:3 gave uniform size with the spherical morphology of nano metals. Acetylene gas pyrolised on different metal particles gives the different morphology of CNMs. Pyrolysis over Ni, Fe and Co metal particles at 750°C in hydrogen atmosphere gives SCNFs, CNTs, or CNFs respectively.

ACKNOWLEDGEMENTS

Financial support by the University Grants Commission (F. No. 32-248/2006 (SR)) is gratefully acknowledged. We are also thankful to Dr. Naresh Chandra (Principal), my colleagues of Department of Chemistry & all the Nano Technology Laboratory researchers, for their support.

REFERENCES

[1] Henglein, *Chem. Rev.* 89, 1861, 1989.

[2] G. Schmid, *Chem. Rev.* 92, 1709, 1992.

[3] N. R. Rao, G.U. Kulkarni, P.J. Thomas, P.P. Edwards, *Chem. Soc. Rev.* 29, 27, 2000.

[4] C. Templeton, W.P. Wuelfing, R.W. Murray, *Acc. Chem. Res.* 33, 27,2000.

[5] G. Markovich, P. Collier, S.E. Henrichs, F. Remacle, R.D. Levine, J.R. Heath, *Acc. Chem. Res.* 32, 415,1999.

[6] L.N. Lewis, *Chem. Rev.* 93, 2693, 1993.

[7] R. Elghanian, J.J. Storho, R.C. Mucic, R.L. Letsinger, C.A. Mirkin, *Science* 277, 1078, 1997.

[8] S.W. Koch, A. Knorr, *Science* 293, 2217, 2001.

[9] W. L. Wilson, P. F. Szajowski, L. E. Brus, *Science* 262, 1242,1993.

[10] B. Murray, D. J. Noris, M. G. Bawendi, *J. Am. Chem. Soc* 115, 1993.

[11] Sandesh Jaybhaye, Maheshwar Sharon, Madhuri Sharon and L.N. Singh, *Int. J. of SRINMC* Vol. 36 (2), 37-42, 2006.

SYNTHESIS AND CHARACTERIZATION OF Ni DOPED ZnS NANOPARTICLES

J. Yuvaloshini, G. Ravi[1] and R.A. Shanmugavadivu[2]

Department of Physics, Karaikudi Institute of Technology, Karaikudi
[1]School of Physics, Alagappa University, Karaikudi
[2]Department of Physics, Rajadoraisingam Govt. Arts College, Sivagangai
E-mail: rsvadivu@rediffmail.com

ABSTRACT

Ni doped ZnS nanoparticles were prepared by Hydrothermal method. At first, the nanoparticles were synthesised by the reaction of zinc chloride $ZnCl_2$, thiourea $SC(NH_2)_2$ with Ethylenediaminetetraacetic acid (EDTA) as complexing agent. 0.01M of nickel chloride $NiCl_2 \cdot 6H_2O$ is doped and the sample was annealed at 100 ºC. Optical and Structural properties of the sample were studied by UV-VIS absorption spectrum and X-Ray Diffraction, hexagonal structure of the sample was identified. The resultant composition was confirmed by EDAX spectrum. Scanning Electron Micrograph shows the cluster like structure.

INTRODUCTION

The II-VI semiconductors such as ZnS, CdS, ZnO, CdTe are the foremost promising materials and much in demand for optical and optoelectronic applications [1-3]. Among these semiconductors ZnS has wide band-gap energy of 3.68 eV for bulk cubic phase and 3.80 eV for bulk hexagonal phase [4]. Due to large band gap, ZnS is an essential phosphor host lattice material for electroluminescent devices.

One of the applications of ZnS is in reflective coatings. A reflective coating is an engineered formulation of pigment, binder, solvent and additives that mix to create a specific product with

V. Rajendran, K. Saminathan, P. Paramasivam and K.E. Geckeler (eds.)
Nano Materials Synthesis and Characterisation, pp. 65-70 (2012).

its own reflective characteristics [5]. ZnS particles so obtained are unstable to visible light and get decolourized. To avoid the decolourization, doping of ZnS with cobalt is done [6]. But due to this doping a large absorption edge in the range of 650nm to 700 nm is developed in ZnS, so it cannot be used as white reflective pigment in the industry. But the above mentioned problems can be solved by (a) creating a novel ZnS pigment with unique crystal structure, morphology and surface chemistry, (b) preventing the decolourization of ZnS by doping it with dopant which does not produce absorption edge in the visible region and can also control the morphology, (c) functionalizing ZnS surface with surfactant such that it can control the morphology of particles as well as avoid its decolourization. This part of research work includes synthesis of Ni-doped ZnS nanoparticles and the study of structural and optical properties.

EXPERIMENTAL DETAILS

In order to synthesize the Ni doped ZnS nanoparticles, stock solutions of $ZnCl_2$ (1 M) and $SC(NH_2)_2$ (0.2M) was prepared in 50ml de-ionized double distilled water under stirring with Ethylenediaminetetraacetic acid EDTA as complexing agent. Then 0.01M of $Nicl_2 \cdot 6H_2O$ was prepared in 50ml de-ionized double distilled water and was added to the mother solution. To this stock solution 14ml of freshly prepared NaOH (2M) solution was added under continuous stirring in order to obtain the pH value of 11. These solutions was transferred into teflon lined sealed stainless steel autoclaves and maintained at temperature in the range of 95°C for 4h to 5h under autogenous pressure. It was then allowed to cool naturally to room temperature. After the reaction was complete, the resulting pale yellow nanocrystalline powders were washed with ethanol for several times to remove the other ionic impurities, if any. The filtered powder was then dried in a laboratory oven at 100°C for 5h.

The synthesized samples were characterized for their structure by x-ray diffraction (Rigaku D max-C) with Cu Kα radiation. The optical band gap Eg was estimated from the UV-VIS-NIR diffuse reflectance spectroscopic (UV-Vis-NIR DRS) studies in a wavelength range from 190nm to 1200nm with JASCO V-570 spectrophotometer. The samples for this study were used in the form of powder and pure ZnS was used as the reference.

RESULTS AND DISCUSSION

Structural Analysis Using X-ray Diffraction Method

Generally, zinc sulphide exists in two forms, cubic (zinc-blend) and hexagonal (wurtzite). The cubic form is stable at room temperature [7]. Figure 1 shows the XRD patterns of Ni doped ZnS nanoparticles. XRD peaks, shows the nano-crystalline nature. All the peaks in the XRD patterns clearly indicate that the Ni-doped ZnS nanoclusters possess hexagonal wurtzite structured. These results are verified from JCPDS card no. 89-7334.

The average size (D) of Ni doped ZnS was calculated using Debyue-Scherrer formula [8], the calculated sizes of Ni doped ZnS nanoparticles were found to be 3 nm with ZnS:Ni contents of 1.32% respectively.

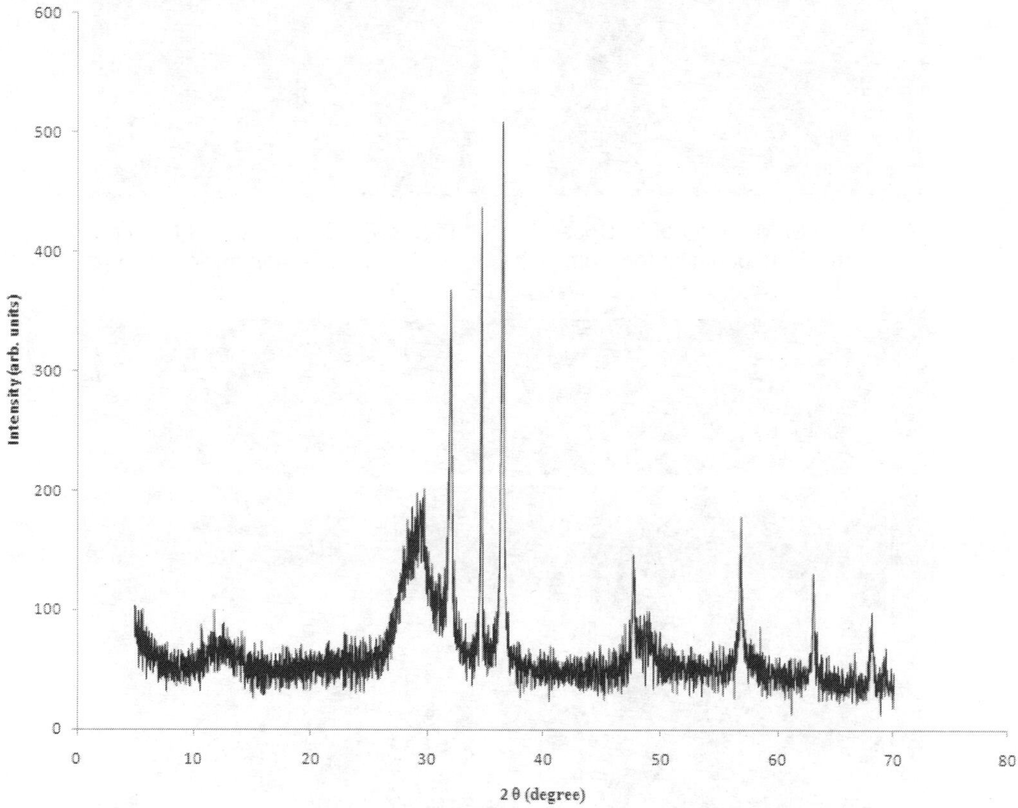

Fig. 1: XRD Pattern of Ni Doped ZnS Nanoparticles

Surface Morphology by Scanning Electron Microscopy (SEM) and Elemental Analysis by Using Energy Dispersive X-Ray Analysis (EDAX)

Scanning electron micrograph of the powder sample is as shown in Figs. 2(a) and 2(b). From the figure it can be seen that the Ni doped ZnS nanoparticles posses uniform grains and is almost homogeneously distributed. Energy dispersion X-ray (EDX) analysis is performed to confirm the atomic composition (%) of constituents' atoms and foreign impurity atoms. Figure 3 shows the EDX spectrum measurement of 100% Ni-doped ZnS nanoparticles. The sample was found to have atomic percentage as 66.88% of Zn, 31.79% of S and 1.32% of Ni, which confirms doping of Ni in the host ZnS.

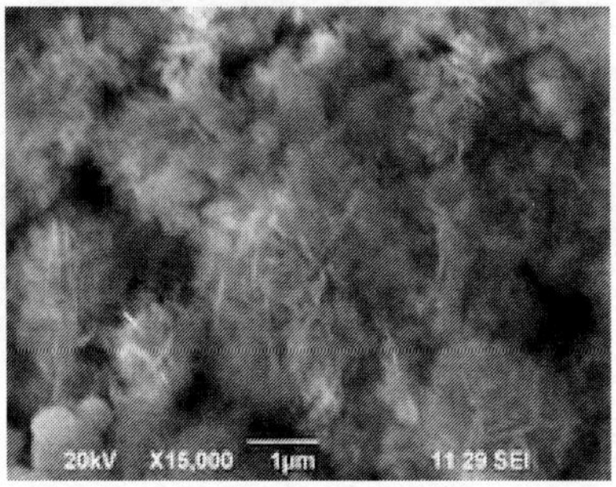

Fig. 2: (a) SEM Image of Ni Doped ZnS Nanoparticles for 1 μm

Fig. 2: (b) SEM Image of Ni Doped ZnS Nanoparticles for 5 μm

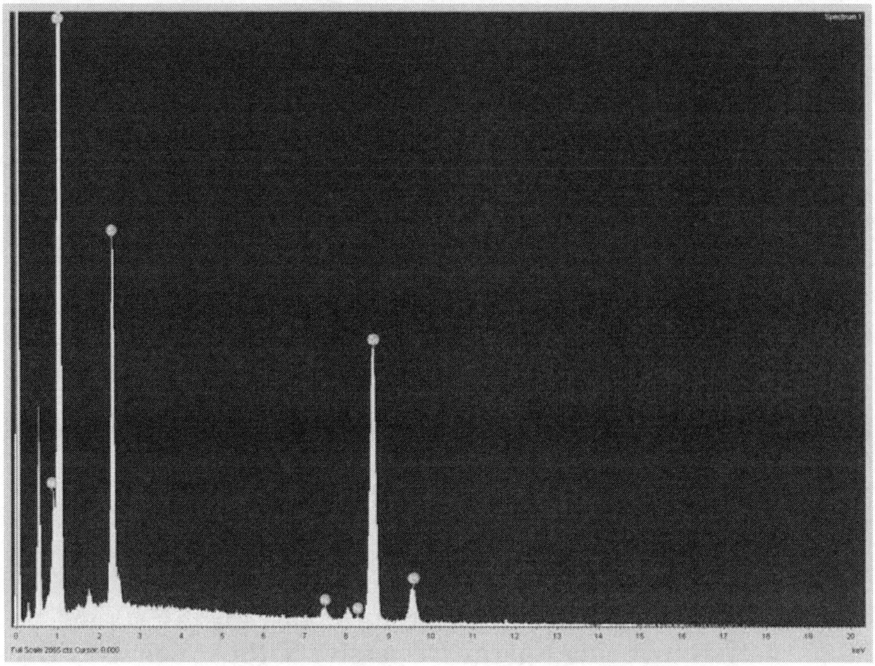

Fig. 3: SEM-EDX Spectrum of Ni Doped ZnS Nanoparticles

UV-Vis Spectroscopy

Figure 4 shows the UV-Vis absorption spectrum of the synthesised of Ni doped ZnS nanoparticles has been recorded to measure their band-gap. The spectrum shows absorption edge of the

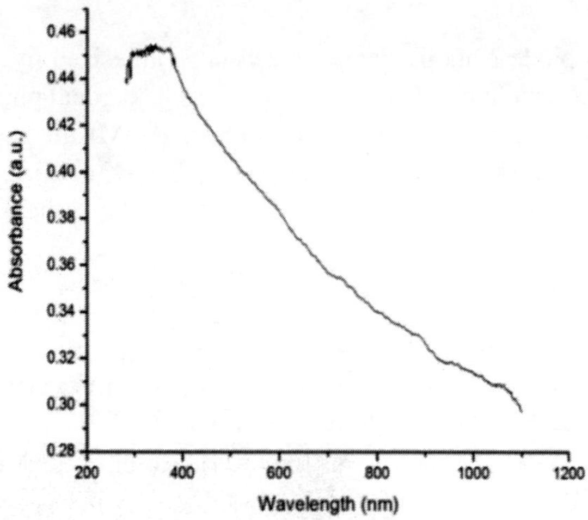

Fig. 4: Plot of Absorption versus Wavelength of Ni Doped ZnS Nanoparticles

nanoparticles in the range - 300 to 320 nm, indicating that these nanoparticles being blue-shifted as compared to bulk ZnS for which the peak is at 260 nm [9]. The blue shift in the absorption edge is due to the quantum confinement of the excitions present in the sample, resulting in a more discrete energy spectrum of the individual nanoparticles. In this research work on doping Ni in ZnS, the blue shift further decreases, which might be due to the fact Ni forming new energy levels in the ZnS energy band. Using Tauc's relation [10], from the Fig. 5 the band-gap of Ni doped ZnS nanoparticles has been calculated and found to be 3.93 eV.

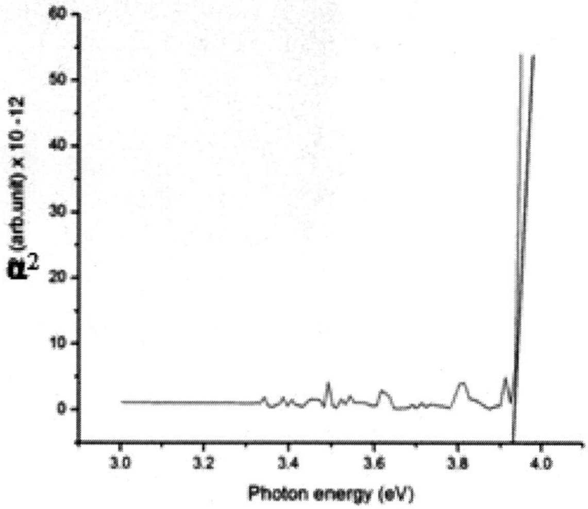

Fig. 5: Plot of Square of Absorption Coefficient α^2 versus Photon Energy of Ni Doped ZnS Nanoparticles

CONCLUSION

Nickel doped Zinc Sulphide Nanoparticles have been synthesized by hydrothermal method. X-ray diffraction measurement confirms the structure as hexagonal phase having particle size in the range of 3 nm. Scanning Electron Microscope (SEM) images clearly indicates the formation of nanoclusters. Compositional analysis by EDAX confirms that the sample with clear peaks of Zinc (Zn), Sulphur (S) and Nickel (Ni) are around the nominal composition.

REFERENCES

[1] A. Tiwari, S. A. Khan, R. S. Kher, *Adv. Appl. Sci. Res.* 2, 105 (2011).

[2] M. Godlewski, E. Guziewicz, K. Kopalko, G. Luka, M. I. Lukasiewicz, T. Krajewski, B. S. Witkowski, S. Gieraltowska, Low Temp. Phys. 37, 235 (2011).

[3] R. E. Pimpinella, A. M. Mintairov, X. Liu, T. H. Kosel, J. L. Merz, J. K. Furdyna, M. Dobrowolska, *J. Vac. Sci. Technol. B* 29, 1 (2011).

[4] N. Saravanan, G. B. Tec, S. Y. P Yap, K. M. Cheong, *J. Mater. Sci. Mater. Electron.* 19, 1206 (2008).

[5] R. Talbert, Paint Technology Handbook, Taylor & Francis Group, London (2008).

[6] G. Buxbaum, G. Pfaff, Industrial Inorganic Pigment, 3rd edn. WILEY-VCH, Weinheim (2005).

[7] L.I. Berger, Semicond. Master. 1997 (1997).

[8] F. Atay, V. Bilgin, I. Akyuz, S. Kose, Mater, *Sci.Semicond. Proc.* 6, 197 (2003).

[9] B. Tripathi, Y. K. Vijay, S. Wate, F. Singh, D.K. Avasthi, *Solid-State Electron* 51, 81 (2007).

[10] Z. Jindal, N. K.Verma, *J. Mater. Sci.* 43, 6539 (2008).

FATTY ACID DIMER PRODUCTION FROM MELON SEED OIL

Udom Mark Akpan

National Agency for Science and Engineering Infrastructure, Idu Industrial Area, Abuja, FCT. Nigeria
E-mail: udymark@yahoo.com

ABSTRACT

Alkyd Resins used for the production of paints, lacquer, etc., in surface coating industries, ordinarily, are prepared from Phthalic Acid and Glycerol for their respective uses. Recently, research on possible substitutes for these acids are still on-going. Fatty acid content of a more readily available seed oil (melon) in our locality appears more promising for the production of Dimers which are further used for the production of Alkyd Resins. To this end, the seed oil was extracted using the cold system, characterized and refined. The refined oil was later thermally polymerized at various temperatures in inert atmosphere of nitrogen using sulphur as catalyst in Alkyd kettle. Dimers were eluted from the products of this thermal polymerisation using Chromatographic plates and their quantities at different temperatures were found to be adequate for large scale Dimers production. Optimal yield of dimer was obtained at temperature 360°C at time 30 minutes of dimersation. Properties such as: the molecular weight, refractive index, acid value, iodine value and colour of the dimer fatty acid were analysed and found to correspond to the standard dimer acids in commercial use.

INTRODUCTION

Melon a vegetable, grown in the tropics contain in their seeds oil which are predominantly glycerides or triesters of glycerol, with one unit of glycerol and three units of fatty acids. The

V. Rajendran, K. Saminathan, P. Paramasivam and K.E. Geckeler (eds.)
Nano Materials Synthesis and Characterisation, pp. 71-76 (2012).

fatty acid content of melon (*Colocynthis vulgaris*) seeds contains 12.8% palmitic acid (C_{16}:0), 17% stearic acid (C_{18}:0), 18% Oleic acid (C_{18}:1) and 54% linoleic acid (C_{18}:2). By this composition, melon seed oil is found to be a semi drying oil which puts melon oil in the category of important Industrial raw materials. The term Dimer Fatty Acids (DFA) is applied to the dicarboxylic acids formed as a product of thermal polymerization of two or more C_{18} unsaturated fatty acids such as oleic and linoleic acids. Commercial DFA products are mixtures of C_{36} dibasic acids containing some trimers (C_{54}) and higher oligomers. Small amounts of monomer acids (C_{18} and saturated, unsaturated and structurally modified fatty acids)[6] is also present in the product. During thermal polymerization, The amount of unsaturation decreases rapidly, and their density and viscosity increase.[7] Under this condition, the polyunsaturated C_{18} fatty acids present in the oil is dimerized to form mixtures of C_{36} dibasic acids with some trimers (C_{54}) and higher oligomers. Small amounts of monomeric fatty acids which include C_{18} saturated, unsaturated and structurally modified fatty acids are also present. Although catalyst is not necessary, the use is to lower the temperature and time necessary for dimerisation to occur as the reaction passes through the process of isomerisation in the polyunsaturated fatty acids. Therefore, sulphur in this project was used as an isomerisation catalyst and the mechanism through which this occur is represented in the scheme below:

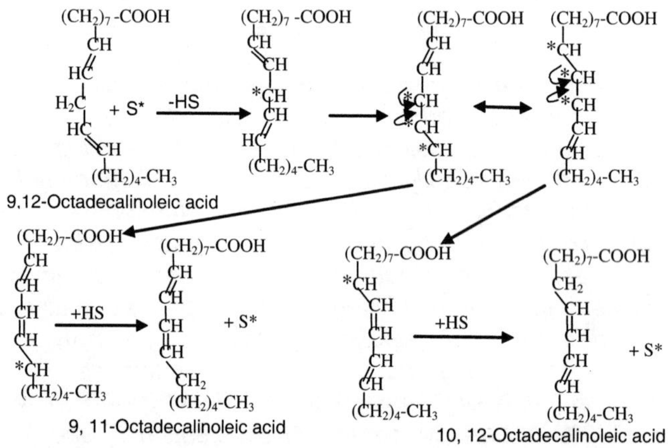

Mechanism of Sulphur catalysed isomerisation of linoleic acid moiety

The DFA are extremely used in adhesives, ink and surface coating formulation, reactive and non-reactive polyamides, corrosion inhibitors (basically, trimers), and metal working lubricants[55].

EXPERIMENTAL

Materials Used

Melon seed (*Colocynthis vulgaris*): Melon seeds were bought from Nsukka main market. The seeds were winnowed and dried at about 105°C. They were then ground coarsely, in a mechanical grinder and stored in a clean polyethylene bag.

Reagents: Petroleum ether (40–60°C), potassium hydroxide (KOH), diethyl ether, methanol, toluene, sulphur, carbon tertrachloride, phenolpthalene, Wij's solution, sodium thiosulphate, potassium iodide, iodine trichloride, starch, fuller's earth, nitrogen gas.

Equipment: Vaccum pump, 4-necked alkyd resin kettle, heating mantle, mechanical stirrer, PZO Abbe refractometer, Gardner-Delta colour comparator, mechanical grinder, chromatographic column with tap, Ferranti portable viscometer.

Oil Extraction

Petroleum ether was used in a chromatographic column to elute the oil from ground melon seeds. The oil/solvent solution was filtered and redistilled, weighed and stored in an airtight container.

Refining the Crude Melon Seed Oil Extract

The following refining processes were carried out on the oil. These include demunification, deacidification, and decolourisation.

Dimerisation of Melon Seed Oil (MSO)

200.0g of the melon seed oil (MSO) was accurately weighed into a 4-necked resin Kettle fitted with mechanical stirrer, thermometer, nitrogen inlet tube and a condenser. 1.0g (0.5%w/w) of monoclinic β-sulphur (yellow crystalline sulphur) was weighed and added to the oil and the mixture heated on the mantle. A steady stream of nitrogen was released from the cylinder into the system while stirring continues and it was heated gradually to 300°C at the first instance of dimerisation.

The temperature was maintained at 300°C for about ten minutes before the first sample was withdrawn using a 20ml pipette fixed onto its filler. Subsequent withdrawals were made at 5 minutes intervals and ten samples were collected in 45 minutes. The samples were kept in clean, well labeled sample bottles with rubber tight fitted caps.

The remains of the dimerisation products in the resin kettle was later transferred into a different sample container and labeled 'final crude product' at the specified temperature. The resin kettle was then cleaned-up for subsequent use.

Same procedure was adopted to dimerise the oil at five other temperature values of 310, 320, 330, 340 and 350°C respectively.

Characterisation of the Oil and Dimer Fatty Acid (DFA)

The physicochemical properties of the refined melon seed oil and the DFA produced were later determined which include: acid value[9], iodine value, molecular weight, colour, viscosity, specific gravity, the refractive index and saponification value.

RESULTS AND DISCUSSION

The Physicochemical properties of Melon Seed Oil (MSO) used for this research work is given below:

Percentage yield	47.90%
Colour (Gardner Scale)	1
Relative density (at 28°C)	0.912
Absolute Viscosity	0.3192 poise
Refractive index	1.4680
Molecular weight	448.65 g/mol
Acid value	0.3186 MgKOH/gm
Iodine Value	111.825
Saponification value	194.246

This Characteristics shows that the oil is semi-drying and within the range of high molecular weight glycerides[1,5]. The low acid value recorded was obtained after alkali refining which gave rise to the high relative density[2,12]of the refined MSO. The high molecular weight of the refined oil also explains the reason for the low saponification value which reflects a large percentage of C_{16} and C_{18} in the oil[5].

Characteristic of the Crude DFA

Some of the physicochemical properties of the DFA produced, amongst others, are discussed below which are relevant for comparative analysis with the commercial DFA.

Acid Value (AV)

The presence of 1% phosphatide increases the acid value by 0.30 – 0.33 mg KOH[10]. The non-hydrated phospholipids which are difficult to remove during degumming process are hydrolysed during dimerisation. This leads to the formation of polar lipids together with other decomposition products which increased the acid value of the DFA with time and temperature of dimerisation as shown in the Plots inside:

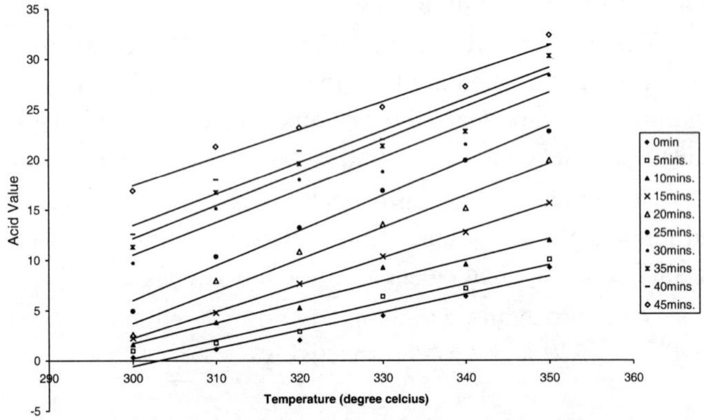

Fig. 2. Plot of Acid Value Against Temperature of Dimerisation

MOLECULAR WEIGHT

The increase in molecular weight of the DFA with respect to temperature of dimerisation as shown in the Plot below is appropriate since the dimerisation process is a bimolecular additive reaction of the unsaturated fatty acids[13].

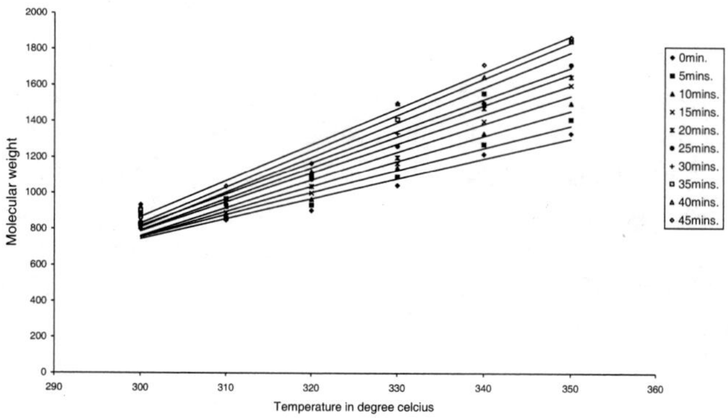

Fig. 3: Plot of Molecular Weight vs Temperature of Dimerisation

Percentage Yield of DFA

Crude DFA usually contains some monomeric and trimeric products together with other thermally decomposed products and the fraction of interest (dimers). Fractions of the samples considered most dimerised (based on their molecular weight, refractive indices and colour) were separated by Thin Layer Chromatography (TLC) followed by cryoscopic determination of their molecular weight and the results given as shown in Table 1.

Table 1: TLC Separation of DFA Samples Into its Components

Sample	*% Yield of Bands*			*Molecular Weight of Bands*		
	1st	*2nd*	*3rd*	*1st*	*2nd*	*3rd*
T_{45}^{310}	96.67	2.67	NI	740.74	1333.33	–
T_{45}^{320}	96.00	3.33	Trace	714.29	1263.16	–
T_{45}^{330}	82.00	16.67	1.33	823.53	1400.00	1647.06
T_{45}^{340}	63.33	34.00	2.67	689.66	1217.39	1750.00
T_{30}^{350}	47.33	48.67	4.00	769.23	1272.73	1866.67
T_{40}^{350}	45.33	46.00	8.67	714.29	1333.33	2000.00
T_{45}^{350}	39.33	43.33	17.33	833.33	1411.76	1866.67

+Sample T_A^B: A = Time of dimerisation, B = temperature of dimerisation
NI = Not Identified, M = Monomer, D = Dimer and T = Trimer

From the table,1st, 2nd, and 3rd fractions denotes Monomer, Dimer, and Trimer respectively. Two basic bands were obtained (monomer and dimer) on separation; and the third band (trimer) was obtained from 330°C as tiny spots. The highest percentage yield of 48.67% dimer was obtained at 350°C after 30 minutes of dimerisation. This therefore marks the OPTIMUM YIELD of the DFA.

CONCLUSION

The dimerisation of Melon seed oil (MSO) is temperature dependent which lead to increases in the acid value, refractive indices, and the molecular weights of the dimer fatty acids. The rate of dimerisation and the percentage yield of dimers are affected by time and temperature of dimerisation.

The optimum parameters in the dimerisation of MSO using sulphur catalyst are temperature of 350°C dimerised for 30 minutes using 0.5% catalyst in an inert atmosphere of nitrogen. The highest yield under the condition is 48.67%.

REFERENCES

[1] Fox, T.E. The Chemistry and Technology of fats and oils. 2nd ed. Mcmillan Publishers, NY P55, 1984.

[2] Chappelow, C. C., Acid Value'. *Encyclopedia of Chemistry.* 3rd Ed. Van Notra Reinhold company, NY. PP 12 – 13, 1964.

[3] Ibemesi, J.A. Vegetable oils as industrial Raw materials: the Nigerian Perspective, Snaap press. Enugu. pp 7–44, 1992. Investigation of Network formation in Drying oils byDilute Solution viscometry. *J. Am. Oil Chem. Soc.* 66: 974–978, 1989.

[4] Strong, F.M. and Koch, G.H. Biochemistry labouratory manual, 2nd ed. Mcmillan press, Nigeria, 1975.

[5] Kroschwitz, Jacqueline I. (Ed). *Concise Encyclopedia of PolymerScience and Engineering,* John Wiley and Sons Inc, 1990.

[6] Wexler, Herman, 'Polymerisation of Drying oils' in *Chemical Reviews,* 64:6. *American Chemical Society*, N.W. S Washington 6 D.C, 1964.

[7] Beat, Meyer. 'Elemental Sulphur' in *Chemical Reviews*, Vol. 68. *American Chemical Society*, 1968.

[8] Howdhury, Diptikalyan and Mukterji, B.K.. *Effect of Temperature on Polymerisation and Isomerisation of dehydrated castor oil* (Univ. Coll. Sci. Technol. Calcutta) *Proc. Nath. Inst.* Sci. India. pp 199–203. 12505, 1951.

[9] Rheklin, V.P., and Semenov, F.A. Trudy Nsesoyuz. Nauch Issgledovatel Inst. Zhirov 59 No. 18, 1958.

[10] ASTM D 180 – 67. Annual Book of ASTM standard. Part 20 Philadelphia.

[11] Girgis, P. and Said F. 'Lesser known Nigerian oils and fats-1, Characteristics of Melon seed oils'. *J. Sc.Ed. Agric.* 19 ed. pp 615 – 16. 1968.

[12] Anderson, L.V and Porter, James V. Relation of Glyceride oil composition to polymerization. In *Industrial and Engineering Chemistry. M*inneapolis, 1946.

ADHESION OF THE CHEMICAL BATH DEPOSITED POLYTHIOPHENE THIN FILMS

Sandip V. Kamat, Vaishali S. Patil, Rahul B. Patil[1], Vijaya Puri[2] and R.K. Puri[3]

Sinhgad Institute of Science & Technology, Narhe, Pune
[1]Sanjeevan Engineering and Technology Institute, Panhala
[2] Department of Physics, Shivaji University, Kolhapur
[3]Vacuum Techniques & Thin Film Lab., USIC, Shivaji University, Kolhapur
E-mail: rkp.usic@gmail.com

ABSTRACT

The polythiophene thin films were prepared by a well known chemical bath deposition technique. The deposited thin films were characterized for structural morphological properties and the adhesion of these thin films were measured by Direct Pull Off (DPO) method, the effect of oxidant concentration on these thin films also studied.

The FTIR spectra of chemically deposited polythiophene thin films shows the absorption peak at 836 cm^{-1} which represents C–S stretching vibrations, shifts to 869 cm^{-1} as the oxidant concentration increases. The band at 666 cm^{-1} representing C–S–C ring deformation becomes sharper and appears with a shoulder peak due to increase in oxidant concentration. The adhesion of the thin films was found to be increased with increase in oxidant concentration from 34.8×103 N/m2 to 41.5 ×103 N/m^2.

INTRODUCTION

Polythiophene (PTh) and its derivatives have been studied for various applications in optoelectronics like light emitting diodes (LED) [1], field effect transistors (FET) [2], super capacitors [3], photovoltaic and photoconductive devices and optical modulator devices, etc.

V. Rajendran, K. Saminathan, P. Paramasivam and K.E. Geckeler (eds.)
Nanomaterials Synthesis and Characterisation, pp. 77-82 (2012).

[4]. Polythiophenes have also been exploited in sensor applications [5]. Polythiophene exhibit superior properties such as low density ease of fabrication [6] as compared to conventional conductors. In addition the conductivity of these materials can be varied from conducting to insulating range by using different techniques of the polymerization. The synthesis conditions and the nature of the dopant anion play a vital role in the polymerization process as they strongly affect the various properties of the material [7]. Most of the studies available on the polythiophene thin films are deposited by electrochemical polymerization techniques and to the authors' knowledge there are no reports available on the chemical bath deposited polythiophene thin films. The Chemical bath deposition (CBD) method appears most suitable for the integration in large scale fabrication process. The main advantages of the CBD method are its low cost, no requirement of sophisticated instrument, low processing temperature and non-polluting properties [8]. The adhesion of the thin film to the substrate is an important parameter, which can affect the reliability and the performance of the device. The present article reports the effect of oxidant concentration on the adhesion of the chemical bath deposited polythiophene thin films.

EXPERIMENTAL

Thin Film Deposition

Thiophene (AR grade Merck) was purified by distillation before prior use. Iron chloride $FeCl_3$ (Sd-Fine), Chloroform (CH_3Cl), Methanol (CH_3OH) were used for the preparation of polythiophene thin films. Monomer solution was prepared by dissolving 0.1 M of thiophene in chloroform, the oxidant solution was prepared in a glass beaker with 0.1 M concentration of $FeCl_3$ in chloroform the ratio of monomer to oxidant was kept 1 : 1. The substrates were immersed in the bath at room temperature at constant stirring. The monomer solution was added drop wise in the oxidant solution the reaction being carried out under N_2 atmosphere. During the precipitation, heterogeneous reaction occurred and the deposition of polythiophene took place on the substrate. The substrates coated with polythiophene thin films were removed after a time interval of 1 hour from the bath, washed with methanol followed by chloroform and acetone repeatly to remove residual oxidant and unreacted monomers, dried in air and preserved in an airtight container. The oxidant concentration was varied to 0.2 M, 0.3 M, 0.4 M and 0.5 M and again the experiment was repeated to study the effect of oxidant concentration on the properties of polythiophene thin films. For oxidant concentration greater than 0.5 M, deposition of thin film did not take place.

Characterization Techniques

FTIR spectra of polythiophene powder and thin films were recorded by using a Perkin–Elmer's Spectrum Spectrometer in the range of 400–4000 cm^{-1}. For FTIR, these films were scratched from glass substrate and mixed with KBr. The adhesion of the polythiophene thin films was measured by direct pull off (DPO) technique.

RESULTS AND DISCUSSION

FTIR Spectroscopy

The FTIR spectra of chemical bath deposited polythiophene thin films are shown in Fig. 1. The absorption peak corresponds to 3408 cm^{-1} in all the PTh thin films is the humidity due to KBr [9]. The peak at 2921 cm^{-1} and 2845cm^{-1} corresponds to C-H symmetric and C-H asymmetric stretching vibrations respectively. The bands at 1628 cm^{-1} and 1384 cm^{-1} represents the C=C symmetric and C=C asymmetric vibrating modes in the thiophene ring [5]. The band at 1134cm^{-1} represents C-H in plane bending [9]. The weaker feature in the range 1030 to 1018cm^{-1} corresponds to (C-C stretching + C-H wagging components) [5]. As the concentration of the oxidant increases the peak corresponding to 1134 cm^{-1} becomes sharper representing the C-H bending vibrations, the band at 836 cm-1 which represents C-S stretching vibrations [9], shifts to 869 cm^{-1} as the oxidant concentration increases. The band at 666 cm^{-1} representing C-S-C ring deformation [9] becomes sharper and appears with a shoulder peak due to increase in oxidant concentration. While the band at 781 cm^{-1} corresponds to the C-Hα out of plane bending in mono substituted thiophene rings [10] becoming sharper due to increase in oxidant concentration. The increase in concentration of $FeCl_3$ increases the rate of polymerization process [11] due to which there may be formation of short chain oligomers or back bone polymers, this may be one of the reason for presence of bending vibrations in FTIR.

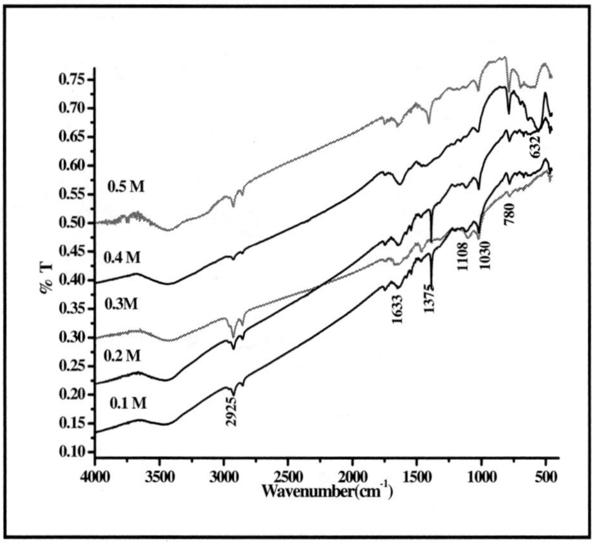

Fig. 1: FTIR Spectra of PTh Thin Films with Various Oxidant ($FeCl_3$) Concentrations

Surface Morphology

Figure 2 shows the surface morphology of chemical bath deposited PTh thin films. It can be seen clearly from the figure that the oxidant concentration strongly affects the surface morphology of PTh thin film. Initially when the oxidant concentration is equal to monomer concentration, i.e. 1:1 ratio the surface morphology shows starting of agglomeration of polymer grains, when the ratio of monomer and oxidant is 1 : 2 then there are presence of more lone pair free electrons of $FeCl_3$ to attack on the weak bonds of monomer due to which the polymerization process becomes rapid due to which the formation of polymer chain takes place. As the oxidant concentration is increased to 0.3 M then the polymerization is observed, giving nano sheet like structure. If the monomer concentration increases beyond this the reaction time reduces and it gives fibers like structure, if the monomer concentration increases to 0.5 M a well grown polymer chain consisting of spherical grains is observed. If the concentration of oxidant is increased beyond 0.5 M then the polymerization process takes place so fast and the precipitate forms very quickly and no deposition takes place on the glass.

Fig. 2: FESEM Images of PTh thin Films with Various Oxidant Concentrations

Adhesion

The adhesion of chemical bath deposited PTh thin films with the substrate is measured by direct Pull Off method.

Figure 3 shows the graph of adhesion vs $FeCl_3$ concentration in mole and it is seen that the adhesion is increased with an increase in the oxidant concentration from 34.8×10^3 N/m^2 to 41.5×10^3 N/m^2 The Path thin films prepared with the lowest concentration of $FeCl_3$ shows lowest adhesion, the increase in the film thickness may be responsible for this rise in the adhesion.

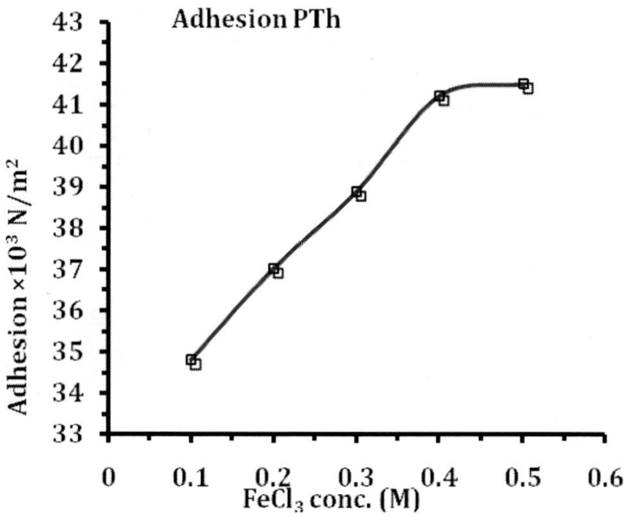

Fig. 3: Adhesion of PTh thin Films Prepared with Different Oxidant Concentrations

CONCLUSIONS

The FTIR spectra confirm that the polythiophene thin films were successfully deposited by chemical bath deposition method on the glass substrates. The variation of oxidant concentration strongly affects adhesion of the films. These thin films show different surface morphologies due to change in the oxidant concentration. The thickness and adhesion of the film increased with increase in oxidant concentration.

ACKNOWLEDGEMENT

One of the authors Prof. Mrs. Vijaya Puri gratefully acknowledges the award of Research Scientist from University Grant Commission (UGC), India.

REFERENCES

[1] S. H. Ahn, M.Z. Czae, E.R. Kim, H. Lee, *Macro Molecules* 34 (2001) 2522-2527

[2] R. Daik, S. Maniam Maly. *J. Chem.* 9 (2007) 16-21.

[3] F. Bloisi, A. Cassinese, R. Papa, L. Vicari, V. Califano., *Thin Solid Films*. 516 (2008) 1594-1598.

[4] X. Ma, G. Li, H. Xu. M. Wang, H. Chen, *Thin Solid Films* 515(2006) 2700-2704.

[5] M. R. Karim, K. T. Lim, C. J. Lee, M. S. Lee., *Syn. Met.* 157(2007) 1008-1012.

[6] X. Zhao, *J. of Mater. Sci.* 40 (2005) 3423-3428.

[7] O.Yu Posudievskii, N.V. Konoshchuk, A.L. Kukla, A.S. Pavlyuchenko, G.V. Linyuchev, V.D. Pokhodenko, *Thero. and Exp. Chem.*, 42 (2006) 339-345.

[8] A.M. Chaparro, M.A. Martinez, C. Guillen, R. Bayon, M.T. Gutierrez, J. Herrero, *Thin Solid Films*, 361, (2000), 177-180.

[9] A. Gok, M. Omastova, A.G. Yavuz, *Syn. Met.*, 157 (2007) 23-29.

[10] V. C. Goncalvis, D.T. Balogh, *Eur. Pol. J.*, 42, (2006), 3303-3310.

[11] T. Olinga, B. Francois, *Syn. Met.*, 69, (1995), 297-298.

STUDY OF AMBIENT AIR AGEING EFFECT ON VAPOR CHOPPED POLYANILINE (AQUEOUS) THIN FILM

J.B. Yadav, A.A. Jatratkar, R.K. Puri and Vijaya Puri[1]

Vacuum Techniques & Thin Film Lab., USIC, Shivaji University, Kolhapur
[1]Department of Physics, Shivaji University, Kolhapur
E-mail: jbyadavphy@gmail.com

ABSTRACT

We present here ambient air ageing effect on vacuum evaporated polyaniline (PANI) thin film, wherein PANI powder synthesized in aqueous medium by oxidation polymerization pathway was used as a source material. The effect of ambient air ageing (30 days at room temperature) on film was studied by checking optical, mechanical and optical waveguide transmission loss properties The vapor chopping technique was successfully used for improving film properties as well as reducing the ambient air ageing effect. The vapor chopped film was observed with low optical transmission loss, refractive index, intrinsic stress and high adhesion as compared to as deposited film. The air ageing was caused to increase in refractive index, optical transmission loss and decrease in adhesion and intrinsic stress but that effect was found to be negligible in vapor chopped PANI thin film. Low ageing effect on optical transmission loss (2.45-3.69 dB/cm.) was observed in vapor chopped film as compared to non-chopped film (3.18-5.08 dB/cm).

INTRODUCTION

In optical communication system, light is used to transport information between different users. To manipulate the optical signals on local scale, devices like optical source, beam splitter, modulator, amplifier and detector are needed. Such devices are integrated on single substrate

V. Rajendran, K. Saminathan, P. Paramasivam and K.E. Geckeler (eds.)
Nano Materials Synthesis and Characterisation, pp. 83-88 (2012).

and technology is called as integrated optics. Planner optical waveguide used for connecting these optical components is the main part of the integrated optics. To fabricate these optical components wide range of material system is required. The stringent requirement of the today's technology is to design all these optical components on a substrate by using a single material.

Polyaniline is the one of the most popular conducting polymer. It has an ability to tailor its properties as per requirement of optical components in integrated optical circuit [1,2]. It is obtained in emeraldine state, pernigraniline state and lucoemeraldine state of different physical and chemical properties, which depends on the oxidation and reduction states of PANI. The oxidation and reduction state of the PANI mostly depends on synthesis process, wherein pH of the reactant solution plays crucial role on reaction mechanism, reaction time and also on chemical structure of outputted PANI [3,4]. Hence it needs to study the synthesis PANI in different mediums. PANI synthesized in acidic medium and deposited by vacuum evaporation has already been studied [5].

In present paper PANI is synthesized in aqueous medium without using acidic solution, which is used as a source in vacuum evaporation technique for thin film deposition. The structural properties of the film have already been reported in previous paper [6]. The effect of ambient air ageing on non-chopped and vapor chopped film on the refractive index, mechanical properties and optical waveguide transmission loss of PANI thin film is reported.

EXPERIMENTAL

PANI powder was synthesized by oxidation polymerization, wherein aniline monomer and oxidizing agent ammonium peroxidisulphate (APS) was prepared in aqueous medium[6]. This synthesized PANI powder was used for vacuum evaporation (resistive heating using tantalum boat at 10^{-4} Torr) on glass substrate.

The vapor chopping technique consists of circular metallic vane of 10 cm diameter with V-cut (155^{0}C) shape. This circular metallic vane is connected to the shaft of 6 V DC motor having broad base. The variable power supply of 0-6 V (DC) was used to very the voltage by means of which the speed of rotation could be controlled and hence the chopping speed. The structural, optical and mechanical properties of these films have been reported in previous paper [5].

For air ageing study, the film were kept in air atmosphere for 30 days at room temperature and studied their optical, mechanical and transmission loss properties after 10 days interval. The optical transmission loss was measured by prism coupling technique [7]. The setup consists of He-Ne laser (632.8 nm), spectrometer, polarizer, two special right angled prism of refractive index 1.717, prism mount and detector. The prisms were kept at 1.5 cm from each other. Only the lowest mode m = 0 has been studied. Precise adjustment of angle of incidence for obtaining m = 0 mode was done. The input light intensity was measured at the region of the coupling spot where actual launching into waveguide takes place. The output intensity was measured in terms of current through a photo diode.

Thickness of the film was ~150 nm measured by Tolansky interferometer method [8]. The refractive index was measured by Abele's method [9]. The adhesion was measured by direct pull off method and intrinsic stress by interferometer technique [10].

RESULTS

Optical Properties

The refractive index (RI) and Optical Transmission loss (TL) are the optical properties of the film that of non-chopped and vapor chopped film are given in Table 1. From Table 1, it seems that the RI and optical TL value increases in both the films with air ageing time. But these changes are very negligible in vapor chopped film as compared to non-chopped film.

Table 1: Refractive index and Transmission Loss of Non-chopped and Vapor Chopped thin Films after Ageing.

Ageing Time (days)	*Refractive Index*		*Optical Transmission Loss (dB/cm)*	
	Non-chopped	*Vapor Chopped*	*Nonchopped*	*Vapor Chopped*
Fresh (0)	1.843 ± 0.005	1.808 ± 0.005	3.41 ± 0.04	3.07 ± 0.03
10	1.862 ± 0.006	1.825 ± 0.004	3.52 ± 0.03	3.23 ± 0.05
20	1.881 ± 0.004	1.835 ± 0.005	3.59 ± 0.05	3.27 ± 0.04
30	1.901 ± 0.005	1.838 ± 0.004	3.82 ± 0.04	3.32 ± 0.03

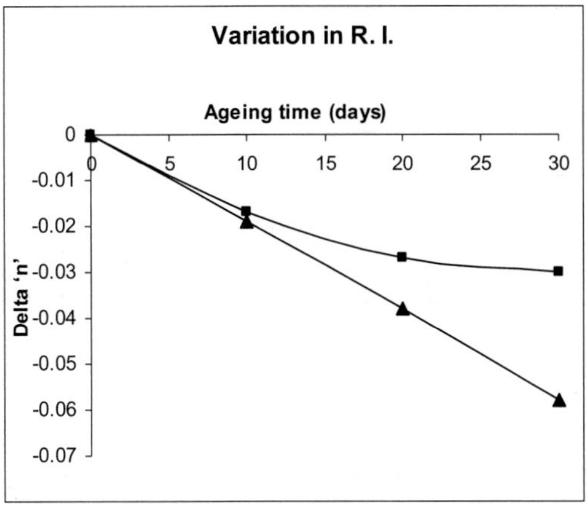

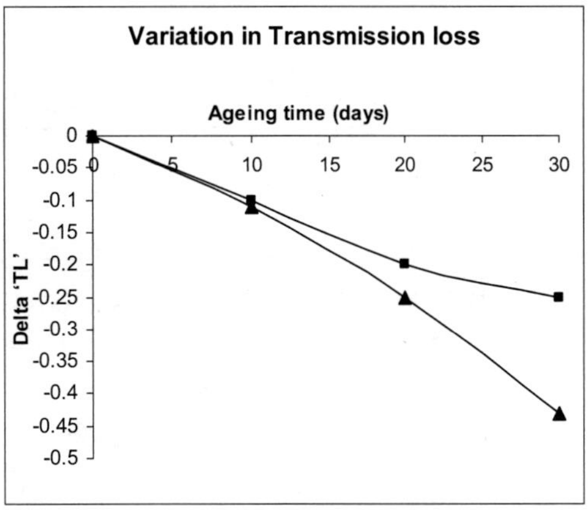

Fig. 1: (A) Graph of Variation in RI (Delta 'n') vs Ageing Time of Both Non-chopped (▲) and Vapor Chopped Films (■).

(B) Graph of Variation in TL (Delta 'TL') Values vs Ageing Time of Both Non-Chopped (▲) and Vapor Chopped Films (■).

The trend of variation in RI and TL value with ageing time is an important aspect to study. It gives the information about how the film properties are influenced in air atmosphere. The variation in RI (Delta 'n') and TL (Delta 'TL') were calculated by using formula 1 and plotted with air ageing time in figure 1;

(Delta 'n') or (Delta 'TL') = (RI or TL of fresh films)

– (RI or TL of air aged films) ... (1)

The negative Delta 'n' and Delta 'TL' values indicates that the RI and TL values increases in air aged film and hence graph observes in fourth quadrant. From figure1, it observes that the Delta graph of both the parameters show similar trend. The nonchopped film show linear increase in delta values, whereas in vapor chopped film, delta value increases gradually and remain constant after 20 days. The vapor chopped film show comparatively low air ageing impact on RI and TL values.

Mechanical Properties

Adhesion and intrinsic stress are the mechanical properties of the film, which of non-chopped and vapor chopped film are shown in Table 2. From table it seems that the vapor chopped film shows high adhesion and low intrinsic stress value as compared to nonchopped film. It also observes that both mechanical properties show decrease in values for both non-chopped and vapor chopped films.

Table 2: Adhesion and Intrinsic Stress of Non-chopped and Vapor Chopped PANI Thin Film after Ageing

Time (days)	*Adhesion $X10^4$ (N/m^2)*		*Intrinsic Stress ($X\ 10^8\ N/m^2$)*	
	Nonchopped	*Vapor Chopped*	*Nonchopped*	*Vapor Chopped*
As deposited	192 ± 4	223 ± 3	10.47 ± 0.08	8.58 ± 0.06
10	187 ± 5	219 ± 4	10.35 ± 0.06	8.49 ± 0.06
20	183 ± 3	215 ± 5	10.20 ± 0.07	8.43 ± 0.04
30	178 ± 4	213 ± 2	10.02 ± 0.08	8.38 ± 0.05

The trend of variation in adhesion (Delta 'A') and intrinsic stress (Delta 'S') value with ageing time is calculated by using the formula 2 and plotted with ageing time in Fig. 2.

Delta 'A' or Delta 'S' = (Adhesion or Intrinsic stress of fresh films)

– (Adhesion or Intrinsic stress of air aged films) ... (2)

From Fig. 2, it clearly seems that both the films show positive Delta 'A' and Delta 'S' values, which indicates the decrease in adhesion as well as intrinsic stress values with air ageing time. The Delta 'A' and Delta 'S' value increases linearly in non-chopped film, whereas the vapor chopped film shows gradually increase in Delta 'A' value up to initial 20 days and thereafter show very negligible change. The Delta 'S' value in vapor chopped film also increases gradually but this change is very negligible.

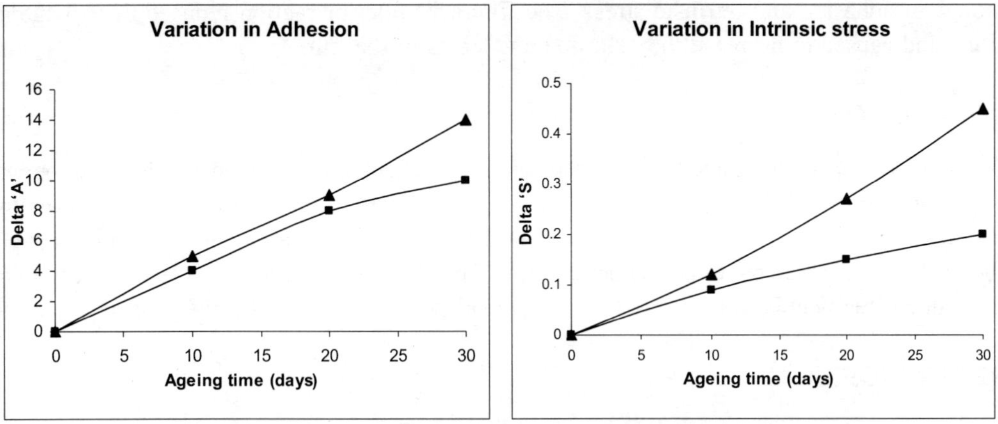

Fig. 2: (A) Graph of Variation in Adhesion (Delta 'A') Value Vs Ageing Time of Both Non-Chopped (▲) and Vapor Chopped Films (■).

(B) Graph of Variation in Intrinsic Stress (Delta 'S') Value Vs Ageing Time of Both Non-Chopped (▲) and Vapor Chopped films (■)

DISCUSSION

At the time of thinking for commercialization of any product, it is very important to study the durability of that product. Hence, we have studied durability of the PANI film for 30 days by keeping the film in air atmosphere. We checked optical and mechanical properties of the film at 10 days time interval. Vapor chopped film showed really good stability as compared to Non-chopped film. The answer of this conclusion is in growth mechanism. In vacuum evaporation for polymer films, it is very difficult to evaporate chain segment more than five or six aromatic rings but in thin films they observes as super molecules of high molecular weight. The evaporated chain segments when reach on the substrate, they undergo cross-linking and thus increases the molecular weight [11]. But some of the active oligomers remain as it is in the film. These active oligomers interact with the environmental active gases which cause to change the film properties.

Vapor chopping technique gives more time to evaporated oligomers to interact with each other and settle uniformly on the substrate which reduces voids and achieve highly compact film structure. Thus, the film gets minimum surface area open to the environment and also less number of active oligomers to interact. Hence the vapor chopped film shows small change in film properties as compared to non-chopped film. The refractive index and transmission loss values increased with ageing time. This is because of the oxidation of PANI film from leucoemeraldine to emeraldine state in air atmosphere that shifted the absorption peak of the film and caused to increase RI and TL of the film[12]. Same reasoning is also applied for change in adhesion and intrinsic stress. The film substrate interaction reduced due to ageing effect and caused to decrease the adhesion and stress of the film.

CONCLUSIONS

PANI powder was synthesized in acidic medium and vacuum evaporated for thin film optical waveguide and studied air ageing effect on their properties. Both the films showed air ageing effect on both optical and mechanical properties. But the impact was found negligible in vapor chopped film as compared to non-chopped films. Thus the economic and simple vapor chopping technique is really effective for producing high quality films for optical waveguide applications.

REFERENCES

[1] J. Chung, B. Choi and H. H. Lee., *App. Phys. Lett.* 74(1999) 3645, D. M. Delongchamp, P.T. Hammond. *Chem. Mater.* 16 (2004)4799.Z.

[2] Tan, E. Zhou, Y. Yang, Y. He, C. Yang, Y. Lee., *Eur. Polym. J.* 43(2007) 855.

[3] T. Abdiryim, Z. Xiao-Gang, R. Jamal, *Mater. Chem. Phys.* 90 (2005) 367.

[4] E.N. Konyushenko, J. Stejskal, I. Sedenkova, M. Trchova, I. Sapurina and M. Cieslar, *Poly. Int.* 55 (2006) 31.

[5] J.B. Yadav, R.B. Patil, R.K. Puri, Vijaya Puri, *J. Non Cryst. Solids* 353 (2007) 4691.

[6] J.B. Yadav, R.K. Puri, Vijaya Puri, *App. Surf. Sci.* 253 (2007) 8474.

[7] P.V. Patil, R.K. Puri, Vijaya Puri, J.Opt. 27 (1998)33.

[8] S. Tolansky, Surface Microtopography, Interscience Publ., New York, 1960.

[9] F. Abeles, *J. Phys. Radium* 19 (1958) 327.

[10] P.V. Patil, R.K. Puri, Vijaya Puri, *Mat. Chem. Phys* 49 (1997) 156.

[11] K. Uvdal, M. Logdlund, P. Dannetun, L. Bertilsson, S. Stafstrom, W.R. Salaneek, A.G. MacDirmid, A. Ray, E.M. Scherr, T. Hjerberg, *A.J. Epstein, Synth. Met.* 29 (1989) E451.

[12] T.R. Dilligham, D.M. Cornelison and E. Bullock,. *J. Vac. Sci. Technol. A* 12 (4), (1994) 2436.

ATMOSPHERIC PRESSURE GROWTH OF Cu-DOPED SnO_2 NANOWIRES AND THEIR PROPERTIES

ANIMA JOHARI, ANOOPSHI JOHARI, M.C. BHATNAGAR[1] AND MANISH SHARMA

Centre for Applied Research in Electronics, Indian Institute of Technology, New Delhi
[1]Department of Physics, Indian Institute of Technology, New Delhi
E-mail: mukesh@physics.iitd.ac.in

ABSTRACT

In present work, one-dimensional nanostructure of Cu-doped Tin oxide (SnO_2) was synthesized by using thermal evaporation method in a tubular furnace under Nitrogen (N_2) ambience. The growth was carried out at atmospheric pressure. SEM images reveals the growth of wire-like nanostructures of Cu-doped SnO_2 on Au (100nm)/Si substrate. The XRD analysis confirms that the synthesized SnO_2 nanowires have tetragonal rutile structure with polycrystalline nature and X-ray diffraction pattern also showed that Cu gets incorporated into the SnO_2 lattice. EDX spectra confirm the doping of Cu into SnO_2 nanowires and and atomic fraction of Cu in nanowires is 0.977 at%. The Vapor Liquid Solid (VLS) growth mechanism for Cu-doped SnO_2 nanowires was also confirmed by EDX spectra. The optical properties of as grown Cu-doped SnO_2 nanowires were studied by using UV-vis spectra which concludes the band gap of about 3.7 eV.

INTRODUCTION

Tin oxide (SnO_2) is N-type, wide band gap semiconductor which is widely used for chemical sensing. Among all metal oxides semiconductors, tin oxide has attracted more attention because of its potential applications in electronics, gas sensor [1] and optoelectronics [2].

V. Rajendran, K. Saminathan, P. Paramasivam and K.E. Geckeler (eds.)
Nano Materials Synthesis and Characterisation, pp. 89-94 (2012).

The 1D nanostructures of SnO_2 can be synthesized by various growth techniques such as laser ablation, thermal evaporation [3] and rapid oxidation. Among these, thermal evaporation is the most popular method for the growth of SnO_2 nanostructures due to its process simplicity, easy-to-handle equipment, low cost infrastructure and high through put. As is well known that doping in semiconductors with selective elements offers an effective approach to adjust their optical, electrical and gas sensing properties, to make them suitable for practical applications [4]. SnO_2 nanostructures can be doped by various metals like Cu [5], Sb [6] and Ru [7] and these doped SnO_2 nanostructures can be used as gas sensors. As reported, CuO-doped SnO_2 thin films [8] as well as nanostructures [9] can also be used as gas sensor for H_2S.

In present work, we have synthesized Cu-doped SnO_2 nanostructures by thermal evaporation method. The grown nanostructures were characterized with different characterization techniques for analyzing their structural morphology, crystallographic nature and optical properties.

EXPERIMENTAL WORK

For the growth of SnO_2 nanostructures, silicon (Si) was used as substrate. The Si substrate was atomically cleaned and gold (Au) catalyst thin film (~ 50 nm) was then deposited on the Si substrate in radio-frequency (RF) sputtering chamber at the pressure of 1×10^{-6} mbar and 100 watt power. For the synthesis of Cu-doped SnO_2 nanostructures, mixture of high purity SnO, graphite and CuO powders was used as source powder and Au catalyst deposited silicon (Au/Si) was used as substrate.

During synthesis, source powder was placed in maximum temperature zone and substrates were placed at a distance away from source materials in a conventional horizontal tube of chemical vapor transport system. The growth was carried out at atmospheric pressure. The furnace temperature was increased to a temperature of 1050°C in the presence of nitrogen gas flow and this temperature was maintained for 1 hour. After that the furnace was cooled down to room temperature and a light white layer was found on the surface of the substrates.

The morphologies and dimensions of the as-synthesized products were characterized by scanning electron microscopy (SEM) (ZEISS, EVO-50 operated at 20 kV) and transmission electron microscope (TEM) (Philips CM12). The phase of the as-synthesized products were identified by X-ray diffraction (XRD) with Cu Ka radiation (λ = 0.1542 nm) from 20° to 80° (2θ). Energy Dispersive X-Ray Spectroscopy (EDX) (RONTEC instrument) was used for analyzing the elemental composition of products. The band gap was investigated using UV–vis spectroscopy (Shimadru UV-2401PC).

RESULTS AND DISCUSSION

SEM image in Fig. 1 shows the morphology of as-grown Cu-doped SnO_2 nanostructures on Au/Si substrate. It can be seen from the images that the growth of Cu-doped SnO_2 nanowire takes place throughout the Au/Si substrate with uniform density. The diameter of Cu-doped SnO_2 nanowires was in the range of 50 nm to 100 nm and their length was about 70 ìm.

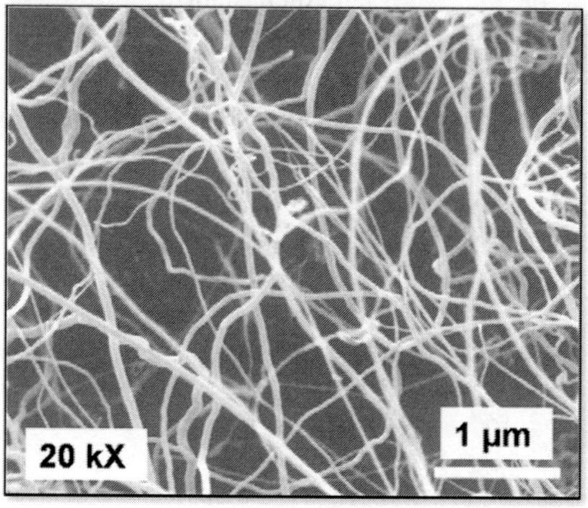

Fig. 1: SEM View of Cu-doped SnO_2 Nanowires

Figure 2 shows EDX spectra of Cu-doped SnO_2 nanostructures grown on Au/Si substrate. The EDX analysis of Cu-doped SnO_2 nanowires indicates the presence of Sn, O, Cu and Au elements, regardless of position in the SnO_2 nanowires. The impurities such as C in EDX spectra are due to carbon coating during sample preparation for EDX analysis. The EDX showed that Cu has been doped in the SnO_2 nanowires and atomic fraction of Cu in nanowires was about 0.977 at%.

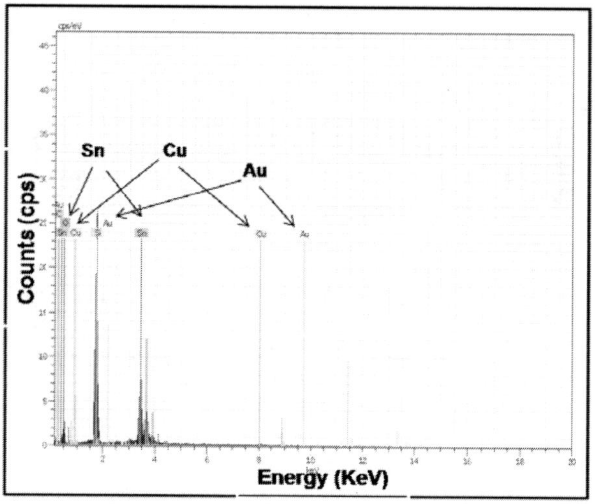

Fig. 2: EDX Spectra of Cu-doped SnO_2 Nanowires.

The XRD pattern (Fig. 3) reveals the overall crystal structure and phase purity of Cu-doped SnO_2 nanowires. All diffraction peaks of Cu- doped nanowires could be indexed to the tetragonal rutile structure of SnO_2 with lattice constants of a = 4.740 Å and c = 3.190 Å as reported in data file (JCPDS 77-0450).

No peaks pertaining to Cu or CuO was visible since the amount of Cu in the nanowires was about very less (~0.977%) in comparision to Sn amount, as confirmed by EDX spectra.

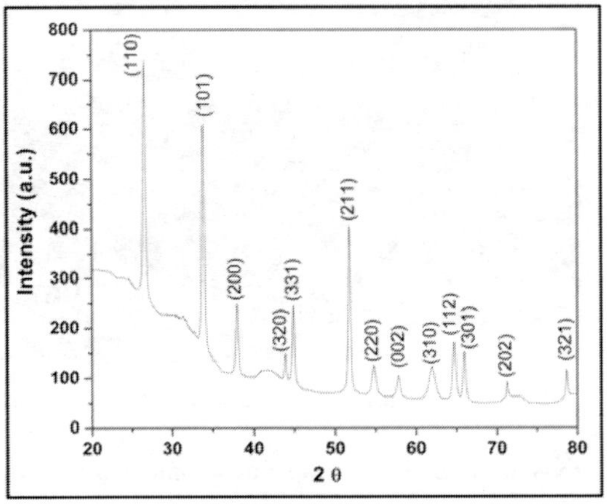

Fig. 3: X-Ray Patterns of Cu-doped SnO_2 Nanowires

The room temperature UV luminescence from Cu-doped SnO_2 nanowires is shown in Fig. 4. The as-grown Cu-doped SnO_2 nanowires were dispersed in ethanol and this solution

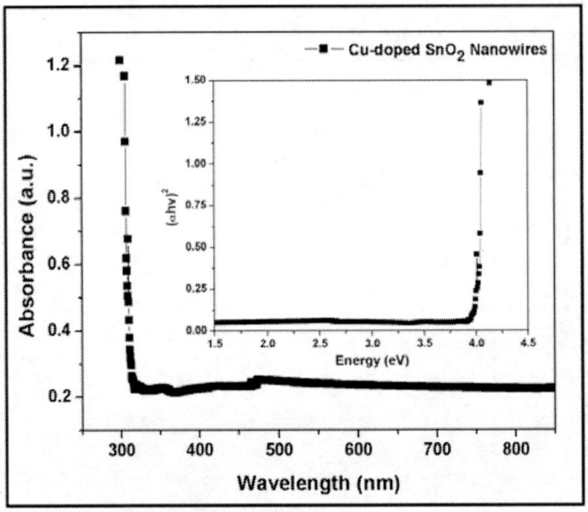

Fig. 4: UV-vis Absorption Spectra for Cu-doped SnO_2 Nanowires

was excited by 325 nm laser and UV-Vis spectra was then recorded in absorption mode. The calculated energy band gap of Cu-doped SnO_2 nanowires was about 3.7 eV. This energy band gap was less than pure SnO_2 nanowires (~ 3.99 eV). Similar decrease in the band gap of SnO_2 has been reported for doping with other transition metals like Co and In [10,11].

CONCLUSION

In summary, 1D Cu-doped SnO_2 nanostructures have been successfully synthesized on Au/Si substrates with a wire like morphology by using thermal evaporation method. The Cu-doped SnO_2 nanowires show polycrystalline nature with tetragonal rutile structure. The EDX analysis confirms the Cu doping in SnO_2 nanowires and also confirms VLS growth of the nanowires.

Optical band gap of Cu-doped SnO_2 nanowires was concluded to be about 3.7 eV. As-grown Cu-doped SnO_2 nanostructures can be used as a sensing element in gas sensor for H_2S gas.

ACKNOWLEDGEMENT

The authors gratefully acknowledge use of Central SEM Facility and XRD facility of IIT Delhi. The authors thank Dr. Ritu Srivastava for UV-vis measurements in National Physical Laboratory Delhi.

REFERENCES

[1] Kolmakov, Zhang, Y.X., Cheng, G.S., Moskovits, M., Detection of CO and O_2 using Tin Oxide Nanowire Sensors, *Adv. Mater.*, 15, 997, 2003.

[2] Ferrere, S., Zaban, A., Gregg, B.A., Dye Sensitization of Nanocrystalline Tin Oxide by Perylene Derivatives, *J. Phys. Chem. B*, 101, 4490, 1997.

[3] Chen, Y., Cui, X., Zhang, K., Pan, D., Zhang, S., Bulk-quantity Synthesis and Self-catalytic VLS Growth of SnO_2 Nanowires by Lower-temperature Evaporation, *Chem. Phys. Lett.*, 369, 16–20, 2003.

[4] Geng, B., Wang, G., Jiang, Z., Xie, T., Sun, S., Synthesis and Optical Properties of S-doped ZnO nanowires, *Appl. Phys. Lett.,* 83, 4791,2003.

[5] Kumar,V., Sen, S., Muthe, K.P., Gaur, N.K., Gupta, S.K., Copper doped SnO_2 nanowires as highly sensitive H_2S gas sensor, *Sensors and Actuators B*, 138, 587–590, 2009.

[6] Wan, Q., Wang, T.H., Single-crystalline Sb-doped SnO_2 nanowires: synthesis and gas sensor application, *Chem. Commun.*, 30, 3841–3843, 2005.

[7] Ramgir, N.S., Mulla, I.S., Vijayamohanan, K.P., A room temperature nitric oxide sensor actualized from Ru-doped SnO_2 nanowires, *Sens. Actuators B: Chem*, 107, 708–715, 2005.

[8] Khanna, A., Kumar, R., Bhatti, S.S., CuO-doped SnO_2 thin films as hydrogen sulfide gas sensor, *Appl. Phys. Lett.*, 82, 4388–4390, 2003.

[9] Wan, Q., Huang, J., Xie, Z., Wang, T., Dattoli, E.N., High sensitivity of CuO modified SnO_2 nanoribbons to H_2S at room temperature, *Appl. Phys. Lett.,* 92, 102101–102103, 2008.

[10] Bouaine, A., Brihi, N., Schmerber, G., Ulhaq-Bouillet, C., Colis, S., Structural, optical, and magnetic properties of Co-doped SnO_2 powders synthesized by the coprecipitation technique, *J. Phys. Chem.* C, 111, 2924–2928, 2007.

[11] Drake, C., Seal, S., Band gap energy modifications observed in trivalent In substituted nanocrystalline SnO_2, *Appl. Phys. Lett.*, 90, 233117-233123, 2007.

CHARACTERIZATION AND ELECTROMAGNETIC STUDIES OF NANO-SIZED GAMMA FERRITE

A.M. Bhavikatti, Subhash Kulkarni[1], Arunkumar Lagashetty[2] and Mahadevi B.[3]

Department of Electronics and Communication Engineering, Dr. MGR University, Chennai
[1]JPNCE, Mahaboobnagar
[2]Department of Chemistry, Appa Institute of Engineering Technology, Gulbarga
[3]Department of E&E, PDA College of Engineering, Gulbarga
E-mail: arvindbhavikatti@gmail.com

ABSTRACT

The increased interest of the researchers in nano objects is due to the discovery of unusual physical and chemical properties of these objects, which is related to the so-called 'quantum size effects'. When materials are reduced to the nanometer size, they unveil new and exciting behaviors which are expected to become the basis for a new generation of electronic and power devices. Nano particles would exhibit unexpected and strange physical and chemical properties compared to conventional or classical materials. For example, copper is a good conductor, while the nano-copper is an insulator.

A key reason for the change in the physical and chemical properties of small particles as their size decreases is the increased fraction of the 'surface' atoms, which occur under conditions, differing from those of the bulk atoms. Currently, unique physical properties of nano particles are under intensive research.

The present article reports some of the interesting and important electromagnetic studies of nano-structured Gamma ferrite. Here, the nano-sized Gamma ferrite (ã-Fe_2O_3) was synthesized by using microwave route. The sample was characterized by X-ray diffraction and SEM. From the XRD it is observed that the structure can be indexed as a monophasic metal oxide cubic spinel structure with no extra lines corresponding to any other crystallographical phase or unreacted ingredient.

V. Rajendran, K. Saminathan, P. Paramasivam and K.E. Geckeler (eds.)
Nano Materials Synthesis and Characterisation, pp. 95-100 (2012).

The peaks 220, 311, 320, 420, 422, 511 and 440 correspond to spinel phase. The d-spacing values of the sample matches well with standard JCPDS (4-755) files. From the SEM of gamma ferrite particles, is observed that, particles are irregular in shape with high porosity. Some particles are found as agglomerates. The average grain size of the Gamma ferrite is calculated to be around 55nm.

Electrical resistance(R) measurements of the sample revealed a very high resistivity and the results indicated a fall in resistivity with the rise in temperature. The obtained dc conductivity is found to be very low of the order of 10^{-8} S/cm. The complex impedance spectroscopy (CIS) technique was used to study the electrical response of the nanostructured sample in a wide range of frequencies (1 KHz to 1MHz) at different temperatures (0–700^0 C).

INTRODUCTION

Gamma ferric oxide (γ-Fe_20_3) is a ferromagnetic material that finds wide use as magnetic storage media in audio and video recording[1]. Recently efforts are underway to find new applications of these materials as drug delivery agents, as superparamagnetic iron oxide (SPJON) particles for hyperthermia, contrast enhancing agents in magnetic resonance imaging (MRI)[2] and also as biosensors[3]. However, with the advent of nanosciencc and nanotechnology, shape and size induced modifications of various properties like optical, magnetic, structural properties is a rich area of physics where quantum mechanical phenomenon like quantum size effects can be investigated on candidates materials like gamma iron oxide[4]. 1t is also known that size dependent magnetic properties manifest itself in to superparamagnetism and single domain characteristics.

CHARACTERIZATION

The structural, morphological and bonding for the prepared Gamma ferrite is studied. Figure 1 shows the XRD pattern of Gamma ferrite. From the XRD it is observed that the structure

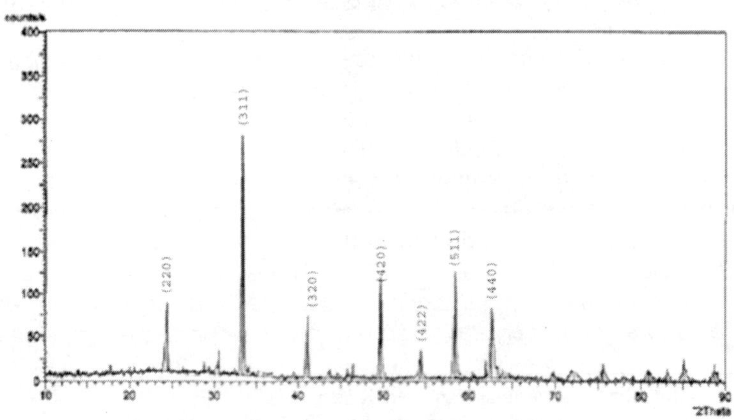

Fig. 1: XRD of Nano-sized Gamma Ferrite

can be indexed as a monophasic metal oxide cubic spinel structure with no extra lines corresponding to any other crystallographical phase or unreacted ingredient. The peaks 220, 311, 320, 420, 422, 511 and 440 correspond to spinel phase. Indexed XRD data of Gamma ferrite sample is given in Table 1. The d-spacing values of the sample matches well with standard JCPDS (4-755) files. Figure 2 shows the SEM of gamma ferrite particles which are in big size. It is observed that, particles are irregular in shape with high porosity. Some particles are found as agglomerates. The average grain size of the Gamma ferrite is around 55nm.

Table 1: Indexed XRD Data of Nano-sized Gamma Ferrite

hkl	d_{obs}	d_{lit}	I_{obs}
220	2.910	2.950	35
311	2.450	2.520	100
320	2.310	2.320	32
420	1.789	1.870	45
422	1.690	1.700	12
511	1.600	1.610	53
440	1.440	1.480	40

d_{obs} = Observed d values, d_{lit} = literature d values, I_{obs} = Observed intensity

Fig 2: SEM of Nano-sized Gamma Ferrite

ELECTROMAGNETIC PARAMETERS

Variation of Resistivity with Temperature

Figure 3 shows the variation of resistivity of Gamma ferrite with different temperatures. For gamma ferrite, high dc resistivity of 1.39×10^{11} Ohm-cm is obtained at a temperature of

175°C and it decreases with increase of temperature according to the Arrhenius equation [5]. The plots show that, the resistivity decreases as the temperature increases. Hence all the samples exhibit semiconducting behavior. Such high values of resistivity are already reported by many investigators [6].

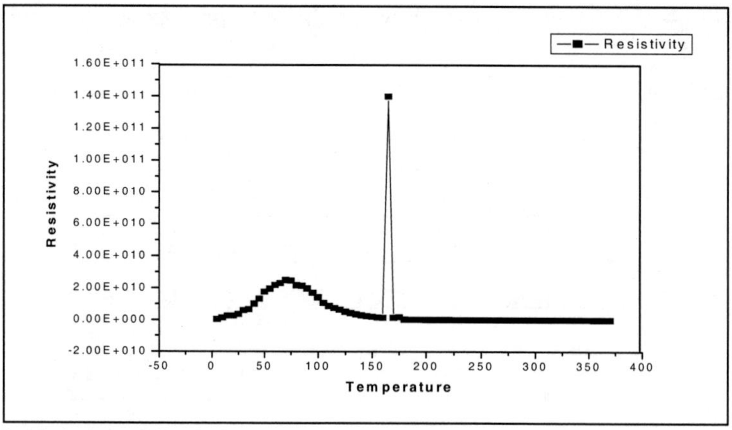

Fig. 3: Variation of Resistivity with Temperature for Nano-sized Gamma Ferrite

Variation of DC Conductivity (σ_{dc}) with Temperature

Figure 4 shows the variation of dc conductivity with temperature for Gamma ferrite. It is observed that σ_{dc} increases with increase in temperature. In the present study, hopping of charge carriers is responsible for electronic conduction [6]. It is observed that the conductivity is very low of the order of 2.4×10^{-9} S/cm around the temperature of 400°C. The minimum conductivity 9.7×10^{-10} S/cm is obtained at a temperature of 340°C.

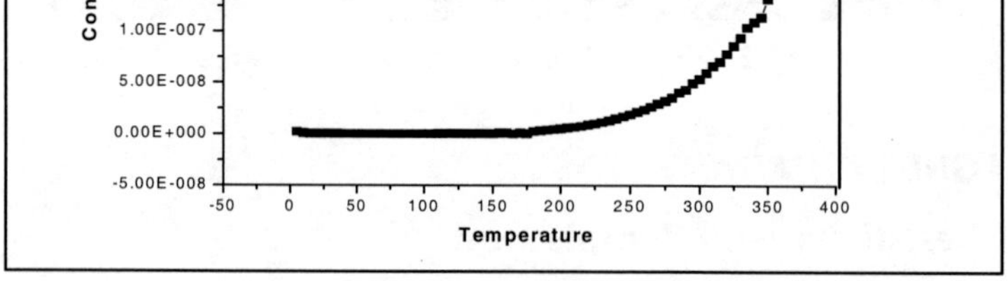

Fig 4: Variation of D.C. Conductivity with Temperature for Nano-sized Gamma Ferrite

Study of Hysteresis Loop Parameters

The hysteresis loop of nano-sized Gamma ferrite is shown in Fig. 5. For the present sample, magnetic properties were evaluated with the help of M-H curve or hysteresis loop. The remanance (M_r) was found to be 3.98 emu/gm, saturation magnetization (M_s) is found to be 18.75 emu/gm and coercivity (H_c) is found to be 338.84 Oe when the applied field is 108×10^2Oe. Ferrite is found to be a soft magnetic material. However, comparing these results with the magnetic properties of nano-sized gamma ferrite prepared by nonaqueous solution approach as reported earlier [7], the values obtained in those cases were (M_s)was found to be 57.71 emu/gm and coercivity (H_c) was found to be 78.11 Oe and remanance (M_r) was found to be 7.326 emu/gm. As the Ms Value obtained in the case of nano-sized gamma particle was high, we can say that, these particles are good for application in magnetic recording.

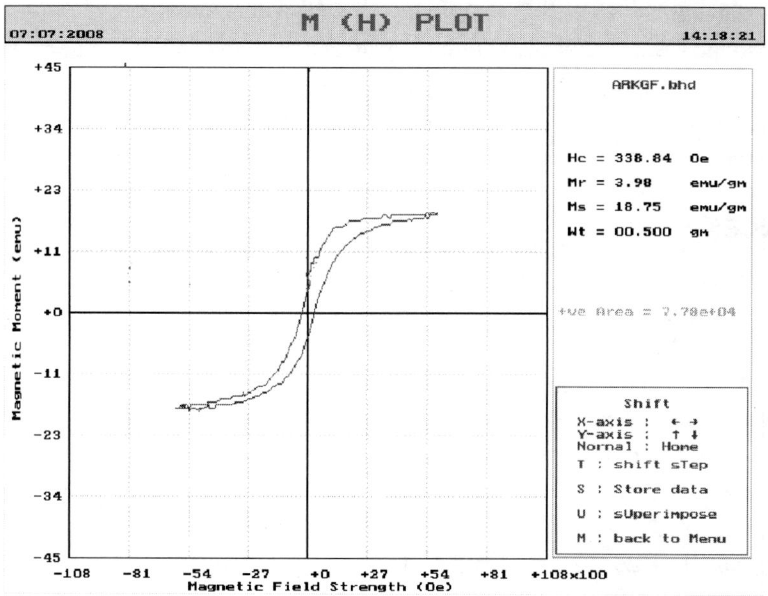

Fig. 5: Hysteresis Loop of Nano-sized Gamma Ferrite

INSTRUMENTATION USED

The x-ray diffraction patterns were obtained at room temperature, using GEOL JDX-8P X-ray diffractometer at a scanning rate of 2° minute using CuK_α radiation of wavelength (λ) 1.54056 Å . Leica Cambridge-440 Scanning Electron Microscope was used here to get SEM image. DC electrical resistivity [R] values of ferrite sample were obtained by using a two probe method with keithley electrometer (Model 2010) in the temperature range from room temperature to

400^{0}C and conductivity was determined. Resistivity[ρ] and DC conductivity [σ_{dc}] values were calculated. Hysteresis loop parameters at room temperature were evaluated for the sample by using a vibrating sample magnetometer (Model DMS 1660 VSM) at a maximum applied field.

CONCLUSIONS

We prepared nano-sized Gamma ferrite material by microwave synthesis process. Some of the structural and electromagnetic properties were investigated. Based on the results obtained, the following conclusions are drawn,

(i) High value of dc resistivity obtained in this ferrite sample makes it suitable for high frequency applications where eddy current losses become prominent.

(ii) Ferrite is found to be a soft magnetic material.

(iii) The d-spacing values of this sample match well with standard JCPDS (4-755) files.

(iv) The average grain size of the Gamma ferrite is calculated and found to be around 55 nm.

(v) It may also be observed that, standard values are obtained after experimentation even though microwave synthesis scheme is used. The graphs obtained in this case are comparable to the existing standard graphs available in literature.

REFERENCES

[1] B.R.V. Narasimhan, S. Prabhakar, P. Manohar, F.D. Gnanam, "Synthesis of gamma ferric oxide by direct thermal decompositionof ferrous carbonate", *Materials Letters*, 52, pp. 295-300, 2002.

[2] Sabino,V.-verdaguer, Maria, del P., Morales, Oscar, Bomati-M., Carmen, B., Xinquing, Z.; Pierre, B., Rigoberto, Perez de A., Jesus, Ruiz-cabcllo, Martin, S., Fransisco, J. Tendillo-Cortijo, Joaquin, F. J. *Phys. D: Appl. Phys.* (2004), 37, pp. 2054.

[3] Panagiotis, D., Nicolas, M., Eamonn, D., Dimitrios, N., Dimitros, P., Nanotechnology (2006), 17, pp. 5019.

[4] Kaehkachi, H.; Ezzir, A. Nogues, M., Tronc, *E. Eur. Phys. J.B.* (2000) 14, pp 681.

[5] B. Viswanathan, VRK Murthy, "Ferrite materials", Narosa Publishing House, New-Delhi.

[6] M.C. Chhantbar, U.N. Trivedi, P.V. Tanna, H.J. Shah, R.P. Vara, H.H. Joshi, K.B. Modi (2004), "Infrared spectral studies of Zn-substituted $CuFeCrO_4$ spinel ferrite system", *Indian J. Phys.* 78A(3), pp 321-326.

[7] B. Parvatheeswarararao, A. Maheshkumar, K.H. Rao, Y.L.N. Murthy, O.F. Caltun, I. Dumitru, L. Spinu (2006), "Synthesis and magnetic studies of Ni-Zn ferrite nanoparticles," *Journal of Optoelectronics and Advanced Materials*, Vol. 8, No. 5, pp. 1703-1705.

CHARACTERIZATIONS ON TETRAKIS (DIMETHYL AMINO) ETHYLENE-TREATED-GRAPHENE

A.P. Alegaonkar[1,3], Arvind Kumar[2], P.S. Alegaonkar[2], S.A. Waghmode[3] and S.K. Pardeshi[1]

[1]*Department of Chemistry, University of Pune, Maharashtra*
[2]*Dept. of Applied Physics, Defence Institute of Advanced Technology, Pune, Maharashtra*
[3]*Department of Chemistry, Abasaheb Garware College, Pune, Maharashtra*
E-mail: skpar@chem.unipune.ac.in; shkhawaghmode@gmail.com

ABSTRACT

Characterization studies were carried out on the Graphene after treating it with tetrakis(dimethylamino) ethylene (TDAE). Initially, as synthesized–Graphene samples were dispersed and sonicated in the TDAE solution. The as synthesized- and TDAE-treated-Graphene samples were subjected to Fourier Transform Infrared Spectroscopy (FTIR), X-ray Diffraction (XRD) and Vibrating Sample Magnetometer (VSM). The FTIR measurements, showed the emergence of new bands, for the TDAE–treated–Graphene, at 3433 cm^{-1}, 1450–1250 cm^{-1} and 983 cm^{-1} associated with the fragments/radicals of TDAE. The XRD measurements, showed the characteristic Graphene peak at $2\theta \sim 13.11°$. The XRD pattern also revealed that, for the TDAE–treated–Graphene, the peak is shifted from 13.11° to 12.25° which could be due to the change in the interlayer δ–spacing by introducing TDAE. The VSM measurements were carried out at the AC magnetic field from 0 to 15 kOe on both the samples, at room temperature. No hysteresis pattern was observed for the as synthesized–Graphene. Whereas, for the TDAE-treated-Graphene hysteresis was obtained with a remanent magnetization, M_r, of ~ 0.46 emu/gm and coercive field, H_c of ~ 2.172 Oe, recorded at 300 K. Our analysis revealed that, the magnetization, in the TDAE-treated-Graphene, could be due to the spin exchange overlap of unhybridized p_z electrons and the fragments of TDAE via charge transfer mechanism at the Graphene edges or network within. The details are presented.

V. Rajendran, K. Saminathan, P. Paramasivam and K.E. Geckeler (eds.)
Nano Materials Synthesis and Characterisation, pp. 101-106 (2012).

INTRODUCTION

Graphene [1], a two dimensional monoatomic thick building block of the carbon allotrope, has emerged as an exotic material of this century. Graphene could be synthesized in the single-, bi-, and few-layer format. Graphene based materials have significant impact on the electronic parameters such as room temperature quantum Hall effect, ambipolar electric field effect with ballistic charge transport, tunable band gap, optoelectronic devices, chemical sensors, nanocomposites with high elastic modulus, energy storage systems and various types of solar cells [2]. Furthermore, there are few reports on the magnetic properties of carbon based nanomaterials such as nanographite particles and nanoribbons [3]. Such carbon nanomaterials exhibit unusual magnetic properties like switching characteristics and spin glass arising from the nonlocal bonding of the edge state electrons. The bi-layer Graphene is predicted to be ferromagnetic, whereas, the hydrogenated nanographite is reported to show spontaneous magnetism at room temperature. Various measurements such as high field magnetization, electron spin resonance (ESR) and nuclear magnetic resonance (NMR) and pressure dependence of magnetization were carried out on Carbon-TDAE system. However, the mechanism of ferromagnetism in Carbon-TDAE is still unclear [4]. Thus the focus on magnetism of metal-free materials is important to both basic science and technology, with application ranging from nanoscale electronics to cancer treatment.

The purpose of this communication is to investigate the structure–property relationship in the tetrakis (dimethylamino) ethylene (TDAE)-treated-Graphene. The TDAE-treated–Graphene consists of fragments/radicals of TDAE, such as, tetrakisdimethylamino-1,2-dioxetane (TDMD), tetramethylurea (TMU), tetramethyloxamide (TMO), tetramethylhydrazine (THM), bis (dimethylamino) methane (BMAM), dimethylamine (DMA). These fragments could be responsible to induce local magnetic ordering due to charge transfer between them and unhybridized p_z electrons of Graphene at room temperature.

EXPERIMENTAL

Initially, the readily available Graphene was cleaned by admixing it with the isopropyl alcohol (IPA). The solution was sonicated at room temperature for a period of 30 min. Following this, the solvent was filtered using PTFE microfilter and the obtained powder was dried under atmospheric conditions. In order to obtain a stable suspension, Graphene was dispersed in the tetrakis (dimethylamino) ethylene (TDAE) and sonicated for a period of 45 min, at a volume proportion of 1:10. After sonication, the suspension was filtered and dried under air ambient. In this way, the obtained TDAE-treated-Graphene powder samples were kept ready for further characterization.

The as synthesized-Graphene and TDAE-treated-Graphene samples were mixed with KBr and ground to prepare the pellets. These pellets were subjected to Fourier Transform Infrared (FTIR) spectroscopy using Perkin–Elmer Spectrometer over wavenumber range

4000-400 cm^{-1} with resolution of 4 cm^{-1}. Variations in the intensity of absorption as a function of wavenumber were recorded for the as synthesized-, and the TDAE-treated-Graphene samples. For X-ray powder Diffraction (XRD) technique, data was collected on D_8 Advanced Bruker Diffractometer by using Cu $K_á$ (36 kV, 36 mA) radiation and secondary beam graphite monochromater. Initially, the as synthesized-Graphene sample was mounted on the XRD table and spectra was recorded in the 2 theta (2θ) range from 10° to 50° in steps of 0.02° and counting time 2 s per step. Variation in the Bragg scattered X-ray Intensity as afunction of specular reflection angle, 2θ, were collected for the as synthesized- and a few TDAE-treated-Graphene samples. Furthermore, the as synthesized- and TDAE-treated-Graphene samples were subjected to Vibrating Sample Magnetometer (VSM) technique at room temperature and variation in the Magnetization (M) as a function of applied AC magnetic field (H) were studied. The samples of weight ~5 µg was loaded in a cylindrical shaped plastic container and four such containers were prepared and subjected to VSM for each set. The magnetic field (H) was varied from 0 to ± 1.5 kOe for a sweeping period of ~15 min time with 0.1 s time constant. The resolution of the system is 0.1 emu.

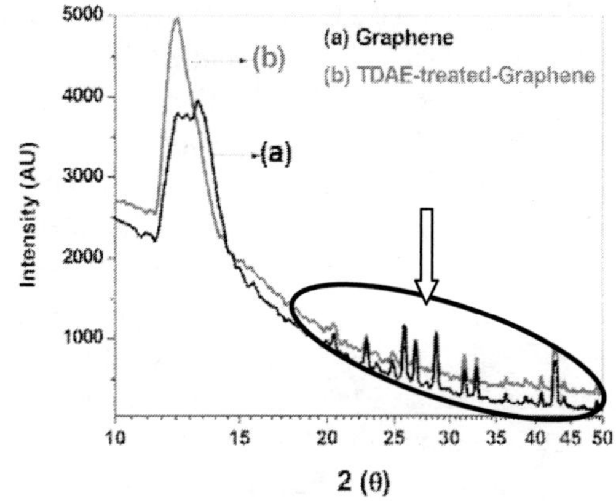

Fig. 1: FTIR Spectrum for (a) As Synthesized-Graphene, and (b) TDAE-Treated-Graphene.

RESULTS AND DISCUSSION

The TDAE is an organic material which shows ferromagnetic property via electron charge transfer mechanism when attached to the Graphene. This is due to unique electron donating property of TDAE [4]. Figure 1. shows the recorded FTIR spectrum for (a) as synthesized–Graphene, and (b) TDAE-treated-Graphene. In the profile (a), mainly three bands were observed, below 3000 cm^{-1}. The absence of major bands above 3000 cm^{-1} indicates that the as synthesized-Graphene contains no hydroxyl (–OH) and amino (–NH_2) group. Moreover, the set of bands between 1500–1600 cm^{-1} could be attributed to the conjugated double bond C=C and single bond, i.e. C–C. The recorded FTIR profile (b), for the TDAE-treated-Graphene, is distinctly different than that of the profile (a). The TDAE has 108 fundamental vibration modes more than half of which consists of C–H bond stretching and bending within 8 methyl fragments. The complete assignment of them is a difficult task therefore we have focused our attention mainly on C–N bonds stretching mode where different charge states are expected from different fragments.

The emergence of broad band, for TDAE-treated-Graphene, at ~3433–3335 cm^{-1} (indicated by the open circle) is assigned for –N–H amines/amides, associated with one of the fragments of the TDAE, i.e. dimethylamine (DMA). The small bands appeared at 3000–2650 cm^{-1} is –C–H stretch, whereas, the bands at 3350–3310 cm^{-1} is for N-heterocyclic compounds. This could easily be reconfirmed with the –C=N–H mode that appeared at ~ 1641cm^{-1}. It also explains R–NH_2 and Ar–NH_2 in plane bending between 1599–1641 cm^{-1} and could be attributed to the fragment of tetramethylurea (TMU). The small peak appearing at ~ 983 cm^{-1} (indicated by an arrow) could be attributed to the out of plane bending corresponding to the CH_2 twisting. The shakeup feature appeared at ~ 1599 cm^{-1}, as a shoulder to main band, is attributed to the fragment of the tetramethyloxamide (TMO). This feature explains the Ar–NH– bonding with aromatic ring and confirmed by the band appearing at ~ 1599 cm^{-1}. The bands appearing at 1450–1350 cm^{-1} is the stretching of C–N bonds and $N_2C=CN_2$ fragments, while peaks at 1372–1382 cm^{-1} contain the C–N bonds stretching within the N–CH_3 fragments and could be attributed to the existence of fragments of tetrakisdimethylamino-1, 2-dioxetane (TDMD), and tetramethylhydrazine (THM). Further, the aromatic C–N stretching could also be explained using the stretching bands appeared between 1350–1265 cm^{-1} (as indicated by the circle in Fig. 1). These are due to the C–C character with C–N stretch. The disubstitution of amines to sp^2 bonded carbon network could be confirmed by observing two bands closely appearing at 1641 cm^{-1} and 1599 cm^{-1}. Moreover, the bands appeared at 1448 cm^{-1} and 1408 cm^{-1} are stretch vibrations assigned to the geminal dimethyl group. The band appearing at 1350 cm^{-1} is for –CO–CH_3 on the zig-zag carbon of Graphene which is attributed to asymmetric bending mode appeared at 1448 cm^{-1} attached with the bis (dimethylamino) methane (BMAM). From above discussions, the FTIR analysis revealed that, the fragments of TDAE are attached to the zig-zag chains of Graphene edges or the honeycomb network. These sites play important role in charge transfer mechanism via exchange interactions for the magnetic ordering for the TDAE-treated–Graphene samples.

Figure 2 shows XRD pattern for (a) as synthesized-Graphene and (b) TDAE–treated-Graphene. In Fig. 2(a), the peak at 2θ ~ 13.11° is assigned to the monolayer of atomically thin Graphene. The XRD peaks appeared in the range from 2θ ~ 20°–45° (indicated in the Fig. 2. by an arrow and circle) are attributed to the nanogaphitic (NG), Graphene like nanocarbon phases (GNC) that are present in the as synthesized-Graphene. This indicates that, the Graphene samples consist of NG as well as GNC in addition to the Graphene monolayers. Further, for the TDAE-treated-Graphene, (Fig. 2(b)) the peak associated with the Graphene is observed to be shifted from 13.11° to 12.25°. Interestingly, for the TDAE

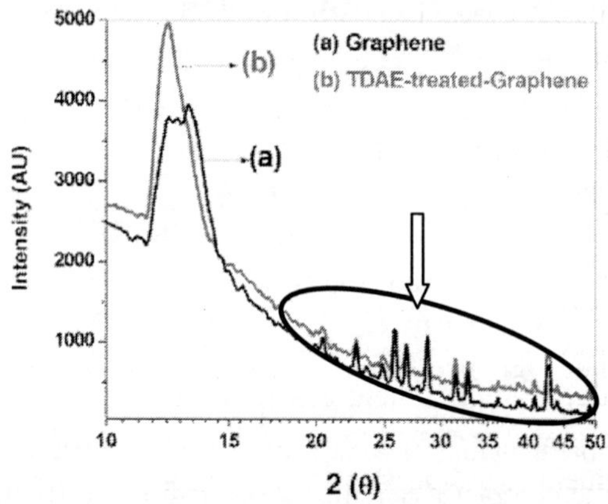

Fig. 2: XRD Pattern for (a) As Synthesized–Graphene, and (b) TDAE-treated-Graphene.

treated-Graphene samples, the peaks in the range of 2θ ~ 20°–45° corresponding to NG and GNC also exists. However, their positions are almost unshifted with respect to their position identified for the as synthesized–Graphene samples. This indicates that, the NG and GNC phases are unaltered by the TDAE treatment. Hence the main contribution could be from the Graphene monolayers selectively, in which TDAE gets doped. Further, the downshift in peak position of the main peak shows the increase in the interlayer d–spacing of the carbon honeycomb network from 7.21A° to 7.95A°, after treating it with TDAE.

Figure 3 shows VSM spectrum for the TDAE-treated-Graphene. No hysteresis pattern was recorded for all the as synthesized-Graphene samples. Hence, the M–H curve for the as synthesized-Graphene is not shown here. The result at ±1.5kOe field, indicates that, there is hysteresis with a remanent magnetization for the TDAE-treated-Graphene. We recorded remanent magnetization per sample, M_r, of ~ 0.46 emu/gm and coercive field, H_c of ~ 2.172 Oe at 300 K. The hysteresis in the magnetization verses the applied external magnetic field was evident over the range of 2.2 kOe. The effective magnetic moment in these samples was determined since the weight of each sample (~ 5.1 μg) and composition of the TDAE-treated-Graphene composite were known. Thus, experimental fact for spontaneous magnetization response in the TDAE-treated-Graphene is obtained at room temperature. Magnetism in carbon basically requires unpaired electrons with atoms carrying a net magnetic moment that can be coupled via exchange with other electrons. This coupling can be between itinerant or localized electrons. Specifically, the magnetic properties of carbon-TDAE molecule depends on the intermolecular interactions and are assumed to be itinerant ferromagnetic in nature [5]. The radicals/fragments of TDAE, as identified by FTIR spectroscopy, are the single electron dopants. These sites on the quasi 2D Graphene network could act as the spin ½ Heisenberg ferromagnet. And the spin ordering is possible via charge transfer mechanism when applied to the external magnetic field even at room temperature. Thus, for the TDAE-treated-Graphene, the radicals which are absorbed or intercalated could possibly influence the magnetic properties, probably due to the overlap of edge state Graphene p_z electrons to TDAE fragments. Thus, the absorption of different itinerant charge donor guest molecules on Graphene could give rise to reversible, low/high spin switching phenomena which depends upon nature of the guest species [6].

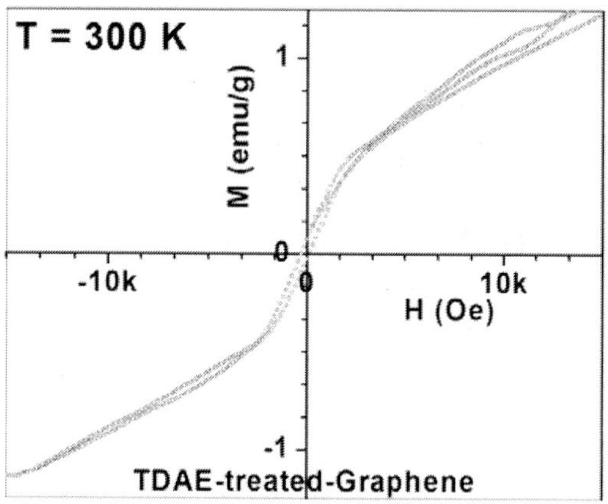

Fig. 3.: VSM Spectrum Recorded for the TDAE–Treated–Graphene at 300 K

CONCLUSIONS

Thus, the analysis of our characterization leads to the conclusion that, the fragments of TDAE attached to the Graphene, shows magnetization at room temperature. The FTIR studies, showed emergence of new bands at 3433 cm^{-1}, 1450–1250 cm^{-1} and 983 cm^{-1}. These bands were assigned to the charge transfer fragments of tetrakisdimethylamino-1,2-dioxetane (TDMD), tetramethylurea (TMU), tetramethyloxamide (TMO), tetramethylhydrazine (THM), bis (dimethylamino) methane (BMAM), dimethylamine (DMA). The analysis of XRD revealed that, the characteristic Graphene peak recorded at $2\theta \sim 13.11°$ was shifted to $2\theta \sim 12.25^0$ for the TDAE-treated-Graphene. The shift is attributed to the deformation of the interlayer d-spacing by introducing the fragments of TDAE. In VSM analysis, no hysteresis pattern was observed for the as synthesized–Graphene, whereas, on TDAE treatment, the hysteresis was obtained with a remanent magnetization per sample, M_r, of ~ 0.46 emu/gm and coercive field, H_c of ~ 2.172 Oe recorded at 300 K. Our study reveals that, the magnetic ordering, at room temperature, for the TDAE-treated-Graphene could be associated with the exchange overlap of unhybridized $p\text{-}_z$ electrons to the fragments of TDAE via charge transfer mechanism at the Graphene edges or within the monolayer network.

ACKNOWLEDGEMENTS

Authors are thankful to Local Organizing Committee for accepting the paper in prestigious MAM-12 conference. APA, SKP is thankful to Head, Dept. of Chemistry, University of Pune, APA and SAW is thankful to Head, Dept. of Chemistry, Garware College, Pune. PSA and AK is thankful to Head, Dept. of Applied Physics and the Vice Chancellor, DIAT, Pune for their help and support. All authors are thankful to CNNC, Korea.

REFERENCES

[1] Novoselov K.S., Geim A.K., Morozov S.V., Jiang D., Katsnelson M.I., et al., Two- dimensional gas of massless Dirac fermions in graphene, *Science*, 306, 666–669, 2004.

[2] Singh V., Joung D., Zhai L., et al., Graphene based materials: Past, present, and future *Prog. Mat. Sc.,* 56, 1178–1271, 2011.

[3] Andersson O.E., Prasad, B.L.V., Sato, H., Enoki, et al., Structure and electronic properties of graphite nanoparticles, *Phys. Rev. B.*, 58, 16387–13,2007.

[4] Marakova, T., and Palacio F., Carbon–Based Magnetism, 1st ed., Elsevier, Netherlands, 2006.

[5] Winberg, H.E., Carnahan, The Chemiluminescence of Tetrakis (dimethylamino) ethylene *J. Am. Chem. Soc.*, 87, 2054–56, 1965.

[6] Heller C. A. and Fletcher K.A., Chemiluminescence quenching terms, *J. Phys. Chem.*, 69, 3313–17, 1965.

STRUCTURAL PROPERTIES OF COBALT SUBSTITUTED BARIUM HEXAFERRITE NANOPARTICLES PREPARED BY A THERMAL TREATMENT METHOD

C.C. Chauhan, R.B. Jotania[1] and K.R. Jotania[2]

Department of Physics, Institute of Technology, Nirma University, Ahmedabad, Gujarat

[1] *Material Science Research Laboratory, Department of Physics, University School of Sciences, Gujarat University, Ahmedabad, Gujarat*

[2] *Department of Physcis, Faculty of Science, The M.S.University of Baroda, Vadodara, Gujarat*

E-mail: rbjotania@gmail.com

ABSTRACT

A series of M-type hexagonal ferrites with the composition $BaCo_xFe_{12-x}O_{19}$ (x = 2.0, 3.0) were synthesized using a simple heat treatment method. The aqueous solution containing metal nitrates and polyvinyle pyrrolidone (PVP), as an capping agent, was used to prepare M-type barium hexferrite nano particles. The prepared hexaferrite particles were calcinated at microwave frequency (2.45 GHz, power 900W, 5 min.) as well as at 650°C temperature. The structural properties of the samples were investigated by using X-ray diffraction (XRD), Fourier transform infrared spectroscopy (FTIR) and scanning electron microscopy (SEM). XRD pattern shows the pure M-phase. The particle size of the powder prepared by this method ranges from 20 nm to 27nm. The thermal properties of the sample were investigated by thermogravimetric Analysis (TGA), which confirms the stability of the sample prepared by the heat treatment method.

V. Rajendran, K. Saminathan, P. Paramasivam and K.E. Geckeler (eds.)

Nano Materials Synthesis and Characterisation, pp. 107-112 (2012).

INTRODUCTION

Hexaferrites are classified into six fundamental types: M, W, Y, X, Z and U on the basis of their chemical composition and crystal structure [1,2]. The crystal structure of these different types of hexaferrites is very complex and can be considered as a superposition of T, R and S blocks along the hexagonal *c*-axis [3]. The M-type barium hexaferrite is a very useful material as absorbers of electromagnetic radiation in the microwave region [4]. The molecular structure of M-type ferrite is made up of one R and one S block with an overlap of hexagonally and cubically packed layers. The crystal structure of $BaFe_{12}O_{19}$ is described as SRS*R*, where R is a three layer block (O_4–BaO_3–O_4) with composition $BaFe_6O_{11}^{2-}$ and S is a spinel like two layer block (O_4–O_4) with composition $Fe_6O_8^{2+}$. Here the asterix (*) means the corresponding block has been turned 180° around the hexagonal *c*-axis. The O^{4-} and BaO^{3-} layers form a close packing. In this structure the smaller iron cations are distributed within five different kinds of sites (Three Octahedral – 12k, 2a, $4f_2$, one tetrahedral – 4f and one pseudo tetrahedral trigonal bi pyramidal – 2b). The ordering of magnetic moments of Fe^{+3} ions and the strong super exchange interaction explain the excellent magnetic behaviour of these materials [5].

Barium hexaferrite is currently of technical interest because of its permanent magnetic property, high uniaxial magneto-crystallilne anisotropy, saturation magnetization, excellent chemical stability and corrosion resistance [6,7]. Recently, ceramic materials are classified by their high resistivity and high permeability, so numerous products of them are used in industry, technology, electronic devices, communication equipments, high density magnetic recording media, materials for permanent magnets and microwave devices[2,8-12].

In present work, barium hexaferrite nanoparticles were synthesized from an aqueous solution of metal nitrates, polyvinyl pyrrolidone and deionized water using a thermal treatment method. This method does not require the addition of other chemicals as surfactants. This method has an advantage of simplicity, low cost, no byproducts and environmental friendly. The structural and morphological characteristics of the prepared hexaferrite nanoparticles were studied to verify the particle size.

SAMPLE PREPARATION

Barium hexaferrite was synthesized using metal nitrate reagent as precursors, polyvinyl pyrrolidone (PVP) as an capping agent to reduce the agglomeration of the magnetic particles and deionized water was used as the solvent. The aqueous solution of PVP was prepared by dissolving 3 gm of polymer in 100 ml of deionized water at a temperature of 70°C before mixing 1.2 mmol of iron nitrate and 0.1 mmol of barium nitrate into the polymer solution and constantly stirring for 2 hours on a magnetic stirrer until a colourless transparent solution was obtained. The solution without precipitate was obtained before heat treatment and had a pH between 1 to 2. The mixed solution was heated at 80°C to evaporate the water. The resulting orange solid was crushed for 15 min and heated at 650°C for 3 hours for the decomposition of the organic compounds and the crystallization of the nanoparticles. The prepared hexaferrite particles were also calcinated at microwave frequency (2.45 GHz, power 900W, 5 minute).

EXPERIMENTAL TECHNIQUES

The characterization of the prepared hexaferrite sample was conducted using various experimental techniques to verify the particle size. The structure of $BaCo_xFe_{12-x}O_{19}$ (x = 2.0, 3.0) was characterized using a Philips diffractometer (PW 1830) using Cu Ká radiation (λ = 1.5405 Å) with a scan rate of 0.02degree/min. The room temperature infrared(IR) spectra of gel and calcinated samples were recorded in mid IR range (4000 cm^{-1} to 400cm^{-1}) using FTIR spectrophotometer (Bruker Tensor 27 Model). Scanning electron micrographs of prepared barium hexaferrite samples are obtained using a Make-Leo/Lica model Stereo scan 440 scanning electron microscope.

XRD Analysis

Figure 1 shows the XRD patterns of $BaCo_xFe_{12-x}O_{19}$ (x = 2.0, 3.0) calcinated at 650°C for 3 hours and heated at microwave frequency. The XRD pattern shows the pure M-phase. All the samples exhibit quite similar XRD patterns assigned to various calcinations methods. The average particle size of the M-type hexaferrite was calculated from the full width at the half maximum (FWHM) of the XRD patterns, using the well known Scherer formula

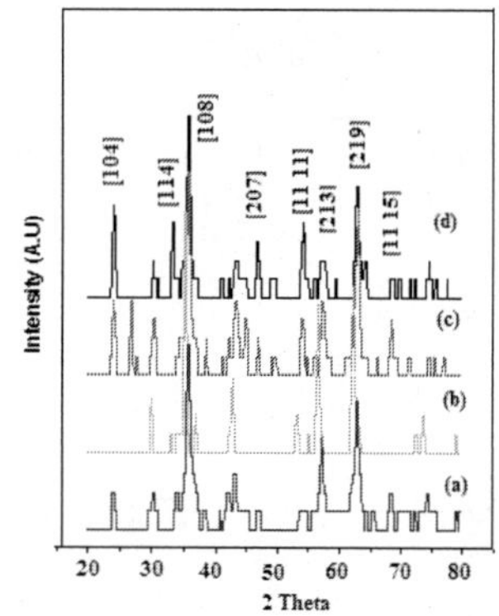

Fig. 1: X-ray Diffractograms of
(a) $BaCo_2Fe_{10}O_{19}$ Calcinated at 650°C,
(b) $BaCo_2Fe_{10}O_{19}$ Heated at Microwave Frequency
(c) $BaCo_3Fe_9O_{19}$ Calcinated at 650°C and
(d) $BaCo_3Fe_9O_{19}$ Heated at Microwave Frequency

$$d = 0.9\frac{\lambda}{\beta Cos\theta} \quad \text{... (1)}$$

where d is the crystallite size (nm), β is the full width of the diffraction line at half of the maximum intensity measured in radians, λ is the X-ray wavelength of Cu Kα = 0.154 nm and θ is the Bragg angle.

Table 1: Structure Parameters *a* and *c* and Cell Volume V for Hexaferrite Samples

$BaCo_xFe_{12-x}O_{19}$ (x = 2.0, 3.0)	*Heat treatment*	*a (nm)*	*c (nm)*	*Cell Volume V (nm)³*
$BaCo_2Fe_{10}O_{19}$	650°C	0.550	2.3456	1.291
$BaCo_2Fe_{10}O_{19}$	Microwave	0.545	2.3780	1.284
$BaCo_3Fe_9O_{19}$	650°C	0.548	2.3423	1.359
$BaCo_3Fe_9O_{19}$	microwave	0.5445	2.3772	1.294

Lattice constants *a* and *c* of hexagonal barium cobalt hexaferrite were calculated using Eq. (2) :

$$\frac{1}{d} = \frac{4}{3}\frac{h^2 + k^2 + l^2}{a^2} + \frac{l^2}{c^2} \qquad ...(2)$$

where *h, k* and *l* are Miller indices, *d* is inter planer distance. Lattice volume of all the samples can be obtained using Eq. (3):

$$V = \frac{\sqrt{3}}{2} a^2 \qquad ... (3)$$

Table 2: Particle Size from Debye Scherrer Formula

$BaCo_xFe_{12-x}O_{19}$ ($x = 2.0, 3.0$)	θ (degree)	Cos θ	β FWHM (rad)	Particle Size d (nm)
$BaCo_2Fe_{10}O_{19}$ (650°C)	17.314	0.954688	0.005152	27.87
$BaCo_2Fe_{10}O_{19}$(micro wave)	17.32	0.954657	0.006185	23.22
$BaCo_3Fe_9O_{19}$ (650°C)	17.171	0.955942	0.00604	23.89
$BaCo_3Fe_9O_{19}$ (microwave)	17.22	0.955175	0.00687	20.89

FTIR Analysis

The room temperature infrared spectra of the samples calcinated by conventional method and microwave heated samples recorded in mid IR range (4000cm^{-1} to 400 cm^{-1}) using FTIR spectrophotometer (Bruker Tensor 27 Model) are shown in figure 2. A few milligrams of the samples were mixed with anhydrous KBr powder and made in the form of pallet for the measurements. The absorption band at 2923.88 cm^{-1} and 2151.3588 cm^{-1} is due to OH vibrations. The most important peaks are at 1427, 1415.5, and 858.26 cm^{-1}, corresponding to the stretching and bending vibrations of C=O, H–C–H, C–H and C–C respectively. It must be mentioned that the absorption peaks of the nitrate anion overlap with vibration of C–H at 1427 cm^{-1}. The absorption band between 513.89 and 468.57 cm^{-1} increases corresponds to the pure hexaferrite nanoparticles, confirming the formation of hexaferrite [13].

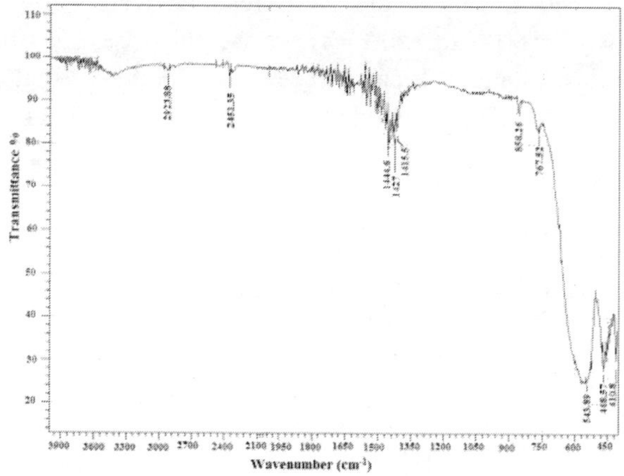

Fig. 2: FTIR Spectra of $BaCo_2Fe_{10}O_{19}$ Hexaferrite Heated at 650°C

SEM Analysis

Figure 3 shows the SEM images of the samples calcinated at various temperatures. The SEM images of the samples shows the agglomerated particles of different shapes and sizes. The platlet stucture of the hexaferrite particles are observed. The particles are found to be porous in nature, indicating that the hexaferrite particles of nano dimensions are formed.

Fig. 3: SEM Images of $BaCo_2Fe_{10}O_{19}$ (a) Calcinated at 650°C (b) Heated at Microwave Frequency, $BaCo_3Fe_9O_{19}$, (a) Calcinated at 650°C, and (d) Heated at Microwave Frequency

Thermogravemetric Analysis (TGA)

TGA curve of the dried precursor was recorded using SII Differential Thermal analyzer Model No. SSC 5100 in the region of 50°C to 700°C with a rate of 10°C/min and is shown in Fig. 4. Thermo gravimetric analysis shows the first step weight loss between 0°C to 100°C, with 11% wt. loss due to the desorption of absorbed water molecules. In the second step, 50% wt. loss is observed between 100ºC to 180ºC. In the third step, the weight loss is about 7%, in the temperature range between 180ºC to 350ºC. These weight losses may be due to combustion and oxidation of organic substances. It can be seen from the Fig. 4 that above 500ºC there is very less weight loss which confirms the thermal stability of the prepared sample synthesized by the heat treatment technique.

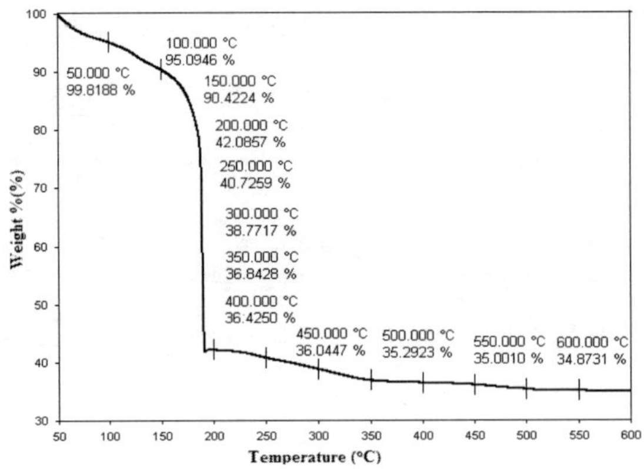

Fig. 4: TGA Curve of $BaCo_3Fe_9O_{19}$ Precursor

REFERENCES

[1] Salunkhe M.Y., Kulkarni, D.K. *J. Magn. Magn. Mater.* 64, 279, 2004.

[2] Smit, J., Wijn, H.P.J., in: Ferrites, Philip's Tech. Library, New York, 1959.

[3] Zhang Haijun, Yao Xi, Zhang Liangying, *Ceram. Int.* 28, 17, 2002.

[4] Eugene, P., et al., *Chem. Sustain. Dev.* 161**,** 10, 2002.

[5] Carp, O., Barjega, R., Segal, E., Brezeanu, M., *Thermo-chem. Acta* 318, 57, 1998.

[6] Pullaar, R.C., Bhattacharya, A.K., Mater. Lett. 537, 57, 2002.

[7] Dho, J., Lee, E.K., Park, J.Y., Hur, N.H., *J. Magn. Magn. Mater.* 164, 285, 2005.

[8] Vishwanathan, B., Murthy, V.R.K., Ferrite Materials; Science and Technology, Narosa publishing house, New Delhi, 1990.

[9] Snelling, E.C., Soft Ferrites/Properties and Appllication; second ed. Butter Worths, London, 1988.

[10] Morisako, A., Matsumoto, M., Naoe, M., *IEEE Trans. Magn.* 24, 3024,1988.

[11] Hylton, T.L., Parker, M.A., Ullah, M., Coffey, K.R., Umphress, R., Haward, J.K., *J. Appl. Phys.* 75, 5960, 1994.

[12] Niesen, T.P., M.R, de Guire, *J. Electroceram.* 6, 16, 2001.

[13] Temuujin, J. J., Aooyama, M., Senna, M., Masuko, T., Ando, C., and Kishi, H. *J. Solid State Chem.* 177, 221, 2004.

EFFECT OF X-RAY IRRADIATION ON NANO-COMPOSITE POLYMER ELECTROLYTES COMPLEXED WITH $LiCF_3SO_3$

C. VIJIL VANI, N. AMMAKUTTY SRIDEVI, S. BALAKUMAR AND X. SAHAYA SHAJAN

Centre for Scientific and Applied Research, School of Basic Engineering and Sciences, PSN College of Engineering and Technology, Melathdiyoor, Tirunelveli, Tamil Nadu
E-mail: hajan89@gmail.com

ABSTRACT

Polyethylene oxide (PEO)/$LiCF_3SO_3$/TiO_2 polymer electrolytes were prepared and the effect of X-ray irradiation on the electrochemical properties was studied. The irradiated membranes are found to show good thermal properties, smooth surface morphology, reduced crystallinity and enhanced ionic conductivity. The room temperature conductivity after irradiation was found to be 4.05×10^{-4} Scm^{-1}. The reorganising nature of PEO below 60°C is blocked due to X-ray irradiation. The details are presented herein.

INTRODUCTION

Solid polymer electrolytes used in batteries have attracted much attention because of their flexibility, formation into thin films of desirable shape and size and especially for their safety issues [1]. Polyethylene oxide (PEO), though highly explored remains a polymer of great interest due to its high lithium ion solvating property, ease of formation into thin films and its good electrochemical characteristics. But it suffers from poor ionic conductivity of nearly 10^{-7} Scm^{-1} at room temperature in its pure form. Among various methods suggested to improve the room temperature conductivity, addition of salt into the polymer matrix [2, 3], addition of nano sized inorganic fillers into the polymer electrolytes [4,5] and irradiation of polymer

V. Rajendran, K. Saminathan, P. Paramasivam and K.E. Geckeler (eds.)
Nano Materials Synthesis and Characterisation, pp. 113-118 (2012).

electrolytes are found to be most suitable methods for enhanced electrical, thermal and mechanical properties.

X-ray irradiated polymers show loss of crystallinity, loss of mass or chemical modification [6, 7]. Though they are similar to electron beam in many aspects they are found to be more penetrating than electron beam, at the same time they offer better dose uniformity. Addition of the filler particles is effective for developing a cross linked structure with a smaller irradiation dose.

In the present study, we have prepared polymer electrolytes with different concentrations of PEO, lithium triflate and TiO_2. The films are then subjected to X-ray irradiation of various dosages. As far as we are aware of, there are no reports available in the literature on x-ray irradiated polymer electrolytes. The complexation behavior of the polymer electrolytes were confirmed by FTIR analysis. The electrochemical analysis of the polymer electrolytes were studied by AC impedance analysis. Morphological and thermal studies on the electrolytes were carried out using SEM and DSC analysis.

EXPERIMENTAL

Preparation of Sample

PEO, lithium triflate and nano-TiO_2 (all chemicals were purchased from Sigma Aldrich) were dried under vacuum for 2 days at 40°C in order to remove any moisture content in them. PEO, lithium salt and nano- titania were dissolved in suitable solvent and stirred overnight to get a homogeneous solution. The resulting solution was cast on a glass plate and the solvent was allowed to evaporate in air at room temperature in a dust free environment. Further the films were pressed to thin membranes of 100 μm thickness using membrane hot press. The temperature and pressure of membrane hot press is maintained at 90 ˚C and 5 torr respectively. The temperature of hot press is sufficient enough to remove any traces of the solvent present, if any. In the present study, the weight percentage of lithium salt is kept constant at 5wt% and filler concentration is varied from 3wt% to 5%. Each sample was irradiated using soft X-ray source with dosages 1.5, 3 and 4.5 Gy.

Materials Characterization

The FTIR spectra were recorded using JASCO (Japan) FTIR-4100type A-spectrometer using ATRPRO 450-S accessory, in the region of 400-4000 cm^{-1}, with a signal average of 20 scans at a resolution of 4 cm^{-1}. The conductivity studies were carried out by Zahner IM6 ac impedance analyzer, Germany with amplitude of 1 V in the frequency range of 50 Hz-100 KHz. The phase transition temperature, nature of phase transition, thermal stability, melting point and degree of crystallinity were determined using DSC (DSC Q20,V24.2, build 107) in a temperature range of ambient to 100˚C at a heating rate of 10°C per minute by ramp method. Surface morphology was analyzed using S-3400 scanning electron microscope.

RESULTS AND DISCUSSIONS

FT-IR studies

The FT-IR spectra of the electrolyte before and after irradiation are given in Fig. 1.

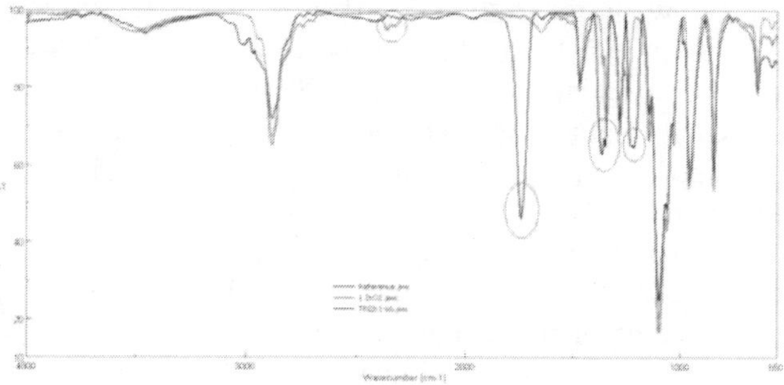

Fig. 1: FT-IR showing the chemical changes after irradiation

The effect of irradiation on the electrolyte can be understood by comparing the FT-IR spectra recorded before and after irradiation. The spectrum before irradiation is used to investigate the complex formation. The C-O-C stretching region which normally occurs at 1100 cm^{-1} could be used to analyze polymer –salt interaction [8]. The up shift noted in this region, i.e. at 1200 cm^{-1} suggests a strong interaction of lithium salt with ether oxygen. The bands occurring at 2900cm^{-1}, 1450 cm^{-1}, 1320 cm$^{\hat{u}1}$ and 1250 cm^{-1} are assigned to C-H vibrations of methyl group which corresponds to the polymer. The broadening of the C-O-C interaction band may be due to the overlapping of the stretching mode of SO_3. The band at 650 cm^{-1} corresponds to äSO_3 vibrations [9]. The dissociation of $LiCF_3SO_3$ into Li^+, CF_4, and SO_3 in the polymer matrix gives rise to two strong bands at 875cm^{-1} and 775 cm^{-1} which corresponds to CF_4 vibration.

After irradiation a downward shift with broadening of the peak at 1200 cm^{-1} can be interpreted as the replacement of Li^+ ions by Ti^{2+} ions through Lewis acid base interaction [8]. Irradiation has enhanced the solvation of filler particles into the polymer matrix. The broadening and shift to higher wave number of the peak at 1250 cm^{-1} and formation of new peak at 1700 cm^{-1} corresponding to C = O suggest the chemical changes which is mainly due to intra-chain cross linking and chain scission (produced by the breakage of C-H bonds) in the polymer chain [10]. Figure 1, shows the overlay spectrum of polymer electrolyte before and after irradiation.

Ionic Conductivity

The ionic conductivity of the prepared polymer membranes were determined using the following equation given by [11],

$$\sigma = \frac{t}{R_b A}$$

where *t* and *A* are the thickness (cm) and area (cm^2) of the polymer membrane in contact with the electrodes respectively. Thickness of the film has been measured by means of micrometer screw gauge. R_b represents the bulk resistance of the electrolyte film.

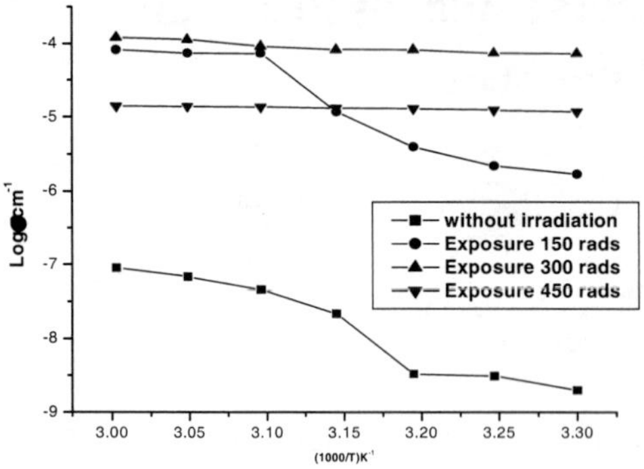

Fig. 2: Arrhenius Plot for the Sample without Filler Particles Before Irradiation and After Irradiation to 3Gy

Ionic conductivity measurements were done before and after irradiation from ambient temperature to 60°C. The temperature dependence of conductivity was found to follow Arrhenius equation. The sample without filler exhibited a conductivity of 9.08×10^{-8} Scm^{-1} at 60°C. As the irradiation dose was increased the conductivity was also found to increase. For 3Gy exposure the conductivity reaches a maximum value of $1.21 \times 10^{-4} Scm^{-1}$ at 60°C. However there is a decrease in conductivity was noted for 4.5Gy dosage. It is expected that irradiation of polymer electrolyte with x-ray has created cross linking in the polymer matrix. But as degree of cross linking increases it results in rigid three dimensional networks, allowing very little ion motion through the matrix [12]. When TiO_2 filler particles were added (3 wt %), the conductivity was found to increase by four orders. This may be due to the liberation of Li^+ ions from ether oxygen of PEO by the replacement with Ti^+ ions which form a strong Lewis acid-base interaction [13]. After irradiation there is no significant increase in conductivity was noted. However the reorganising tendency of the polymer is blocked significantly and the maximum conductivity obtained was 3.5×10^{-4} Scm^{-1}. The value of ionic conductivity was found to remain constant throughout the temperature range considered in the present study. Further increase in irradiation dosages resulted in a decrease in conductivity and this may be due to retardation in polymer chain motion by developing cross linking structure. When the samples with 5% filler particles are exposed to irradiation the conductivity enhances and it may be due to enhancement in dissolution of filler particles into the polymer matrix and thereby enhancing the filler-ether oxygen interaction and hence paving way for lithium ion movement. Figure 2, shows the Arrhenius plot for the sample without filler particles before irradiation and after irradiation to 3Gy.

SEM Analysis

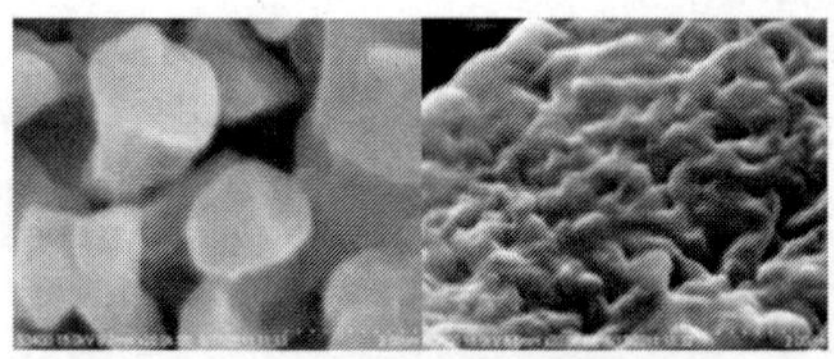

Fig. 3: SEM Image of Sample with 3 wt% Filler before and After Irradiation

SEM is often used to study the compatibility between the various components of the composite electrolytes through the detection of phase separations and interfaces [14]. The compatibility between the polymer matrix and the inorganic fillers has great influence on the mechanical, thermal, ionic conductivity and interfacial properties [15]. The image before irradiation shows a rough morphology. The filler particles and lithium salt are found to distribute over the surface of polymer matrix and not being dissolved into it. A significant improvement is achieved by irradiation. The surface was found smooth and the added particles are found to be completely dissolved in the polymer matrix. The smooth surface morphology is closely related to the reduction of PEO crystallinity via the interaction between PEO segments and added particles which is enhanced by irradiation [16]. Figure 3 shows the SEM images of the sample with 3 wt% filler particles before irradiation and after irradiation (3Gy dosage).

DSC Analysis

DSC analysis is used to study the melting point, phase transition temperature and to calculate the % of crystallinity. Figure 4, shows the DSC curve of the sample with 3 wt% filler particles before irradiation and after irradiation (3Gy dosage). DSC studies on irradiated samples shows an increase in phase transition temperature. This indicates the polymer become more amorphous with exposure to irradiation. This can be attributed to the formation of more stable structure due to cross linking of polymer chains in amorphous region. The broadening of melting peak indicates the reduction in crystallinity with addition of filler and increase in irradiation dosage. Similar results have been reported in the literature earlier [16,17]. The conductivity studies also confirm the observed results.

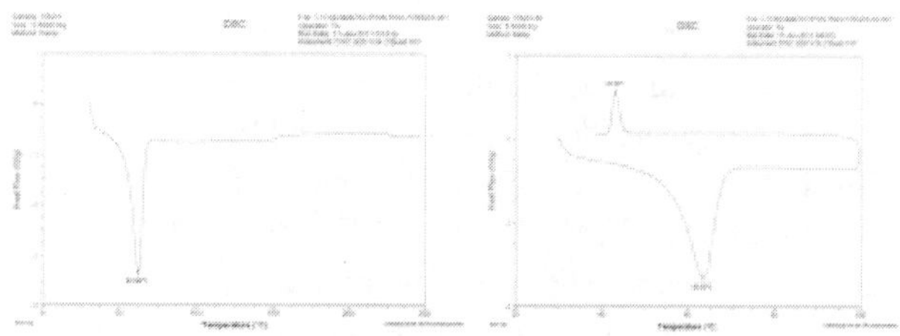

Fig. 4: DSC Curve of the Sample with 3 wt% Filler Particles Before Irradiation and After Irradiation to 3Gy

CONCLUSION

Polymer electrolyte complexed with lithium triflate salt and TiO_2 filler were prepared. The effect of x-ray irradiation on the ionic conductivity of the prepared polymer electrolytes were studied for different irradiation dosages. Conductivity was found to be highly dependent on filler concentration and irradiation dose. The optimum amount of filler and irradiation dose was found to be 3 wt% and 3Gy for the prepared samples. The highest conductivity was found to be 4.05×10^{-4} at ambient temperature. FT-IR confirms the chemical changes caused by irradiation and is in accordance with already reported results. The SEM analysis and DSC analysis confirms the increase in amorphocity in polymer electrolyte which is responsible for the increase in ionic conductivity.

ACKNOWLEDGEMENT

Author (XSS) acknowledge the financial support received from DAE/BRNS for the carrying out this work (Sanction no: 2009/34/40/BRNS/3226 dated 22.2.2010).

REFERENCES

[1] R. Uchiyama et al. Development of dry polymer electrolyte based on polyethylene oxide with co-bridging agent crosslinked by electron beam, *Solid State Ionics*, 180, 205-211, 2005.

[2] Y.G. Andreev, et al., Crystal structures of poly (ethylene oxide) 3:LiBF4 and (diglyme) n:$LiBF_4$ (n = 1,2), *Chem. Mater* 17, 767-772, 2007.

[3] Z. Stoeva, et al. Ionic conductivity in the crystalline polymer electrolytes PEO_6:$LiXF_6$, X = P, As, Sb, *J. Am. Chem. Soc.*, 125, 4619-4626, 2003.

[4] F. Croce, et al., Nanocomposite polymer electrolytes for lithium batteries, *Nature*, 394,456-458, 1998.

[5] Binod Kumara, Stanley et al., Ionic Conductivity of Polymer-Ceramic Composites, *J. Electrochem. Soc.*, 148, A1191-A1195, 2001.

[6] T. Coftey, et al., Characterization of the effects of soft X-ray on polymers, *J. Electron. Spectrosc. Relat. Phenom*, 122, 65-78, 2002.

[7] Y.A. Aggour, Spectroscopic and electrical studies for the influence of ion beam and X-ray irradiation on poly-2-(N-propenamido-2-methylpropanesulfonic acid) and its polymer complex with cobalt chloride, *Polymer Testing*, 20, 879-884, 2001.

[8] W. Wieczorek, et al, Effect of salt concentration on the conductivity of PEO-based composite polymeric electrolytes, *J. Phys. Chem. B*, 102, 8725-8731, 1998.

[9] Manuel Stephan A, et al. Electrochemical studies on nanofiller incorporated poly(vinylidene fluoride-hexafluoropropylene) (PVdF-HFP) composite electrolytes for lithium batteries, *J. Appl Electrochem*, 36, 1091-1097, 2006.

[10] Gladius Lewis, Properties of crosslinked ultra-high-molecular-weight polyethylene, *Biomaterials*, 22, 371-401, 2001.

[11] S. Ramesh, Liew Chiam Wen, Investigation on the effects of addition of SiO_2 nanoparticles on ionic conductivity, FTIR, and thermal properties of nanocomposite PMMA–$LiCF_3SO_3$–SiO_2, *Solid State Ionics*, 16, 255-262, 2010.

[12] Ray, R., Ballabh, T.K., Tarafdar, S., Variation in different properties of polymer electrolytes on gamma irradiation: A computer simulation study. In: proceedings of *12th Asian Conference on Solid State Ionics*, Wuhan, China, May 2010.

[13] F. Croce , et al., Role of the ceramic fillers in enhancing the transport properties of composite polymer electrolytes, *Electrochimica Acta* 46, 2457-2461, 2001.

[14] H.M. Kao, et al. Organic-Inorganic nanocomposite electrolyte derived from triblock copolymers *Chem. Int Ed*, 43 ,980,2004.

[15] Jingyu Xi, Xiaozhen Tang, Nanocomposite polymer electrolyte based on Poly(ethylene oxide) and solid super acid for lithium polymer battery, *Chem. Phys. Lett* , 393, 271-276, 2004.

[16] M.Saheer Akhtar, et al., Effect of electron beam irradiation on the electrochemical properties of heteropolyacid-polyethylene oxide composite electrolyte for dye-sensitized solar cell, Curr. Appl Phys, 10, s161-s164, 2010.

[17] Ribeiro R., et. al. A comparison of ionic conductivity, thermal behaviour and morphology in two polyether–LiI–$LiAl_5O_8$ composite polymer electrolytes, *Electrochim Acta*, 46 , 1679-1686, 2001.

THE EFFECTS OF DEFECTS ON ELECTRON TRANSPORT IN METALLIC SINGLE WALL CARBON NANOTUBES

S. Sivasathya and D. John Thiruvadigal

Department of Physics and Nanotechnology, Centre for Material Sciences and Nanodevices, SRM University, Kattankulathur, Tamil Nadu
E-mail: john.d@ktr.srmuniv.ac.in

ABSTRACT

We report the transport behaviour of an open end metallic single wall carbon nanotube with and without local structural defects using the non equilibrium Green's functions approach together with the density functional theory (DFT). The transmission spectra and the projected density of states for the devices such as SWCNT (3, 3), (4, 4), (5, 5) and (6, 6) are compared with and without defects. In all the cases, we found that the Stone Wales defect has almost negligible impact on the electrical performance compared to the monovacancy defect of the single wall carbon nanotubes at the Fermilevel. The Current-Voltage (I-V) characteristics of the devices are studied using the generalized Landauer - Buttiker formalism under low bias condition. From our results, we concluded that our systems used in various CNT based nanoelectronic devices.

INTRODUCTION

In recent years, electronics based on carbon nanotube has received a lot of attention because of its tremendous potential application such as active components, switches, memory devices, display devices and sensors. Single walled carbon nanotubes (SWCNT) exhibit metallic or semiconducting behavior depending on its diameter and chirality [1,2]. The single wall armchair

---

V. Rajendran, K. Saminathan, P. Paramasivam and K.E. Geckeler (eds.)
Nano Materials Synthesis and Characterisation, pp. 119-124 (2012).

carbon nanotubes are one dimensional conductor with only two open conduction channels. Hence, the conduction of electrons ultimately becomes localized owing to defects in the tube which is inevitable with increasing length due to the interactions between the tube and its environment. Their transport properties have been studied extensively. Early theoretical and experimental work on SWCNT predicted that a special achiral subset of these structures known as armchair nanotubes should be metallic. A coherent electron transport can be maintained through the nanotubes and this is confirmed experimentally by Tans et al. [3]. Recently, ab initio method [4] based on Non equilibrium Green's Function (NEGF) in combination with Density Functional Theory (DFT) allows the determination of electronic transport properties of the systems with hundreds of atoms in presence of a finite voltage accurately. In this paper, we report a study on the effects of defects such as monovacancy and Stone-Wales on the electron transmission in metallic SWCNTs. The I-V characteristics are analyzed and negative Differential Resistance (NDR) effect is observed which is a useful for nano devices like switches, logic cells and memory storage.

COMPUTATIONAL FORMALISM

To analyze the quantum transport through ab initio method, one must deal with open systems within the DFT formalism as the system is either isolated or periodic and can be considered as two semi infinite electrodes coupled via a contact region. Hence, SWCNTs is divided into three regions, namely, left electrode, right electrode and the scattering region in between the electrodes. The scattering region consists of five unit cell and the length of the electrode is 0.492nm. For convergence, a tolerance of 1×10^{-5} of the total energy was used. The electron transport were performed on devices of metallic single wall nanotubes by a fully self consistent NEGF combined with first principles DFT, which is implemented in Atomistic Tool kit (ATK) [5] package (Version 11.8.2). The generalised gradient approximation (GGA) with Perdev Zunger (PZ) pseudo potentials was used in the calculation. A double-ξ polarization (DZP) basis set is employed to describe the localized atomic orbitals and the multigrid poisson-solver must be set with the same boundary conditions. The K-point sampling is $1 \times 1 \times 100$.

In our calculations, the transmission function is calculated by

$$T(E,V) = Tr[\Gamma_L(E,V)G(E,V)\Gamma_R(E,V)G^+(E,V)] \quad \text{... (1)}$$

where $\Gamma_{L/R}$ stands for the coupling matrix between the left/right electrode and scattering region, G and G^+ are the retarded and advanced Green's function.

The I-V is calculated by using Landauer-Buttiker [6] formula, which relates conductance to the transmission probability T (E, V)

$$I(V) = G_0\int_{\mu_L}^{\mu_R} dE\, T(E,V)\{f(E-\mu_L) - f(E-\mu_R)\} \quad \text{... (2)}$$

where $G_0 = {2e^2}/{h}$ quantum unit of conductance, μ_L and μ_R is the chemical potential of left and right electrodes respectively, f (E-$\mu_{L/R}$) *is* the Fermi-Dirac distribution function. The resistance is derived from the current-voltage relation.

RESULTS AND DISCUSSION

The electron transmission coefficients of metallic SWCNTs with and without defects are shown in Fig. 1. In all the cases, the electron transmission coefficient of the defect free NTs is equal to 2 around the Fermi level (0 eV) indicating that there are two transmission channels, Π bonding and Π antibonding state. However, a sudden drop at 0 eV occurs in (4, 4) nanotubes because of weak coupling. The transmission coefficient for Stone Wales defect does not reduce at the Fermi level for all the nanotubes. Hence, the electrical conductance of all metallic nanotubes at the Fermi level is not affected. In the case of monovacancy, the transmission coefficient much reduced due to the formation of dangling bonds where σ bond states are perpendicular to the Π valence bond states which results weak coupling in between them. [7, 8]

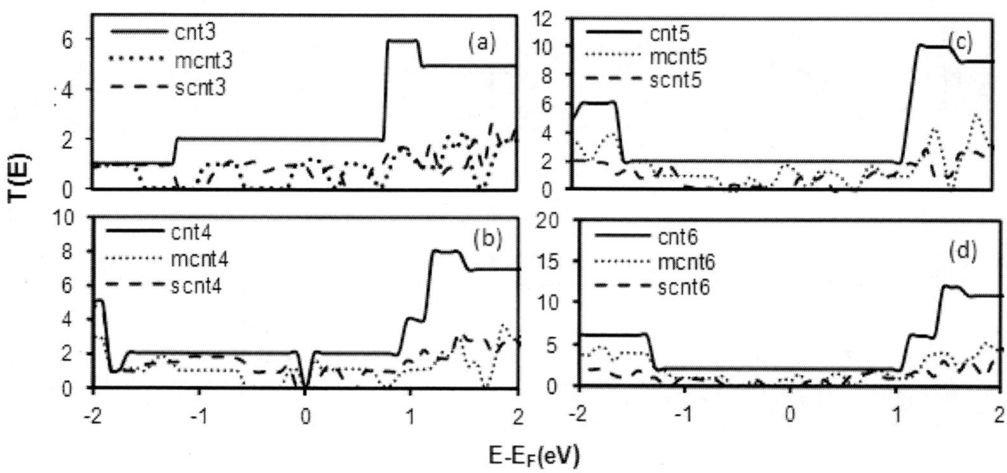

Fig.1: Electron Transmission Spectra of Metallic Single Wall Nanotubes with and Without Defects at Zero Voltage (a) (3, 3) (b) (4, 4) (c) (5, 5) (d) (6, 6). Here 'mcnt' Denotes Monovacancy and 'scnt' Denotes Stone Wales Defects

Under bias condition, the transmission peaks are shifted from Higher Occupied Molecular orbitals (HOMO) to the Lower Unoccupied Molecular Orbitals (LUMO). Above 1 Volts, there is a formation of a symmetric gap in the transmission channels for all the nanotubes but it appears at 0.2 Volts for (4, 4) nanotube indicating suppression of transmission channels in that position.

Based on the calculated electron coefficient, the current-voltage (I-V) relation of the systems can be determined is shown in Fig. 2. Ohmic behavior occurs at low bias voltages (0-1 Volts). However, above 1-2 Volts, the current decreases as the bias voltage increases exhibiting a non Ohmic behavior known as Negative Differential Resistance (NDR) [9, 10] in all the cases. The NDR occurs at 0.2 V for (4, 4) nanotube. Hence, the I-V characteristics of the systems will depend on the leads and the radius of the nanotubes (i.e.) nature of the nanotubes. The monovacancy permits less current than the Stone Wales defect in all the cases except (6, 6) nanotubes since the monovacancy defect is formed near the open ends of the nanotube. In other cases, the monovacancy and Stone Wales defects are in the curvature. These are comparable with diode characteristics, indicating diode like applications of nanotubes.

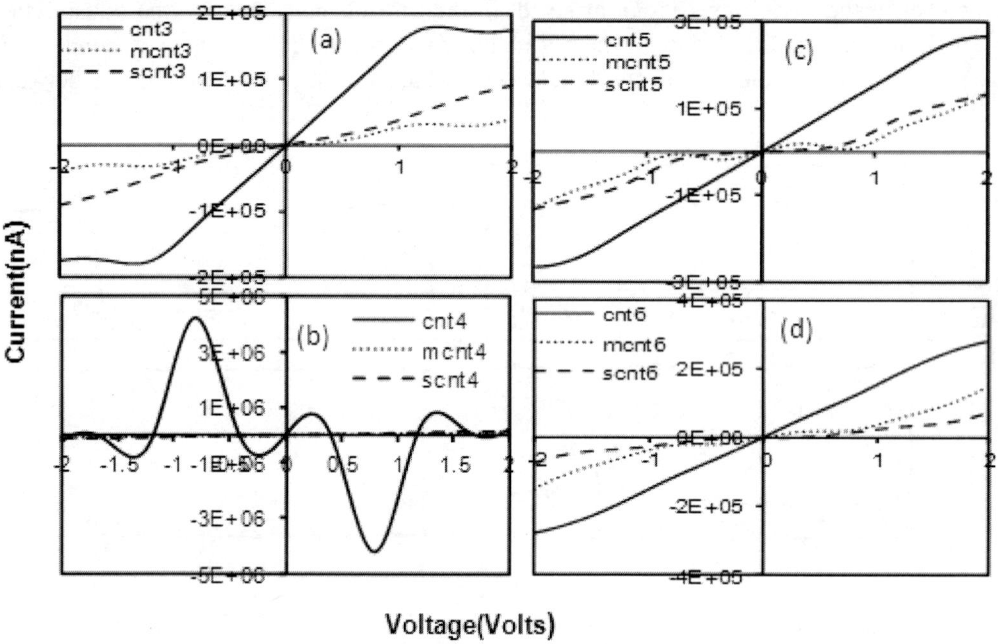

Fig.2: The I-V Characteristics of Metallic Nanotubes with and Without Defects (a) (3, 3) (b) (4, 4) (c) (5, 5) (d) (6, 6). Here 'mcnt' Denotes Monovacancy and 'scnt' Denotes Stone Wales Defects

The total resistance of the systems can be obtained by averaging the values in the 0-0.1 regime. These values for defect free single wall metallic nanotubes are 6.46 kΩ, 0.214 kΩ, 6.46 kΩ, 6.28 kΩ for (3, 3), (4, 4), (5, 5), (6, 6) nanotubes. The resistance for (4, 4) nanotube is very low when compared to other nanotubes. So NDR occurs at very low voltage under bias condition. The other resistance values are in good agreements with the experimental value. The total resistance values for SWCNTs with Stone Wales defect are 279 kΩ, 14.7 kΩ, 13.2 kΩ, 15.6 kΩ and with monovacancy defects are 32.9 kΩ, 31.3 kΩ, 89.4 kΩ, 1180 kΩ. The defects increase the resistance in the nanotubes. The increase in resistance observed in the monovacancy defect induced NT further confirms the formation of weak coupling.

CONCLUSIONS

We have investigated the electron transport properties of metallic SWCNTs with or without defects using NEGF formalism combined with first principles DFT. From our results, we concluded that all the metallic nanotubes have same phenomena under low bias condition and Stone Wales defect has negligible effect on the electrical conductance of SWCNTs at the Fermi level while monovacancy defect significantly reduces the electrical conductance by severely blocking one of the transmission channels. These conclusions demonstrate the effective application of our systems in various CNT based nanoelectronic devices such as memory devices, switches and sensors.

ACKNOWLEDGEMENT

The authors are thankful to the financial support from DST-FIST (Government of India) under Ref. No. SR/FST/PSI – 010/2010.

REFERENCES

[1] S. Ijima, Helical microtubules of graphitic carbon, *Nature*, London, 354, 56-58, 1991.

[2] R. Saito, G. Dresselhaus and M.S. Dresselhaus, Physical properties of carbon nanotubes, Imperial College press, 1998.

[3] S.J. Tans et al., Individual Single wall carbon nanotubes as quantum wires, *Nature* 386, 474-477, 1997.

[4] Kikuji Hirose et al., First-principles calculations in real-space formalism, Imperial College Press, London, 2005.

[5] Atomististix tool kit 11.8.2 version, Quantumwise A/S (*www.quantumwise.com*).

[6] M. Buttiker et al., Generalised many-channel conductance formula with application to small rings, Phys. Rev. B 31(1985) 6207.

[7] Yaroslav V. Shtogun and Lilia M. Woods, Properties of Carbon Nanotubes under External Factors, Carbon Nanotubes, Jose Mauricio Marulanda (Ed.), USA, 2010.

[8] B.C. Pan et al., Formation energies of topological defects in carbon nanotubes, *Phy. Rev.* B 62, 12652-12655, 2000.

[9] J.B. Neaton et al., Electron transport and optical properties of carbon nanostructures from first principles, *Computer physics communications*, 169, 1-8, 2005.

[10] Yingying Zhang et al, Negative differential resistance behavior of silicon monatomic chain encapsulated in carbon nanotubes, *Computational material science*, 62, 87-92, 2012.

MICROWAVE SYNTHESIS AND CHARACTERIZATION OF Co_3O_4 NANOSPHERES BY MICROWAVE REACTION METHOD

G. ANANDHA BABU, M.R. MANIKANDAN AND G. RAVI

Photonic Crystals Lab, School of Physics, Alagappa University, Karaikudi
E-mail: gravicrc@gmail.om. raviganesa@rediffmail.com

ABSTRACT

Metal oxide nanospheres constitute an outstanding class of functional materials with potential applications in almost all fields of technology. Among a wide variety of materials, Co_3O_4 has attracted a great amount of attention and becomes one of the most important functional materials due to its unique application in heterogeneous catalysts, anode materials in Li-ion rechargeable batteries, solid-state sensor, solar energy absorbers, ceramic pigments, and electrochromic devices. The Co_3O_4 nanospheres have been successfully synthesized through a simple microwave route using cobalt acetate, ammonia and citric acid at 180 W for 15 min. The face centered cubic phase structures of Co_3O_4 nanosphere with average grain size of 23 nm were characterized by means of powder X-ray diffraction (XRD). The nanospheres like morphologies of Co_3O_4 were examined by scanning electron microscope (SEM). The functional properties of Co_3O_4 nanospheres were examined by Fourier transform infrared spectroscopy (FTIR).

INTRODUCTION

Nanostructures of transition metal oxides have attracted significant attention in recent years because of its efficient mechanical, electrical, optical, or magnetic properties of nano-structured materials. The intrinsic properties of nanoscale materials are mainly determined by their composition, structure, crystallinity, size and morphology. Among various transition metal oxides,

V. Rajendran, K. Saminathan, P. Paramasivam and K.E. Geckeler (eds.)
Nano Materials Synthesis and Characterisation, pp. 125-130 (2012).

cobalt oxide was found to have the largest potential applications in various fields. Cobalt oxides belong to the family of transition-metal oxides and the most stable phase, Co_3O_4 is an intrinsic p-type semiconductor (direct optical band gaps at 1.48 and 2.19eV). The chemical stability of Co_3O_4 over a wide temperature range and its high mechanical strength (Young's modulus 116-160GPa) render this material a potential candidate for field emitters. So, special attention has been paid to the preparation of Co_3O_4 with different nanostructures to promote its properties [1].

Microwave method finds much importance in the preparation of nanoparticles compared to other synthetic literature methods. Irradiation of microwaves on reaction mixture converts the required product at faster rate and gives the application oriented crystalline product having controlled particle size [2]. The metal oxide obtained by microwave synthetic technique shows nanocrystalline nature and also good morphology. Materials based on cobalt oxides have been extensively investigated because of their potential applications in many technological fields. Co_3O_4 shows the spinal structure based on a cubic close packing array of oxide ions where one-eighth of tetrahedral holes are occupied by Co (II) ions and one-half of the octahedral holes are occupied by Co (III) ions [3].

The present work has demonstrated the preparation of Co_3O_4 nanospheres by microwave route. The structural, morphological and functional properties were characterized by using XRD, SEM, FTIR and EDAX.

EXPERIMENTAL

Materials

Analytical grade reagents cobalt acetate, citric acid and ammonia were purchased from MERCK used for the preparation of cobalt oxide nanoparticles. All the above chemicals were used as purchased without further purification.

Synthesis of Cobalt Oxide Nanospheres

In a typical reaction process for the growth of Co_3O_4, 5 g of cobalt acetate was dissolved in 50 ml distilled water and stirred for 5 min at room temperature. Simultaneously, ammonia was added drop wise into this aqueous cobalt acetate solution under vigorous stirring with control pH = 10. In addition, 0.5 g of citric acid was used as a capping agent. The above solution was taken in a autoclave bottle and placed in a microwave oven. The solution was irradiated by microwaves with the power of 180 W for 15 minutes. The system was then allowed to cool to room temperature. The final products deposited at the bottom of the vessel and the precipitates were washed with distilled water several times and then filtered with water, and subsequently left to dry in hot air oven for 12 hrs.

Finally dried powder was calcined at 600°C. At the end, the cobalt oxide (Co_3O_4) nanospheres can be obtained.

RESULTS AND DISCUSSIONS

X-ray Diffraction (XRD)

Fig. 1 presents the XRD pattern of the Co_3O_4 sample. All the peaks can be indexed to pure cubic phase of Co_3O_4 (Fd3m) with lattice parameter a = 8.0850 Å (JCPDS: 78-1969). No other peaks for impurities can be detected indicating that the final product is of high purity. The spectrum consists of well-defined diffraction peaks indicating that very thin crystalline single phase is formed [4].

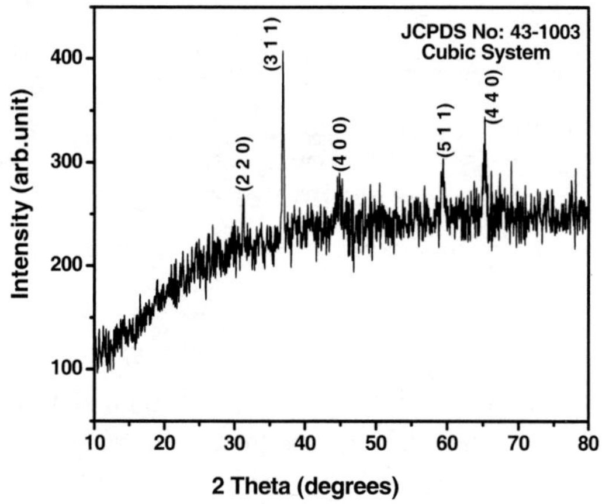

Fig. 1: XRD Pattern for Co_3O_4 Nanospheres

The average particle diameter (D) was calculated by using the Debye-Scherrer formula

$$D = 0.91\lambda/\beta \cos \theta$$

where λ is the X-ray wavelength, β is the FWHM (full width at half maximum intensity) and θ is the Bragg's angle. The average grain size calculated by using Scherer's formula was about 27 nm.

Fourier Transform Infrared Analysis

Chemical bonding information was studied with a Fourier transform infrared (FTIR) spectrometer. Figure 2 shows the FT-IR spectrum for Co_3O_4 nanospheres. Two sharp peaks exists around 660 and 560 cm^{-1} assigned to the ν-CoO of Co_3O_4 appear after calcination at 400°C for 2 h. No other vibration peak exists when the calcination temperature reached 400°C, only the band belonging to Co-O was retained [5].

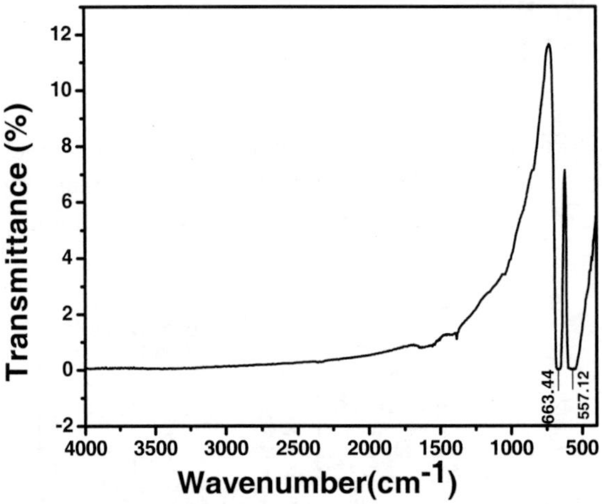

Fig. 2 : FTIR Spectrum for Co_3O_4 Nanospheres

Scanning Electron Microscope and EDAX Analyses

Figure 3 shows a scanning electron microscopy image of Co_3O_4 nanoparticles. The crystallinity of the Co_3O_4 nanoparticles was well developed and the Co_3O_4 grains had nanosphere morphology with two different magnifications. Energy-dispersive X-ray spectroscopy (EDS) has showed that the Co_3O_4 composition is stoichiometric.

Fig. 3: SEM Images of Co_3O_4 Nanospheres

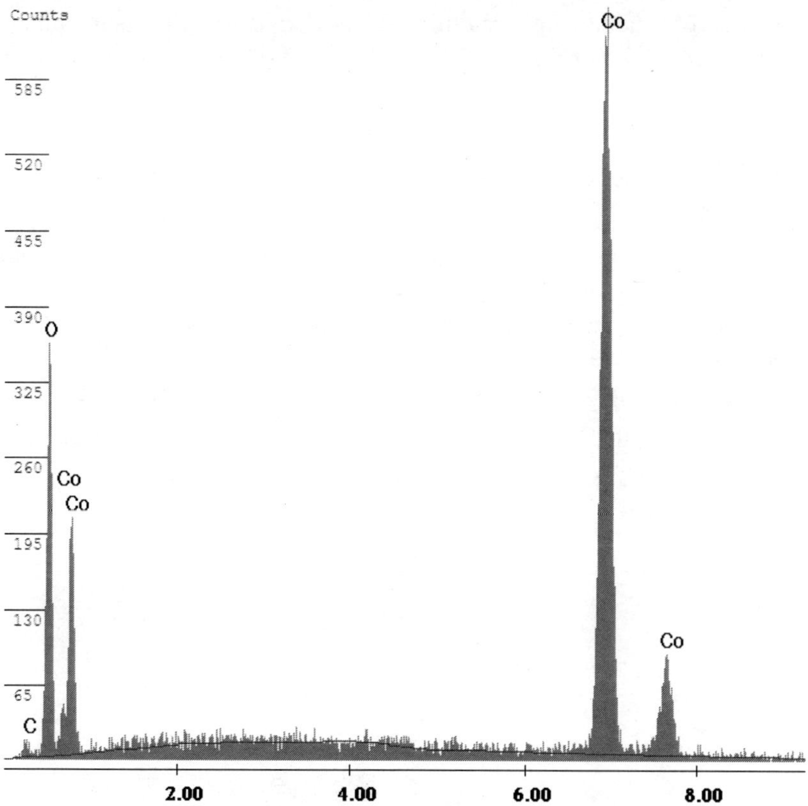

Element	*Atomic Percentage*	*Weight Percentage*
Co	53.59	81.68
O	37.84	15.66
C	8.56	2.66

Fig. 4: EDAX Spectrum of Co_3O_4 Nanospheres and its Composition

CONCLUSION

A pure phase of Co_3O_4 nanospheres have been successfully synthesized using simply microwave route in 15 min. XRD studies indicated that the Co_3O_4 is in a cubic phase with a cell constant of a = 8.085 Å. The SEM images are showed that the obtained Co_3O_4 powder in the form of nanospheres. The functional study confirms the vibration of Co_3O_4 nanospheres. Energy-dispersive X-ray spectroscopy (EDS) confirms that the Co_3O_4 composition is stoichiometric.

ACKNOWLDGEMENT

One of the authors G.Anandha babu gratefully acknowledges the financial support for this study from DST under the scheme of INSPIRE-Fellowship (Grant no: IF110040).

REFERENCES

[1] Masoud Salavati-Niasari, Noshin Mir and Fatemeh Davar, *Journal of Physics and Chemistry of Solids* 70, 847–852, 2009

[2] Bhavikatti, A.M., Kulkarni, S., Lagashetty, A., Characterisation and Electromagnetic studies of nano-sized barium ferrite, *Int.J. Engg. Sci. & Tech.*, 2(11), 6532, 2010.

[3] RuiXu, Jiawei Wang, Qiuyu Li, Guoying Sun, Enbo Wang , Siheng Li et al., *Journal of Solid State Chemistry* 182, 3177–3182, 2009.

[4] Wenbo Yue and Wuzong Zhou, *Progress in Natural Science* 18, 1329–1338, 2008.

[5] Yunshuang Ding, Linping Xu, Chunhu Chen, Xiongfei Shen, and Steven L. Suib, *J. Phys. Chem.* C, 112 (22). 2008.

MICROWAVE SYNTHESIS AND CHARACTERIZATION OF UNDOPED AND Co-DOPED NiO NANOSTRUCTURES

G. ANANDHA BABU, S. VINODHINI, M. THANGARAJ AND G. RAVI

Photonic Crystals Lab, School of Physics, Alagappa University, Karaikudi
E-mail: gravicrc@gmail.om; raviganesa@rediffmail.com

ABSTRACT

Nanostructures which exhibit exotic physical and chemical properties that significantly different from those observed in bulk three-dimensional because of effects in terms of small size effect, surface effect, quantum size effect and macroscopic quantum tunnel effect. The synthesis of NiO nanostructures is of great technological and scientific interest owing to the use of these powders as gas sensors, catalysts, magnetic and electrode materials. In our present work, undoped and Co doped NiO nanostructures were prepared by microwave route using source material (nickel chloride), precipitating agent (ammonia) and cobalt acetate as a doping material. The cubic structure with face centered lattice type was obtained for both undoped and Co-doped NiO nanostructures revealed from X-ray diffraction (XRD). The grain size of cobalt doped NiO nanostructures is in the range of 40-60 nm than undoped NiO with grain size less than 20 nm. The morphology of undoped and Co-doped NiO were analyzed by using scanning electron microscopy. The functional and optical changes of the samples were observed from FTIR and UV-Vis spectroscopy. The influences of Co on structural, morphological and optical properties of NiO nanostructures were investigated.

V. Rajendran, K. Saminathan, P. Paramasivam and K.E. Geckeler (eds.)
Nano Materials Synthesis and Characterisation, pp. 131-136 (2012).

INTRODUCTION

Nanostructures of transition metal oxides have attracted significant attention in recent years because of its efficient mechanical, electrical, optical or magnetic properties. NiO is an intrinsic antiferromagnetic with stable wide band gap of 3.6-4 eV. So, special attention has been paid to the preparation of NiO with addition of dopant to promote its properties [1]. The incorporation of small amounts of monovalent, divalent and trivalent in a NiO lattice generally affects its electronic structure and hence modifies its catalytic, adsorptive and electrical properties. When metal ions are doped into this material, Ni^{2+} vacancies might be induced, leading to semiconducting behavior. Doping may also produce structural defects such as cationic and anionic vacancies, the presence of which markedly affects the diffusion of the NiO lattice ions [2]. Microwave method finds much importance in the preparation of nanoparticles compared to other synthetic literature methods. Irradiation of microwaves on reaction mixture converts the required product at faster rate and gives the application oriented crystalline product having controlled particle size. The present work has demonstrated the effect of cobalt doping on structural, morphological and optical properties of NiO nanostructures using XRD, FTIR, SEM and UV spectroscopy.

EXPERIMENTAL

Materials

Analytical grade reagents nickel chloride hex hydrate, cobalt acetate and ammonia purchased from MERCK were used for the preparation of nickel oxide nanoparticles. All the above chemicals were used as received without further purification.

Synthesis Procedure

In a typical reaction, 1g of $NiCl_2{\cdot}6H_2O$ is added to 50 ml distilled water. Then ammonia was added to the above solution (pH=10). This prepared solution is reacted with microwave oven for 10 minutes with 180 W irradiation. The same procedure is followed for doping of cobalt (10% Wt) in NiO nanoparticles. Here also the pH of the solution is same as in pure NiO. The resulting pale green slurry was rinsed with deionized water and absolute ethanol for several times to remove soluble impurities. The precipitate was dried at 80°C in a hot air oven for 12 h to get the sample of Ni $(OH)_2$ architectures. The as-prepared powder was calcined under air at 400°C for 2 h to obtain NiO nanoparticles from Ni $(OH)_2$ particles.

RESULT AND DISCUSSIONS

X-ray Diffraction (XRD)

The structures of the products are examined by XRD and the results are depicted in Figure 1. All of the diffraction peaks can be ascribed to the cubic NiO structure (FCC phase) with lattice parameter a = 4.194 Å and matched with standard JCPDS card no 89-7130.

All the diffraction peaks of NiO samples exhibits standard peaks at 2θ values 37°, 43°, 63°, 75° and 79° corresponds to (111), (200), (220), (311) and (222) planes respectively [3].

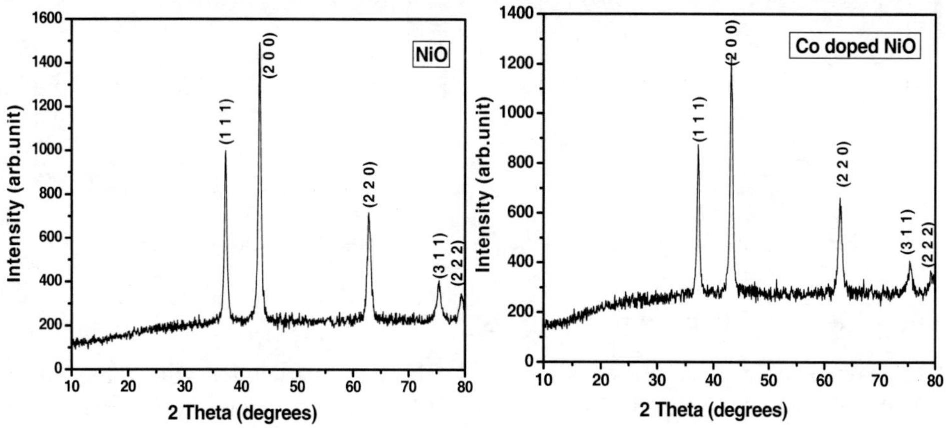

Fig. 1: XRD Pattern for Undoped and Co Doped NiO Nanoparticles

No peaks corresponding to any dopant such as Co metals, CoO and Co_3O_4 are detected within the sensitivity of the XRD scan. Thus, the cubic structure not modified by the addition of Co ion into the NiO matrix indicates that the Co dopant ought to be incorporated into the lattice as substitutional ion [4]. The variation of micro structural properties for undoped and Co doped NiO nanostructures are shown in Table 1. It concluded that the dopant Co mainly affects the micro structural properties of the NiO nanostructures depend on its own properties.

Table 1: Micro Structural Properties of NiO and Co-doped NiO Nanostructures

S. No.	*Sample*	*Crystallite size (nm)* $D = \frac{0.9\lambda}{\beta\cos\theta}$	*Micro strain (ε)* $\epsilon = \beta \, Cos\,\theta / 4$ *lines*$^{-2}$ *m*$^{-4}$	*Dislocation Density* $\delta = 1/d^2$ 10^{15} *lines/m*^{2}
1	NiO	37.20	0.0016	2.1
2	Co-NiO	15.00	0.0023	4.6

Fourier Transform Infrared Analysis

Figure 2 shows the FT-IR spectra with the absorption band appears around 3400–3580 cm^{-1} in undoped and cobalt doped samples due to absorption of OH stretching mode. The band exists near 1600-1660 cm^{-1} (H-O-H bending mode) is related to the evidence of water absorption in the NiO nanostructures. The strong band exists at 437 cm^{-1} for cobalt doped and 420 cm^{-1} for undoped samples is the nature of vibration modes of NiO [5].

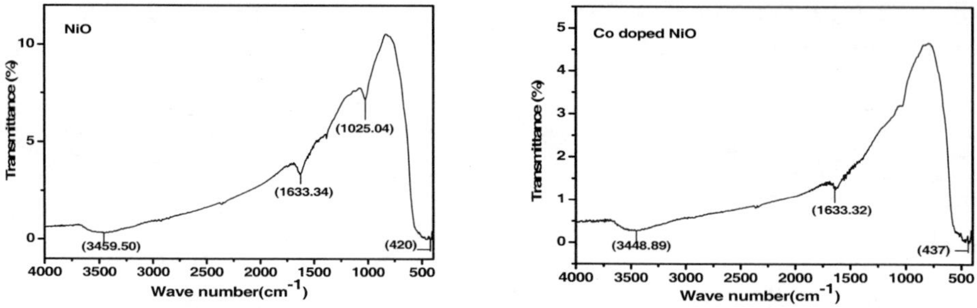

Fig. 2: FTIR Spectra for Undoped and Co-doped NiO Nanostructures

Scanning Electron Microscope Analysis

The unsymmetrical disk shaped structures of NiO nanostructures with three different magnifications for both undoped and cobalt doped NiO nanostructures were obtained from scanning electron microscopy images shown in Fig. 3. Figure 4 shows the EDAX spectrum for undoped and cobalt doped NiO nanostructures. EDAX spectra confirms the presence of cobalt in Co-doped NiO nanoparticles.

Fig. 3: SEM Images of Undoped (a-c) and Codoped (d-f) NiO Nanostructures

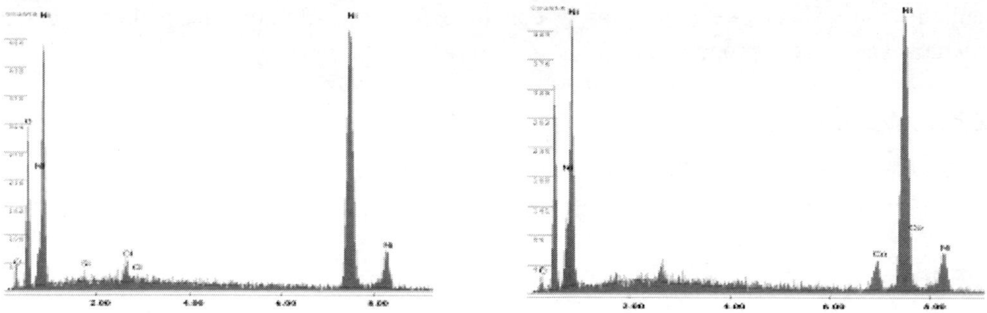

Fig. 4: EDAX Spectra of Undoped and Co Doped NiO Nanostructures

UV-Visible Analysis

Figure 5 shows the optical band gap curve of undoped and Co doped NiO nanostructures is estimated using the following relation

$$\alpha h\nu = A(h\nu - E_g)^n$$

where α is absorption coefficient in cm^{-1}, $h\nu$ is photon energy, E_g is an energy gap, A is energy dependent constant and n is an integer. The lower band gap value for cobalt doped NiO sample showed red shift when compared to bulk NiO due to quantum size effect [7].

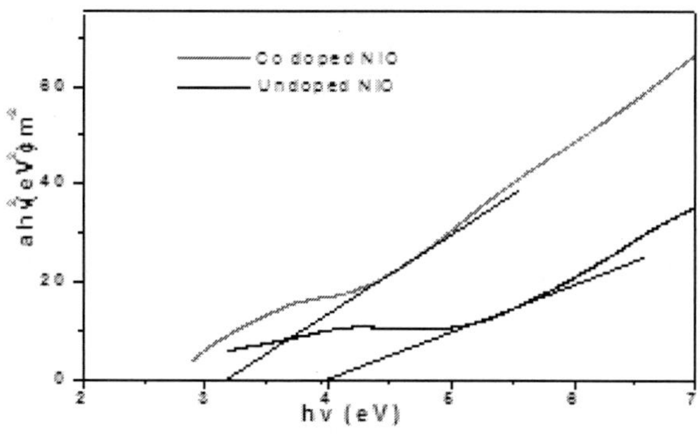

Fig. 5: Plots of hν versus $(\alpha h\nu)^2$ for Undoped and Cobalt Doped NiO Nanostructures

CONCLUSION

Cubic phase NiO nanostructures were not changed and particle size of co-doped sample is less than undoped samples because of the incorporation of cobalt into NiO lattice site. The SEM analysis reveals that unsymmetrical disk like morphology was grown for both undoped and doped NiO nanostructures. The lower band gap value compared to bulk NiO is due to quantum confinement effect. It strongly concludes that the addition of cobalt changes the structural and optical properties.

ACKNOWLDGEMENT

One of the authors G. Anandha babu gratefully acknowledges the financial support for this study from DST under the scheme of INSPIRE-Fellowship (Grant no: IF110040).

REFERENCES

[1] Harraz F.A., Mohamed R.M., Shawky A. and Ibrahim I.A., *Journal of Alloys and Compounds* 508, 133-140, 2010.

[2] Jinda Khemprasit, Suriyasit Kaen-ngam, Bualan Khumpaitool and Phalakorn Kamkhou, *Journal of Magnetism and Magnetic Materials* 323, 2408–2412, 2011.

[3] Guoxing Zhu, Chunyan Xi, Huan Xu, Dan Zheng, Yuanjun Liu et al., *RSC Adv.*, 2, 4236–4241, 2012.

[4] He Jiang, Xiao Fang Liu, Zhi Yu Zou, Zhang Ben Wu, Bo He et al., *Applied Surface Science* 258, 236– 241, 2011.

[5] Yanping Wang, Junwu Zhu, Xujie Yang, Lude Lu and Xin Wang, *Thermochimica Acta* 437, 106-109, 2005.

[6] Mai Y.J., Tu J.P., Xia X.H.,. Gu C.D. and Wang X.L., *Journal of Power Sources* 196, 6388–6393, 2011.

[7] Zhenfeng Zhu, Na Wei, Hui Liu and Zuoli He, *Advanced Powder Technology* 22, 422–426, 2011.

PHOTOELECTROCHEMICAL PROPERTY OF THERMALLY EVAPORATED NANOSTRUCTURE WO_3 FILM

R. SENTHIL KUMAR, T. MAHALINGAM[1] AND G. RAVI[2]

Department of Physics, KIT and KIM Technical Campus, Karaikudi
[1]Karunya University, Coimbatore
[2]Photonic Crystals Lab, School of Physics, Alagappa University, Karaikudi
E-mail: gravicrc@gmail.om; raviganesa@rediffmail.com

ABSTRACT

The photocurrent and dark current measurement of WO_3 have been predominant, because of their interesting photoelectrochemical property. Tungsten oxide (WO_3) nanostructure film were grown on indium tin oxide coated glass substrates by thermal evaporation of WO_3 powder at high temperature using tube furnace without any catalyst and then by subsequent O_2/Ar flow rate. Morphology, phase structure and crystallinity of the film have been characterized by Scanning electron microscope (SEM), X-ray diffraction (XRD) and Raman spectroscopy. These results confirmed that the nanorods with a well-defined morphology of WO_3 phase. The samples were investigated under the photocurrent and dark current in H_2SO_4 aqueous solution as a function of applied potential. WO_3 nanostructures photocurrent was increased with increase of applied potential to the anode. The saturated photocurrent density of tungsten oxide was found to be H $\approx$ 14.4 μA cm^{-2}.

INTRODUCTION

Metal oxides play very important role in many areas of physics, chemistry and materials science. Among the metal oxides Tungsten oxide (WO_3), a very important type of semiconductor material,

V. Rajendran, K. Saminathan, P. Paramasivam and K.E. Geckeler (eds.)
Nano Materials Synthesis and Characterisation, pp. 137-142 (2012).

has been extensively investigated due to its promising physical and chemical properties. By virtue of its outstanding electrochromic, optochromic, and gasochromic properties, in the past few decades WO_3 has been used to manufacture various devices such as flat-panel displays, electrochromic "smart" windows, optical modulation devices, write–read–erase optical devices, gas sensors, and field emission devices.

A number of techniques have been reported for the deposition of WO_3, including pulse laser deposition, thermal evaporation, sputtering, spray pyrolysis, chemical vapor deposition, hydrothermal, and electrodeposition. Among these techniques, the deposition of WO_3 using thermal evaporation technique offers great advantages as they can produce highly crystalline and orthorhombic structures. In addition, the surface morphology of WO_3 is found to be strongly dependent on the substrate temperature and gas flow conditions.

EXPERIMENTAL

Tungsten tri-oxide powder (WO_3) was used as the source material. A quartz tube is lying in a horizontal tube furnace (Thermolyne-4700) without a vacuum system. 2g tungsten tri-oxide powder was sprinkled in an alumina boat and the boat was placed at constant temperature maintain area of the quartz tube. An indium tin oxide (ITO) coated glass substrate was positioned at 18 cm distance from the alumina boat along one side of the quartz tube. The temperature of the tube furnace was increased at a constant rate of 20 °C/min from room temperature to heating temperature of 1100°C. After attaining that source vaporization temperature, the oxygen, argon gas mixture (O_2/Ar) about 50/50 ratio in sccm (standard cubic centimeter) was admitted through another end of the quartz tube. That process condition was maintained for 6 hrs. After this process, the temperature of the tube furnace was decreased at constant rate of 30 °C/min from source vaporization temperature to room temperature. During this process tungsten tri-oxide powder was heated, vaporized, and then deposited on the substrate to form film.

RESULTS AND DISCUSSION

X-ray Diffraction (XRD) Analysis

The X-ray diffraction pattern of WO_3 film on ITO coated glass substrate is shown in figure (1). The pronounced peaks in XRD pattern confirm the orthorhombic structure of WO_3 with lattice constants a = 7.341Å, b = 7.570Å, c = 7.754Å (JCPDS No: 75-2072). According to the JCPDS data the following most prominent plane of WO_3 (002), (020) and (002) to the corresponding 2θ values of 2θ = 22.920, 23.485 and 24.229 respectively are well indexed.

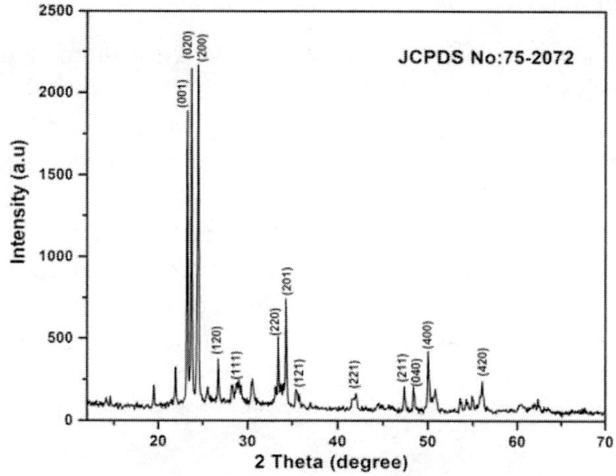

Fig. 1: XRD pattern of WO_3 film deposited on ITO glass substrate

Micro Raman Spectroscopy Analysis

Raman shift spectrum of WO_3 is shown in Figure 2. The spectrum consists of two regions, the (100–500 cm"[1]) and (500–1000 cm^{-1}) for higher and lower wave numbers. Raman spectrum display four distinct bands at 294, 346, 717 and 808 cm^{-1}. The two bands at 294 and 346 cm^{-1} has been assigned to O-W-O bending modes of the bridging oxide, while the 717 and 808 cm/01 bands are the corresponding stretching modes. Consequently, these results give further evidence that the orthorhombic crystals are grown on ITO substrate.

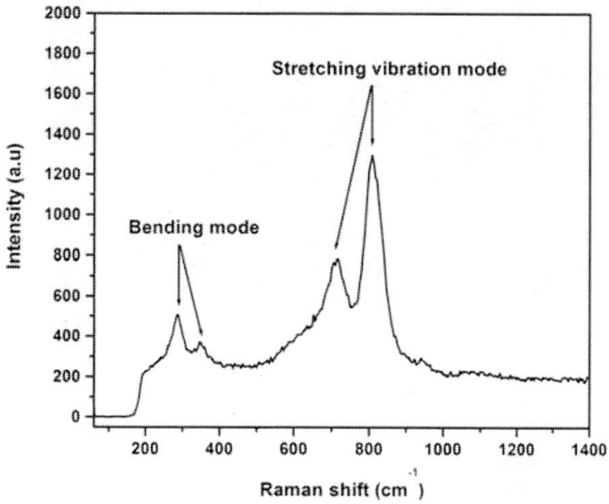

Fig. 2: Laser Raman Spectra of WO_3 Thin Film

Scanning Electron Microscope

The morphology of prepared films both before and after thermal treatment has been investigated by scanning electron microscope (SEM) is shown in fig. 2(a,b). As deposited Film, Fig 3(a) shows that the continuous, finely granulated closed formation. Some conglomerations are formed due to the excess of oxygen flow at the time of processes. The grain size of the film estimated from their SEM images lies well below 50nm and this supports the finding observed from X-ray diffraction pattern. The WO_3 film calcinated for 3 hrs at 45°C is shown in Fig. 3(b). This film consisted of nanorods and nanoparticles superimposed on a porous film. The length and breath of nanorods are 60-100 nm, 50-1300 nm respectively. This indicates that heat treatment induces the morphological evaluation from particle nature to nanorods.

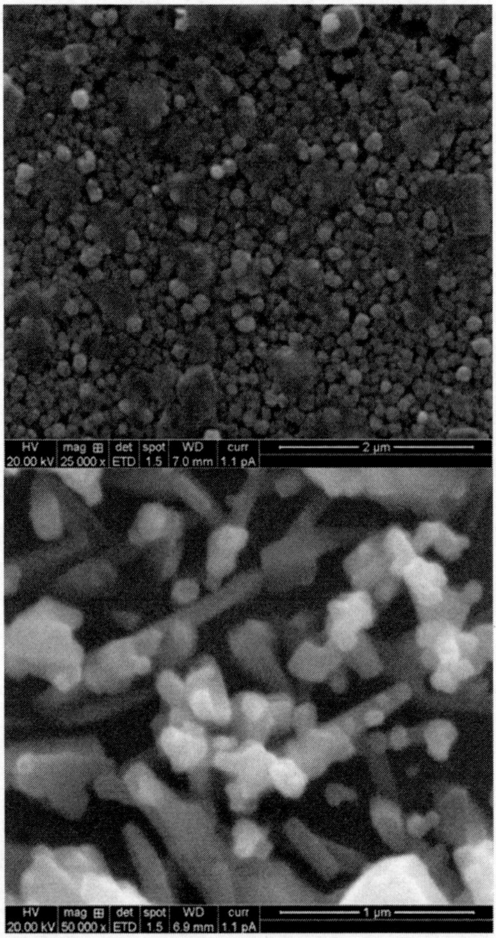

Fig. 3: SEM Image of WO_3 Nanostructure (a) As Deposited (b) Annealed at 450°C

PHOTO–ELECTROCHEMICAL STUDY

The photoelectrochemical measurements were carried out in a potentiostatic system (Autolab PGSTAT30) with the tungsten oxide nanorods as the working electrode, platinum as the counter electrode and a saturated calomel electrode (SCE) as a reference electrode. All potentials are reported versus SCE. An aqueous solution of 0.5 M H_2SO_4 was used as the electrolyte and irradiation was provided by a 1kW Xe arc lamp with an AM1 filter to simulate solar conditions. For monochromatic irradiation experiments, a monochromator was positioned between the source and the photoelectrochemical cell. Linear sweep voltammograms were measured scanning from –1.0 V to + 1.6 V at a scan rate 10 mV s^{-1}.

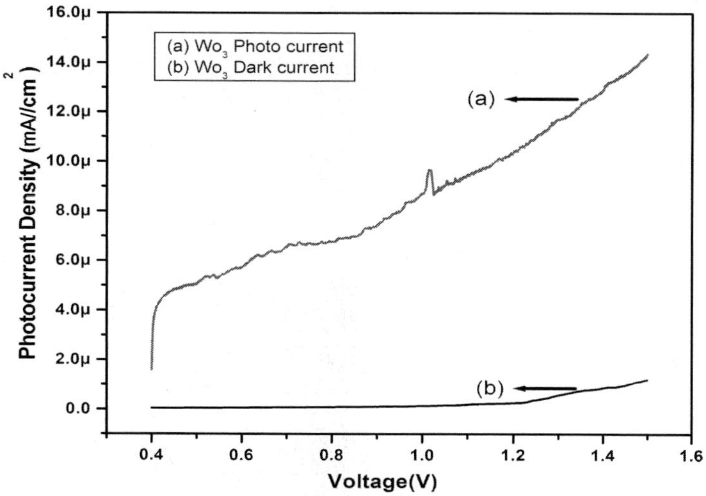

Fig. 4: Photocurrent Measurement of WO_3

The sample under dark conditions is shown in Fig. 4(a). The minimum anodic current is produced in the order of below 1 µA cm^{-2} associated with the coloration of the WO_3 film. Light radiation on the film evidently induced the anodic current increases gradually with the bias potential from 0.45 to 1.45eV. The anodic current was hardly increased in the bias potential range of 1.1 to 1.45eV as shown in Fig. 4(b). The saturated photocurrent density is found to be ≈ 14.4 µA cm^{-2}.

CONCLUSIONS

Tungsten oxide (WO_3) nanorods with an average diameter of 60-100 nm on ITO substrates was prepared by thermal evaporation using tungsten trioxide as a source. Structural, morphology and photoelectrochemical properties were investigated to demonstrate their utility in photoelectrolysis. The WO_3 nanocubes have the highest saturation photocurrent of 14.4 µA/cm^2.

REFERENCES

1. Hyeyoung Kim, Karuppanan Senthil and Kijung Yong, *Materials Chemistry and Physics*, 120, 452–455, 2010.
2. Sung Jong Yoo, Seong Uk Yun, Yung-Eun Sung and Kwang-Soon Ahn, *Journal of Power Sources*, 195, 5422–5425, 2010.
3. S.B. Sadale, S.M. Chaqour, O. Gorochov and M. Neumann-Spallart, *Materials Research Bulletin*, 43, 1472–1479, 2008.
4. Wenzhang Li, Jie Li, Xuan Wang, Jun Ma and Qiyuan Chen, *International Journal of Hydrogen Energy* 35, 13137-13145, 2010.
5. Wenzhang Li, Jie Li, Xuan Wang, Sha Luo, Juan Xiao and Qiyuan Chen, *Electrochimica Acta*, 56, 620–625, (2010).
6. Laura Meda, GabriellaTozzola, AlessandraTacca , GianluigiMarra, StefanoCaramori, Vito Cristino and Carlo Alberto Bignozzi, *Solar Energy Materials & Solar Cells* 94, 788–796, 2010.

Differential Scanning Electroscopic

The Tg value of system B was more than system A and may be explained due to the presence of rigid TGBAPB, offering additional cross links. The Tg values of organo clay reinforced systems C, D and E decreased (Fig. 2), which may be due to catalytic homo polymerization of such systems by the cetyltrimethylammonium ions of the OMMT clay [10] unlikely to produce high crosslink density of the epoxy–amine reaction.

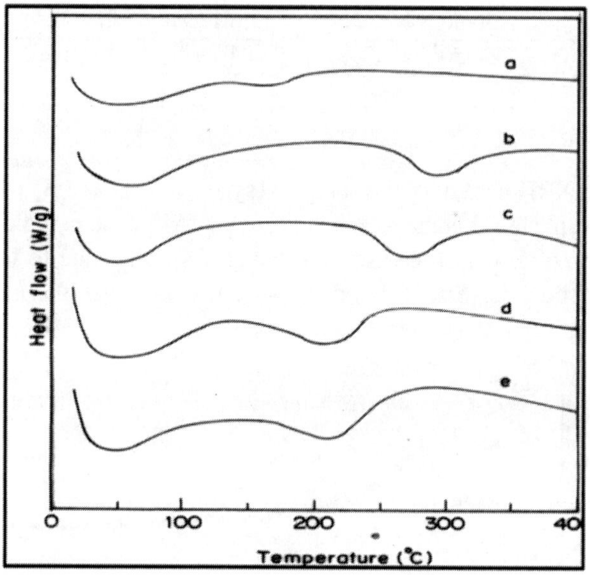

Fig 2: DSC of blend and clay epoxy resins cured with DDM

Mechanical Properties

The tensile, flexural and impact strengths of system B, C, D and E were more than that of system A (Table 1). The order of increase in tensile, flexural and impact strength of the systems studied follows the order E > D > C > B > A. This trend was due to the synergistic effect shown by rigid aromatic rings and flexible ether linkage of TGBAPB for system B. For the C, D, and E systems, the increase can be directly ascribed to the reinforcement of the surface modified OMMT clay offering high strength and stiffness [11].

Table 1: Mechanical Properties of TGBAPB/DGEBA Blend and its Nanocomposites

Sample Dode	*TGBAPB/DGEBA/Clay*	*Tensile Strength (MPa)*	*Flexural Strength (MPa)*	*Impact Strength (Jm^{-1})*
A	0/100/0	63.5 ± 6	110.5 ± 1	97.4 ± 2
B	50/50/0	67.1 ± 2	121.3 ± 3	112.5 ± 3
C	50/50/1%	72.7 ± 1	126.4 ± 2	118.3 ± 4
D	50/50/3%	76.1 ± 6	132.2 ± 4	124.7 ± 2
E	50/50/5%	84.1 ± 5	134.5 ± 2	133.4 ± 3

XRD Analysis

Figure 3.explains the XRD pattern of the samples (systems C, D and E) whose basal reflection plane (001) shifts towards lower angle with increasing OMMT content indicating the intercalated nano composites structure, with increased interlayer *d* spacing due to the presence of alkyl ammonium ions at gallery regions [11] and supports the improved thermal stability observed in C, D and E system by TGA studies.

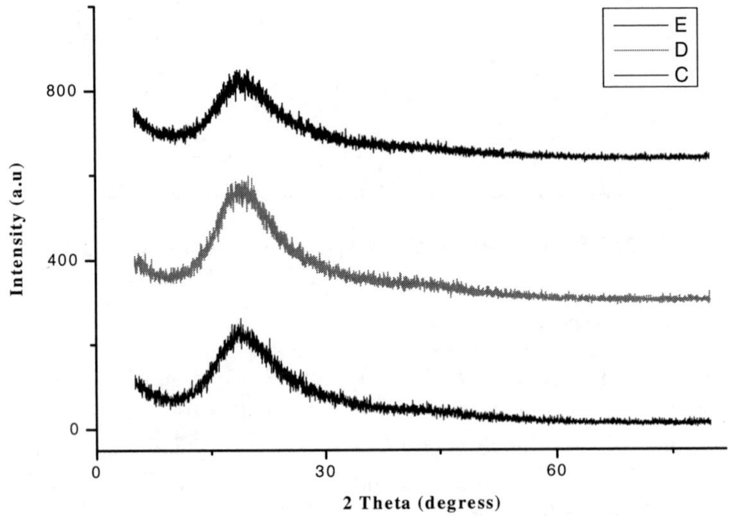

Fig. 3: XRD of Clay Epoxy Resins

MORPHOLOGY OF THE NANOCOMPOSITES

Figure 4(a) and 5(a) Show the SEM and AFM micrographs of the systems B, C, D and E. The surface of B was smooth and thus brittle, while for C, D and E systems, roughness increases

(Figs. 5c & d) indicating a uniform distribution of clay layers (Figs. 4c & d) inducing the toughness [12].

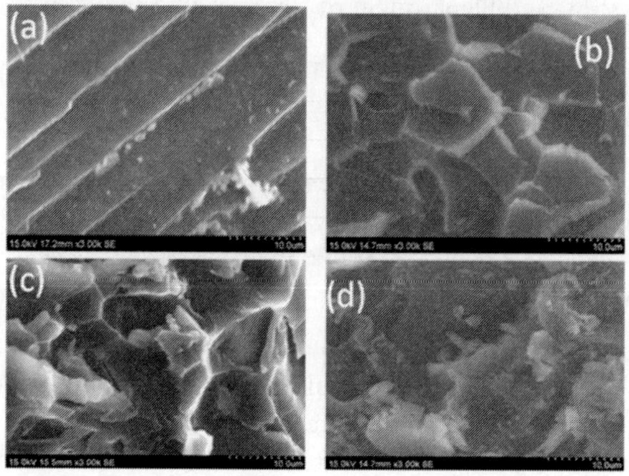

Fig 4: SEM Analysis of System (a) A, (b) B, (c) C and (d) D' Epoxy Resin

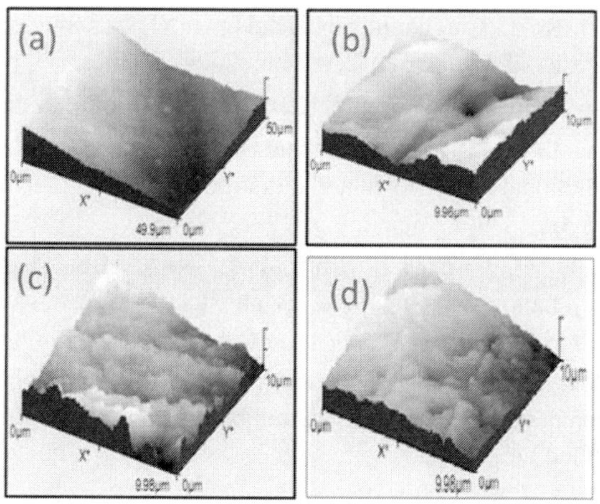

Fig. 5: 3D AFM Image of System (a) A, (b) B, (c) C, and (d) D' Epoxy Resin

CONCLUSIONS

We conclude that TGBAPB may be used as an effective toughening modifier for DGEBA without sacrificing the mechanical properties and this effect was more pronounced with reinforcement of OMMT into the TGBAPB/DGEBA blend.

ACKNOWLEDGEMENTS

One of the authors D. Duraibabu is grateful to UGC for the award of JRF (M.H. No.:6.2.40 UGC Research fellowship) and Department of Chemistry, Anna University, Chennai, to carry out this work.

REFERENCES

[1] Lin, M.S., Liu, C.C., Lee C.T., Toughened interpenetrating polymer network materials based on unsaturated polyester and epoxy, *J. Appl Poly Sci.*, 72, 585–592, 1999.

[2] Pearce, *P.J., Ennis*, B.C., Morris, C.E.M., Tough high temperature aerospace epoxy resin system, Poly Comm, 29, 93–98, 1988.

[3] Choi, J., Yee, A.F., Laine, R.M., Toughening of cubic silsesquioxanxe epoxy nanocomposites using core shell rubber particles, A three component hybrid system, *Macromolecules*, 37, 3267–3276, 2004.

[4] Kinjo, N., Ogata. M., Nishi, N., Kandeda, A., *Adv Polym Sci.*, Berlin, Springer, 1989.

[5] Yanfang, L., Shigang, S., Yanfang, L., Jugang, G., Kinetic of 4,4-diaminodiphenylmethane curing of bisphenol-S epoxy resin, *J Appl Poly Sci* 73, 1799–1803, 1999.

[6] Yee, A.F., Pearson, R.A., Toughening mechanism in elestomer modified epoxies – Part A, mechanical properties, *J Mater Sci*, 21, 2462–2474, 1986.

[7] Ellis, B., Chemistry and Technology of Epoxy Resin, London, Academic Press, 1993.

[8] Musto, P., Martuscelli, E., Rasgosta, G., Russo, P., Scarinzi, G., An interpenetrated system based on a tetrafunctional epoxy resin and a thermo-setting bismaleimide structure properties and correlation, *J Appl Polym Sci*, 69, 1029– 1998.

[9] Park, J.H., Lee, H.M., Chin, I.J., Choi, H.J., Kim, H.K., Kang, W.G., Intercalated polypropylene/clay nanocomposites and its physical characteristics, *J Phys Chem Solids* 69, 1375-1380, 2008.

[10] Becker, O., Varley, R.J., Simon, G.P., Morphology, thermal relaxation and mechanical properties of layered silicate nanocomposites based upon high functionality epoxy resins, *Polymer*, 43, 4365–4373, 2002.

[11] Lee, A., Lichtenhan, J.D., Thermal and viscoelastic property of epoxy clay and hybrid inorganic–organic epoxy nanocomposites, *J. Appl Polym Sci*, 73, 1993–2001, 1999.

[12] Wu, C.L., Zhang, M.Q., Rong, M.Z., Friedrich K., Tensile performance improvement of low nanoparticles filled-polypropylene composites, *Comp Sci tech*, 62, 1327-1340, 2002.

RAMAN STUDY OF COBALT FERRITE NANOPARTICLES

HEMAUNT KUMAR, R.C. SRIVASTAVA, PUNEET NEGI, H.M. AGRAWAL AND ANURAG SINGH[1]

Department of Physics, G.B. Pant University of Ag. & Technology, Pantnagar
[1]Department of Physics Banaras Hindu University, Varanasi
E-mail: hvatsal@gmail.com

ABSTRACT

The nanocrystalline $CoFe_2O_4$ were prepared by nitrate route at the sintering temperature of 300, 500, 700 and 900°C. The X-ray diffraction (XRD) patterns of all the samples show the single phase spinel structure of nanoparticles. The Crystallite sizes are 9, 22, 47 and 61 nm at different sintering temperature respectively. The assigned Raman modes are the characteristics of the spinel structure of the synthesized samples which change slightly with different sintering temperature. The variation in the parameters of various modes follows phonon confinement.

INTRODUCTION

Spinel ferrites are technologically important class of magnetic oxides because of their interesting magnetic properties, high electrical resistivity, low eddy current and dielectric loss. They are used in microwave devices, computer memory chips, etc. Cobalt ferrite ($CoFe_2O_4$) has been the centre of attention due to its large magneto crystalline anisotropy, high coercivity, moderate saturation magnetization, large magnetostrictive coefficient, chemical stability and mechanical hardness [1-3]. The properties of the individual particles as well as their mutual interactions

V. Rajendran, K. Saminathan, P. Paramasivam and K.E. Geckeler (eds.)
Nano Materials Synthesis and Characterisation, pp. 149-154 (2012).

indicate important features of the nanoparticles systems, such as storage capacity and switching field. Raman spectroscopy is one of the powerful techniques, which has often been used to identify the microscopic vibrations caused by the slight structural distortion. So it is very likely that using Raman spectroscopy will provide additional information for reconciling the existing structural disagreement. Raman spectroscopy may provide useful information on the cation migration in spinel $CoFe_2O_4$. Due to its high sensitivity to many lattice effects, such as structure transition lattice distortion charge–lattice and spin–lattice couplings, local cation ordering and magnetic ordering Raman scattering is uniquely suited for probing magnetic oxides.

EXPERIMENTAL

In the present work nanoparticles of $CoFeO_4$ were synthesized by using the chemical route. The stoichiometric amounts of $Fe(NO_3)_3.9H_2O$, and $Co(NO_3)_2.6H_2O$ were dissolved in distilled water. In the aqueous salt solution, citric acid solution was added with the cations to citric acid molar ratio of 1:2. The solution was heated at 85°C under constant magnetic stirring until the solution got converted into a viscous gel, and was allowed to cool at room temperature. The cooled gel was dried in an oven at 100°C for overnight to form the precursor material. The precursor material was sintered at 300, 500 700 and 900°C for 2 hrs to get the nanoparticles of cobalt ferrite of different particle sizes. X-ray diffraction data were collected on Bruker D8 ADVANDED x-ray diffractometer using Cu K_α radiation (λ = 1.54178Å). The micro-Raman system used in this study consisted of a 200mWAr ion laser (488 nm) with a focal spot size of 10–20 μm , a charge-coupled device (CCD) detector (1100×330 pixels) and holographic optics. The spectral range 200–2000 cm^{-1} was examined using an 1800 grooves/mm grating. The spectrometer was regularly calibrated using the Ar emission spectrum. An exposure time of 15 min was used for recording the Raman spectra of the samples. A laser power of 5mW was used in order to avoid sample heating.

RESULTS AND DISCUSSION

Structural Characterization

Figure 1 shows the XRD patterns of the synthesized cobalt ferrite nanoparticles and Fig. 2 shows Rietveld of the sample sintered at 700°C. The diffraction peaks are corresponding to the single phase of the ferrite sample having no impurity peaks. The phase analysis was done with JCPDS card number 03-0864.The crystallite size is determined by Debye Scherrer formula [4] using most intense peak (311)

$$D = \frac{k\lambda}{\beta\cos\theta}$$

where λ is the wavelength of X-ray used, θ is the Bragg's angle, β is the full width at half maximum (FWHM) in radian and 'k' is a constant approximately equal to 0.9. Specific surface area (surface area per unit mass) [5] can be calculated by the following relation

$$S = \frac{6000}{D\rho}$$

The 6000 is called form factor for spherical particles, D is particle diameter and 5 is the x-ray density. The X-ray density ñ was calculated by using the formula:

$$\rho = \frac{8M}{Na^3}$$

where M, N and 'a' are the molecular weight, Avogadro's number and lattice parameter respectively. The factor 8 is due to the fact that there are eight molecules of cobalt ferrite per unit cell. The lattice parameter was calculated with the help of the relation:

$$a = d_{hkl}\sqrt{h^2 + k^2 + l^2}$$

where *h*, *k*, *l* are the Miller indices of the planes having interplanar spacing'd'. All the parameters obtained from the XRD patterns are listed in Table 1. From the XRD pattern it is

Table 1: Calculated XRD Parameters of the Samples

Sintering Temperature	*Particle Size (nm)*	*Lattice Parameter(Å)*	*X-ray Density (gm/cm^3)*	*Specific Surface Area (m^2/gm)*
300 ^{0}C	9 ± 1	8.37 ± 0.01	5.33 ± 0.02	125 ± 1
500 ^{0}C	22 ± 1	8.37 ± 0.01	5.33 ± 0.02	51 ± 1
700 ^{0}C	47 ± 1	8.33 ± 0.01	5.34 ± 0.02	24 ± 1
900 ^{0}C	61 ± 1	8.37 ± 0.01	5.34 ± 0.02	18 ± 1

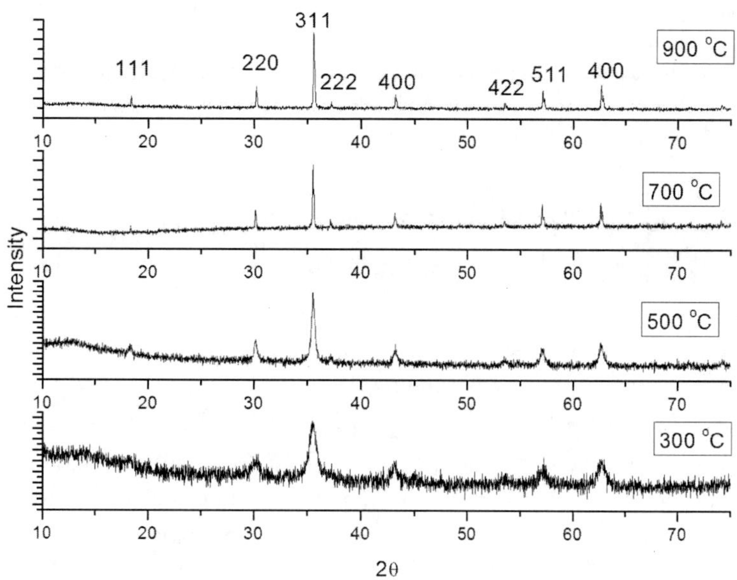

Fig. 1: XRD Patterens at Different Sintering Temperature

clear that peaks becomes sharper with the increase of sintering temperature. This suggests that as the sintering temperature increases the crystallite size also increases. From table 1. it is clear that crystallite size increases significantly from 9 nm to 61 nm as the sintering temperature increases from 300 to 900°C. The sample with the smallest particle size has the largest specific area and the later decreases with increase in sintering temperature. The decrease in specific surface area with sintering temperature can be correlated to the agglomeration of the particles and consequently the densification process may be considered [6].

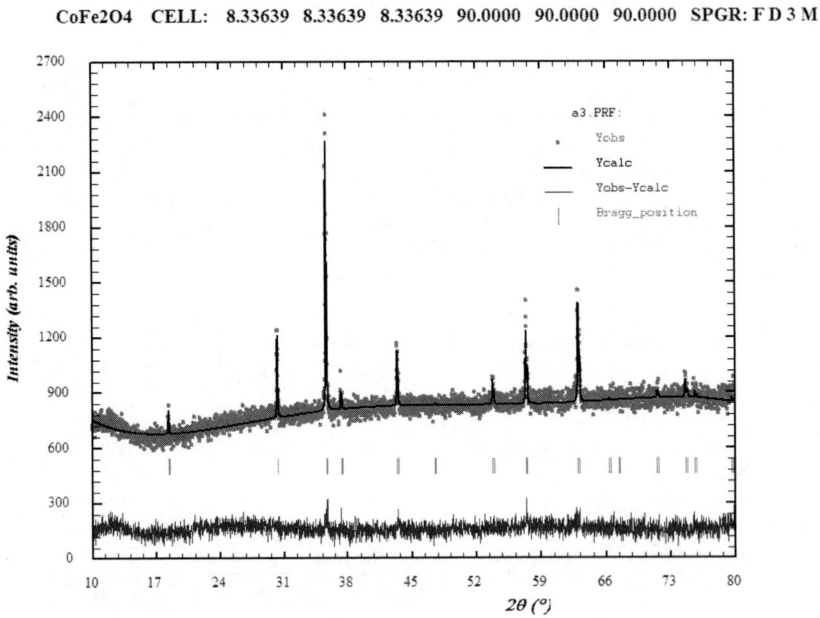

Fig. 2: Rietveld Analysis of the Samples Sintered at 700°C

Raman Analysis

According to the group theory, CoF_2O_4 has a cubic inverse- spinel structure with O^7_h (Fd3m) space group, which gives rise to 39 normal modes.

$$\square = A_{1g} + E_g + F_{1g} + 3F_{2g} + A_{2u} + 2E_u + 4F_{1u} + 2F2u$$

There are five Raman active modes ($A_{1g} + 1E_g + 3F_{2g}$) and all these modes were observed at ambient condition [7]. Figure 3 shows the Raman spectra of cobalt ferrite nanoparticles at sintering temperatures of 300, 500, 700 and 900°C. At 900°C samples showed five peak maxima at 303, 467, 575, 614 and 690 cm^{-1}. Table –2 shows the assignment of modes at different sintering temperature. Assignment of theses modes were carried out in accordance with the literature report [8]. Frequencies above 600 cm^{-1}, i.e. at 614 and 690 cm^{-1} are due to A_{1g}

symmetry involving symmetric stretching of oxygen atom with respect to metal ion at the tetrahedral site. The other low frequency phonon modes are due to metal ion involved in octahedral site, i.e. E_g and F2g (3). These modes correspond to the symmetric and anti-symmetric bending of oxygen atom in M-O bond at octahedral sites [9].The phonon confinement in the materials is effective when the crystallite size of the sample is approximately 20 times the lattice parameter [10]. The variations of parameters are in good agreement with phonon confinement up to the crystallite size of 18 nm [11].

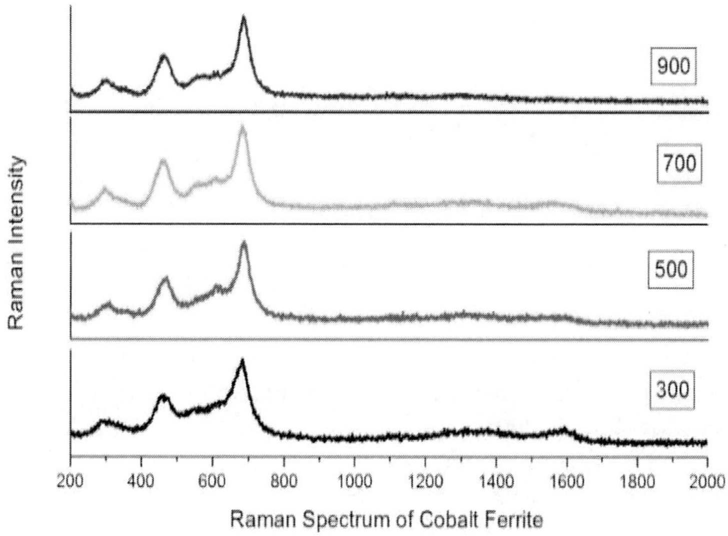

Fig. 3: Raman Spectra of Cobalt Ferrite at Different Sintering Temperature.

Table 2: Assignment of Raman Modes

Sintering Temp.	$A_{1g}(1)(cm^{-1})$	$A_{1g}(2)$ (cm^{-1})	$F_{1g}(2)$ (cm^{-1})	$F_{1g}(1)$ (cm^{-1})	$E_g(cm^{-1})$
300^0C	684 cm^{-1}	--	461 cm^{-1}	552 cm^{-1}	296 cm^{-1}
500^0C	690 cm^{-1}	618 cm^{-1}	474 cm^{-1}	--	310 cm^{-1}
700^0C	686 cm^{-1}	608 cm^{-1}	464 cm^{-1}	--	300 cm^{-1}
900^0C	690 cm^{-1}	614 cm^{-1}	467 cm^{-1}	575 cm^{-1}	303 cm^{-1}

CONCLUSIONS

The XRD and Raman behaviour of nanocrytalline $CoFe_2O_4$ has been studied at different temperature. The particle size increases with increase in sintering temperature. The sample with smallest particle size has the largest specific area and the later decreases with increase in sintering temperature. The assigned Raman modes are the characteristics of the spinel structure of the synthesized samples which change slightly with different sintering temperature.

ACKNOWLEDGMENTS

Hemaunt Kumar is thankful to Dr. Mukul Gupta (UGC DAE CSR Indore Centre) and Dr. R.K. Singh (BHU, Varanasi) for their keen interest and suggestions in XRD and Raman measurements, respectively.

REFERENCES

[1] Pallail V. and Shah D.O., Synthesis of high-coercivity cobalt ferrite particles using water-in-oil microemulsions, *J. Magn. Magn.* Matter 163, 243-248, 1996.

[2] Lijun Zhao, H.Z., Y.X. and F.C. Studies on the magnetism of cobalt ferrite nanocrystals synthesized by hydrothermal method, *J.S. State Chemistry* 181, 245-252, 2008.

[3] R.M.M., M.M.R., F.A.H. and W.S. Structure and magnetic properties of nanocrytalline cobalt ferrite powders synthesized using organic acid precursor methods, J. Magn. Magn. Mat. 322,2058-2064,2010.

[4] B.D. Cullity, Elements of X-ray Diffraction, London: Addison-Wesley, 1978.

[5] Wafaa Bayoumi, Structural and electrical properties of zinc-substituted cobalt ferrite, *J. Mater* Sci 42,8254-8261, 2007.

[6] M. M. El-Okr, M.A. Salem, M.S. Salem, R.M. El-Okr, Ashoush H.M. Talaat, Synthesis of cobalt ferrite nano-particles and their magnetic characterization, *J. Magn. Magn. Mat* 323, 920-926, 2011.

[7] Z. Wang, D. Schiferl, Y. Zhao, H. St. C. O Neil, High pressure Raman spectroscopy of spinel-type ferrite ZnFe2O4, *J. Phys. Chem. Solids*, 64, 2517-2523, 2003.

[8] P.R. Graves, C.Johnston , *J.J. Campaniellom* Raman scattering in spinel structure ferrites, *Materials Research Bulletin* 23, 11, 1651-1660, 1988.

[9] Z.W. Wang, P. Lazor, S.K. Saxena, G. Aritoli, High-Pressure Raman Spectroscopic Study of Spinel ($ZnCr_2O_4$) *J.Solid State Chem.*,165, 2002.

[10] A.K. M.R., T.R. R., in Phonon Confinement in Nanostructur Materials, *Encyclopedia of Nanoscience and Nanotechnology*, American Scientific Publishers: Newyork, 2004, p. 1.

[11] J.P. Singh, R.C. S., H.M., R. Kumar, Micro Raman investigation of nanosized zinc ferrite: effect of crystallite size and fluence of irradiation, *J. Raman spectroscopy* (17 March 2011).

ELECTROCHEMICAL SYNTHESIS OF Cu-BDC METAL ORGANIC FRAMEWORKS: OPTIMIZATION AND ITS CHARACTERISATION STUDIES

R. Senthil Kumar, S. Senthil Kumar[1] and M. Anbu Kulandainathan

Electro Organic Division, CSIR-Central Electrochemical Research Institute, Karaikudi
[1] Electrodics and Electrocatalysis Division, CSIR-Central Electrochemical Research Institute, Karaikudi
E-mail: manbu123@yahoo.com

ABSTRACT

Copper benzenedicarboxylate (Cu-BDC) has been synthesized electrochemically from benzenedicarboxylicacid (BDC) in N,N-Dimethyl formamide solution containing tetrabutylammonium tetrafluoroborate (TBATFB) as electrolyte at room temperature. Copper metal served as both anode and cathode. The reaction condition is optimized by varying the TBATFB concentration, BDC concentration, applied cell potential and reaction duration. The synthesized material is characterized by FT-IR, XRD, SEM, TGA and EDX spectroscopical studies. We obtain crystalline Cu-BDC metal organic framework structure with particle size 0.2-0.5 μm, which is confirmed by XRD and SEM analysis. BDC concentration plays an important role in deciding the shape of the particles (featureless to rods).

INTRODUCTION

Metal-organic frameworks (MOFs) are very promising systems due to the virtually unlimited flexibility in their design. The generation of MOFs enables the tailoring of novel solids with regular porosity from micro to nanopore scale. Nowadays several hundred different types of MOFs are known. The self assembly of metal ions, which act as coordination centres, linked

V. Rajendran, K. Saminathan, P. Paramasivam and K.E. Geckeler (eds.)
Nano Materials Synthesis and Characterisation, pp. 155-160 (2012).

together with a variety of polyatomic organic bridging ligands, results in tailorable nanoporous host crystalline materials as robust solids with high thermal and mechanical stability. These hybrid organic-inorganic compounds exhibit rather low-dense and relatively robust three-dimensional architectures delimiting systems of well-defined cavities or tunnels and they offer new potential applications include molecular magnetism, ionic conductors, sensors, materials for adsorption applications (gas separation and gas storage, in particular hydrogen storage) and heterogeneous catalysis.[1-10] In heterogeneous catalysis, MOFs can exhibit high activity and selectivity in oxidation, esterification, epoxide ring openings, asymmetric hydrogenation, etc.

Metal organic framework material Cu-BDC (BDC = 1,4-benzene dicarboxylate) were synthesized using different routes. Muller et al., have studied the generation of MOFs via electrosynthesis in methanol medium[11] and also developed a series of MOFs using metal ion with various linkers particularly terephthalic acid, naphthalene dicarboxylic acid and benzene tricarboxylic acid.[12,13] Electrochemical methods are easily scalable route with simple procedure and considerably avoid environmental pollution, many organic compounds are being synthesized by such methods.

In this paper, we report the synthesis of Cu-BDC MOF by applying constant D.C potential in the electrolytic solution of tetrabutylammonium tetrafluoroborate (TBATFB) in N,N-Dimethyl formamide. The experiment was carried out at room temperature. In addition to optimization of electrochemical reaction conditions, the effect of TBATFB concentration, BDC concentration and reaction time were also optimised.

EXPERIMENTAL

1,4-Benzene dicarboxylic acid and TBATFB was obtained from Across organics chemicals Ltd, N,N-Dimethyl formamide (HPLC grade) purchased from SRL Ltd and used as such without further purification. Synthesize of Cu-BDC (MOF) by electrolysis under the condition of applied D.C potential between the two copper electrodes in the electrolyte solution, containing 1,4-benzene dicarboxylic acid with tetrabutylammonium tetrafluoroborate(TBATFB) in methanol at room temperature. Interestingly, the addition of TBATFB in electrolyte solution, increase the conductivity of the electrolyte solution and improve the yield of the products. After few minutes of the electrolysis, a blue colour solid started precipitating out from the solution. After 2.5 h, the solid precipitate was filtered off and washed with N,N-Dimethyl formamide for several times to ensure the removal of excess of organic linkers. Afterwards the sample was dried under vacuum at 120°C for 12 h and then activated at 200°C for 2 h.

RESULTS AND DISCUSSION

Figure 1a and b presents the effect of concentration of supporting electrolyte and effect of BDC concentration, when increasing the concentration of supporting electrolyte the yield increase, while considering the 0.02M and 0.05M, there is no more changes, also increasing the concentration of TBATFB and BDC affect the surface morphology. However increasing the applied voltage on the reaction, major change in the amount of yield was observed upto 15 V, which clearly show that while increasing the applied voltage the yield also increases. Fig-1b depicts the yield vs reaction time and current vs reaction time. The yield mainly depends on reaction time but as the reaction time goes on the availability of BDC decreases and thereby current decreases. However, it is worthy to mention that, while increasing the concentration of TBATFB > 0.02M and the applied voltage more than 15 V, the particles size are more featureless. Hence, the optimized experimental conditions are so important to get regular shapes without agglomeration with good yield of crystalline Cu-BDC.

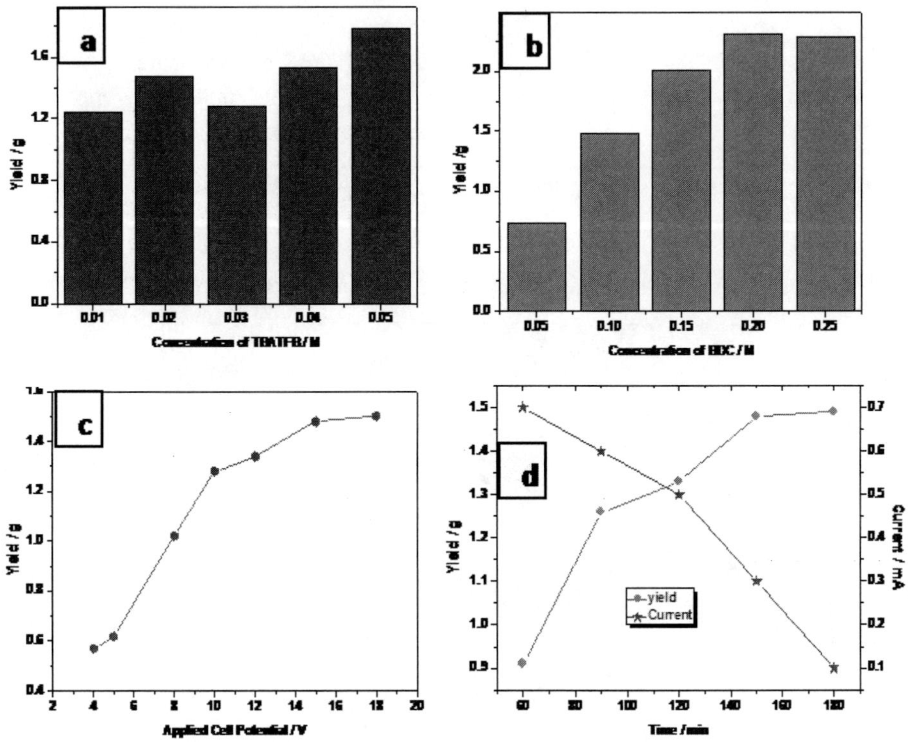

Fig. 1: The Graph Corresponding to (a) Concentration of TBATFB vs yield (b) Concentration of BDC vs Yield (c) Applied cell Potential vs Yield and (d) Reaction time vs Yield and Reaction Time vs Current

In order to confirm the nature of bonding between benzenedicarboxylic acid (BDC) and Cu^{2+}, vibrational spectral analyses was carried out in wavenumber region 400–2100cm^{-1} and are shown in Fig. 2. The band around 1662 cm^{-1} has been assigned to protonated C = O stretching vibration, after complexation which is shifted to deprotonated C = O stretching region at 1631 cm^{-1}, indicates the formation of Cu-BDC. The characteristic vibration of Cu–O and DMF(CO) band were also observed in complex at 671, 1665 cm^{-1} respectively.[14,15]

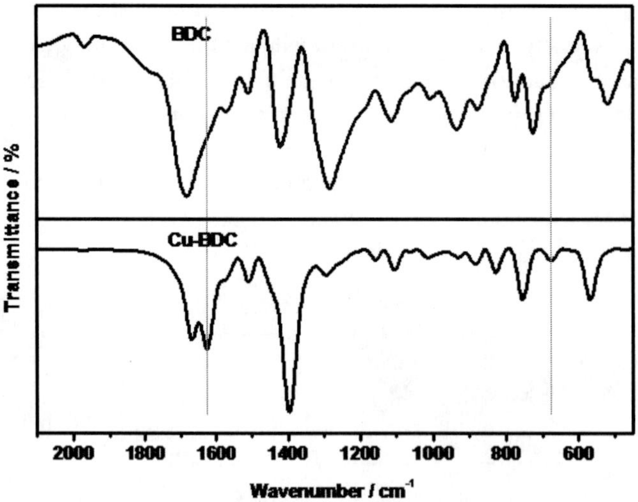

Fig. 2: FT-IR Spectra of (a) 1,4-Benzene Dicarboxylic Acid and (b) Cu-BDC

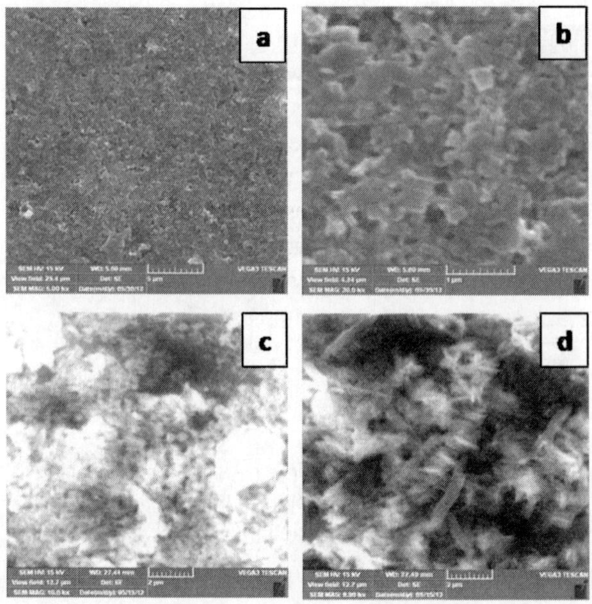

Fig. 3: SEM Images of Synthesized Cu-BDC Samples (a) Lower Magnification Image, and (b) Higher Magnification Image using 0.02 M TBATFB, 0.1 M BDC, (c) 0.04 M TBATFB, 0.1 M BDC and (d) 0.02 M TBATFB, 0.25 M BDC

Scanning electron microscopic images (Figs. 3 a & b) show the synthesized Cu-BDC are formed in featureless shape with particle size 0.2 – 0.5 μm when using 0.02 M TBATFB and 0.1 M BDC. Figure 3c shows that the synthesized material formed in microporous structure. While increasing the BDC concentration to 0.25 M, the rods like structure are formed and are shown in Fig. 3d.

ELEMENTAL ANALYSIS

The chemical composition of electrochemically synthesized CuBDC MOF was determined using energy dispersive X-ray analysis (EDX) as shown in Table 1. The peaks corresponding to Copper, Carbon and Oxygen are clearly seen and also no impurities are present other than these elements.

Table 1: Element Composition Obtained from EDX Analysis for Cu-BDC MOF.

Sl. No.	*Element*	*Weight %*
1	% of C	41.57
2	% of O	25.17
3	% of Cu	33.26

THERMAL STABILITY AND X-RAY DIFFRACTION STUDIES

Thermogravimetric analysis was carried out to obtain the thermal stability information of synthesized Cu-BDC complex. The complex was heated in oxygen atmosphere with heating rate of 10°C/min and the spectrum is shown in Fig. 4a. The results shows that the initial weight loss occur at 200°C corresponds to the removal of solvent molecule from the complex,[15] that results consist with FT-IR analysis. Then the framework remains thermally stable up to 330 °C, and on further heating organic molecules undergo decomposition. The crystalline nature of the synthesized material is confirmed through powder X-ray diffraction pattern shown in Fig. 4b and is consistent with reported literature.

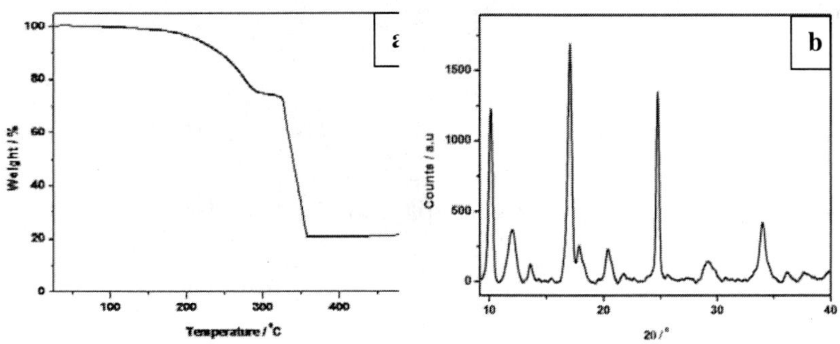

Fig. 4. (a) Thermogravimetric Curve and (b) Powder X-Ray Pattern of Cu-BDC MOF

CONCLUSION

Crystalline copper benzenedicarboxylate (Cu-BDC MOF) was synthesized using electrochemical method. Electrochemical reaction conditions were optimized for the above synthesis along with the effect of TBATFB concentration, BDC concentration, applied cell potential and reaction time. It is also observed that BDC concentration influenced the morphology of the synthesized material beyond 0.20 M. Cu-BDC formation was also confirmed using FT-IR, EDAX and XRD. The thermal stability of synthesized material was found upto 330°C. Further analyses of these crystalline Cu-BDC MOFs, for variety of technological applications particularly in gas adsorption, catalysis and gas separation are in progress.

REFERENCES

[1] Chui, S.S., Y, S.M.F., Lo, Charmant, J.P.H., Orpen, A.G. and Williams, I.D. *Science*, 1999, *283*, 1148.

[2] Hu, A.G, Ngo, H.L. and Lin, W.B. J. *Am. Chem. Soc.* 2003, *125*, 11490.

[3] Kitaura, R. Onoyama, G. Sakamoto, H. Matsuda, R. Noro, S. and Kitagawa, S. *Angew. Chem., Int. Ed.*, 2004, *43*, 2684.

[4] Schlichte, K. Kratzke and T. Kaskel, S. *Micropor. Mesopor. Mater.* 2004, 73, 81.

[5] (a) Czaja, A.U. Trukhan and N. Muller, U. *Chem. Soc. Rev.* 2009, *38*, 1284.
(b) Murray, L.J. Dinca, M. Long, J.R. *Chem. Soc.Rev.* 2009, *38*, 1294.

[6] Harmon, L. Serre, C. Devic, T. Loiseau, T. Millange, F. Ferey, G. and Weireld, G.D. *J. Am. Chem. Soc.* 2009, 131, 8775.

[7] Yang, L. Kinoshita, S., Yamada, T. Kanda, S. Kitagawa, H., Tokunaga, M. Ishimoto, T. Ogura, T. Nagumo, R. Miyamoto, A. Koyama, M. *Angew. Chem. Int. Ed.*, 2010, *122*, 5476.

[8] Xie, Z. Ma, L. Dekrafft, K.E. Jin, A. and Lin, W. *J. Am. Chem. Soc.*, 2010, 132, 922.

[9] Meek Scott, T., Greathouse Jeffery, A. and Allendorf Mark, D. *Adv. Mater.* 2011, 23, 249.

[10] Partha Mahata, Debajit Sarma, Madhu, C. Sundaresen and A. Srinivasan, N. *Dalton Trans.*, 2011, 40, 1952.

[11] Mueller, U., Puetter, H., Hesse, M., Wessel, H. WO 2005/049892, 2005,

[12] Schubert, M., Hesse, M., Mueller, U., Puetter, H. and Tonigold, M. US 7910732 B2, 2011.

[13] Mueller, U., Puetter, H., Hesse, M., Schubert, M., Wessel, H., Huff, J. and Guzmann, M. US 2007/0227898 A1, 2007.

[14] Fathima Beegum, M., Usha Kumari, L., Harikumar, B., Hema Tresa Varghese and Yohannan Panicker, C., *Rasayan J. Chem.* 2008, 1, 117.

[15] Carson, G., Hardcastle, K., Schwartz, J., Liu, X., Hoffmann, C., Gethardt, A. and Tannenbaum, R., *Eur. J. Inorg. Chem.* 2009, 2338.

MICROWAVE STUDIES OF NANO-SIZED $Ba_4Co_2Mn_{(0.2)}Fe_{(35.8)}O_{60}$ MIXED HEXAFERRITE

PRIYANKA KASHID, DEEPTI KULKARNI AND VIJAYA PURI

Thick and Thin Film Research Laboratory, Department of Physics, Shivaji University, Kolhapur
E-mail: Vijaypuri@gmail.com

ABSTRACT

Hexaferrites are the class of ferrites which exhibits high anisotropy, large coercivity and large saturation magnetization. The present paper deals with the investigations on the properties of mixture composed of multiple barium Hexaferrite phases (M type-Y type) synthesized using co-precipitation technique. A Prepared powder was pelletized and sintered at 600, 700 and 800°C. The hexaferrite phase obtained with the synthesized powder was confirmed by X-ray diffraction studies. Effect of sintering temperature on structural and microwave properties was investigated. At lower temperature agglomerated grains are formed due to nanogranules of size 10-30nm.At higher temperature separated grains of size ~60 are observed. Uniform spherical particles were obtained. The microwave transmittance and reflectance of samples was studied by waveguide reflectometer. It has been observed that the microwave transmittance increases with sintering temperature of the sample under test.

INTRODUCTION

Microwave technology is moving up to higher frequencies and higher bandwidths, into the mm wave range, up to 100 GHz. Non conducting materials are essential to ensure total penetration of electromagnetic fields. Ferrimagnetic oxides of iron combine the properties of a magnetic material with that of an electrical insulator. Ferrite elements are widely used in microwave devices, isolators, circulators, phase shifters. The family of hexagonal ferrites can be classified

V. Rajendran, K. Saminathan, P. Paramasivam and K.E. Geckeler (eds.)
Nano Materials Synthesis and Characterisation, pp. 161-166 (2012).

on the basis of their chemical composition and crystal structure. They are subdivided into six fundamental, simplest structural types: M, W, Y, X, Z and U [1-2]. The present paper deals with the investigations on the properties of mixture composed of multiple barium hexaferrite phases (M type-Y type) synthesized using co-precipitation technique. Co-precipitation is a simple and environment friendly method to produce barium ferrite nanoparticles with high intrinsic coercivity [3-4]. It does not require sophisticated instrumentation and low cost starting materials can be used. Very fine particles with high chemical homogeneity and high reactivity can be obtained. In the present paper, we report structural, DC electrical and magnetic properties of Barium hexaferrite nanoparticles. Effect of sintering temperature on structural and microwave properties was investigated.

EXPERIMENTAL

Material Synthesis

$Ba_4Co_2Mn_{(0.2)}Fe_{(35.8)}O_{60}$ mixed hexaferrite was synthesized by a wet chemical route. The starting materials used were barium nitrate, manganese nitrate, ferric nitrate, cobalt nitrate. Stoichiometric amounts of starting materials were dissolved one by one in de-ionized water. Ammonia solution was added slowly in the solution to obtain precipitate and pH was maintained at 9. The solution was stirred for two hours at room temperature. The brown colored precipitate obtained was separated and washed several times with de-ionized water and dried. The powder was pelletized under high pressure. The pellets was sintered in a muffle furnace at 600,700 and 800°C with the rate of 10°C/min in order to obtain sample $Ba_4Co_2Mn_{(0.2)}Fe_{(35.8)}O_{60}$ mixed hexaferrite

Material Characterization

The sintered powder was characterized by X-ray diffraction and SEM. The microwave transmittance and reflectance of pellets sintered at 600,700 and 800°C was studied by waveguide reflectometer.

RESULTS AND DISCUSSION

X-ray Diffraction

Typical XRD patterns of $Ba_4Co_2Mn_{(0.2)}Fe_{(35.8)}O_{60}$ is shown in Fig. (1). The diffraction pattern shows dominant (102) plane which indicates hexagonal crystal system. The diffractogram of $Ba_4Co_2Mn_{(0.2)}Fe_{(35.8)}O_{60}$ shows the presence of M and Y phases. The mixed hexaferrite crystal structure was formed as the combination of M-type ($BaFe_{12}O_{19}$) and Y-type ($Ba_2Co_2Fe_{12}O_{22}$) hexaferrite phases [5] The crystallite size was calculated from highly intense peak using Scherer's formula; $D = 0.9\lambda/\beta \cos \theta$ where D is the crystallite size, β is the broadening of the diffraction peak measured at half of its maximum intensity FWHM and λ is the X-ray wavelength λ = 1.54 A°.

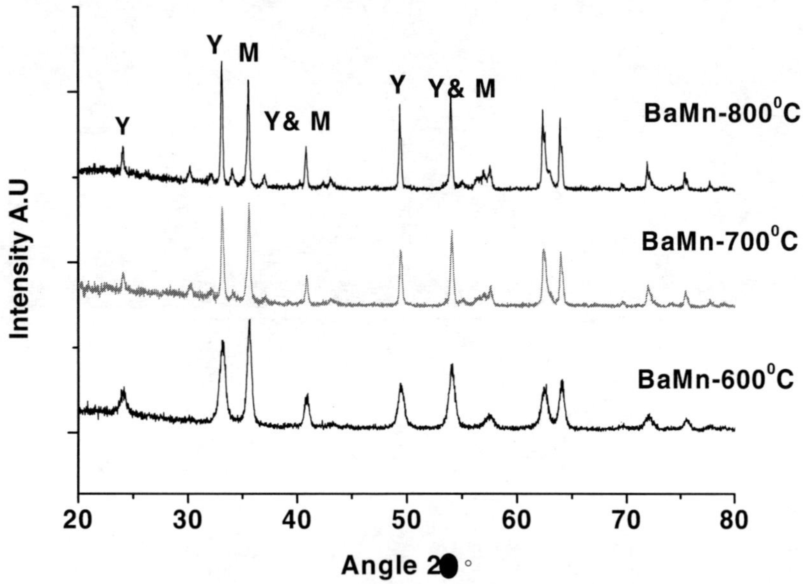

Fig. 1: X-ray Diffraction Patterns of $Ba_4 Co_2 Mn_{(0.2)} Fe_{(35.8)}O_{60}$ Ferrite Sintered at 600,700 and 800°C

Crystallite size obtained using this formula for $Ba_4Co_2Mn_{(0.2)} Fe_{(35.8)}O_{60}$ ferrite sintered at 600,700 and 800^{0}C are 70.8 nm, 83.5nm and 123 nm.It can be observed that crystallite size increases with increase in temperature is shown in Fig. (1)

Scanning Electron Microscopy

The surface morphology of $Ba_4 Co_2 Mn_{(0.2)} Fe_{(35.8\ )}O_{60}$ is shown in Figures 2(a) and 2(b) and 2(c) for samples sintered at 600,700 and 800^{0}C respectively. Fig. 2(a) shows the morphology of ferrite sintered at 600C which indicates the formation of spongy ball like grains of size 300-400nm formed by agglomerated nanogranules of size 5-15 nm. Fig 2(b) shows the morphology of ferrite sintered at 700^{0}C, it can be seen that there is formation of nanogranules of size 10-20nm with the balls starting to distintigrate. Fig. 2(c) illustrates the morphologies of the prepared ferrite calcined at 800^{0}C.This figure indicates that the ferrite particles are non-homogeneous spherical shaped crystals with a wide range of distribution of particle sizes from 50 to 100nm.From the obtained results it can be concluded that particle size increases with increase in temperature. As temperature increases the adjacent ferrite granules coalesce together to form larger grains.

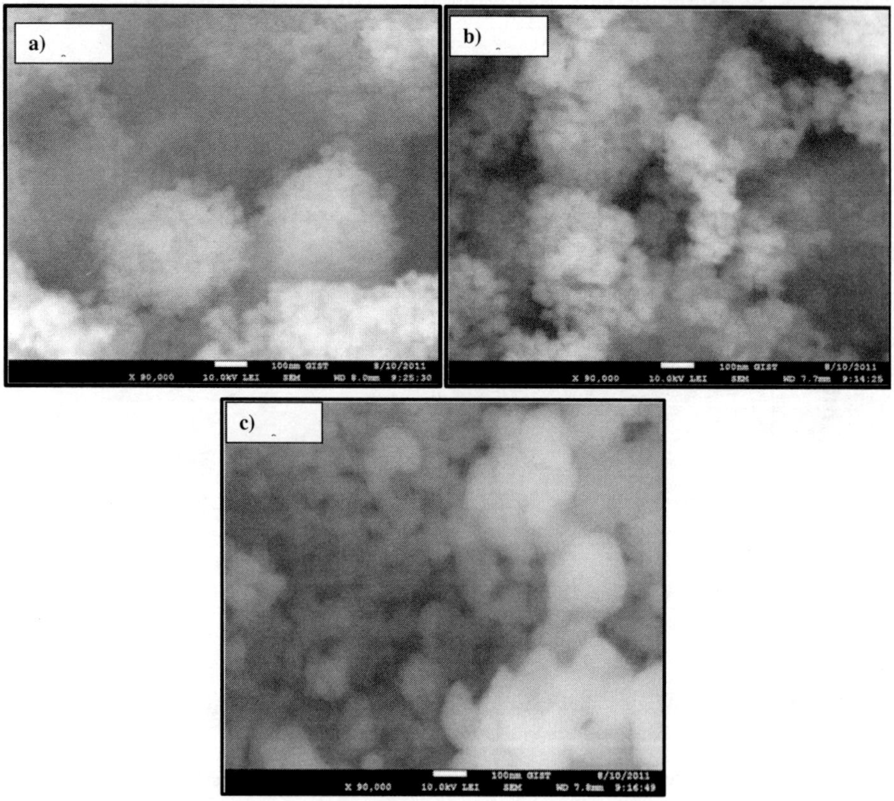

Fig. 2: SEM Images of $Ba_4Co_2Mn_{(0.2)}Fe_{(35.8)}O_{60}$ Ferrite Sintered at 600, 700 and 800^{0}C

Microwave Transmittance and Reflectance

The microwave transmittance and reflectance of $Ba_4Co_2Mn_{(0.2)}Fe_{(35.8)}O_{60}$ pellets sintered at 600°C,700°C and 800°C in X band is shown in Fig. 3 From figure it can be observed that $Ba_4Co_2Mn_{(0.2)}Fe_{(35.8)}O_{60}$ pellets sintered at 600°C,700°C and 800°C shows temperature dependent variation in transmittance with frequency. It can be seen that transmittance increases with increases in temperature to about 70 to 86% at lower frequency.Pellet sintered at 600°C shows transmittance in range of 43% to 82% where as pellets sintered at 700°C and 800°C shows transmittance in the range of 30% to 86% and 24% to 86%. On the other hand $Ba_4Co_2Mn_{(0.2)}Fe_{(35.8)}O_{60}$ pellets sintered at 600°C,700°C and 800°C shows higher reflectance in the frequency range 9.9 GHz to 10.1 GHz Pellet sintered at 600°C shows reflectance in range of 15% to 54% where as pellets sintered at 700°C and 800°C shows reflectance in the range of 17% to 54% and 15% to 62%. Thus it can be concluded that Pellets shows higher transmitting properties.

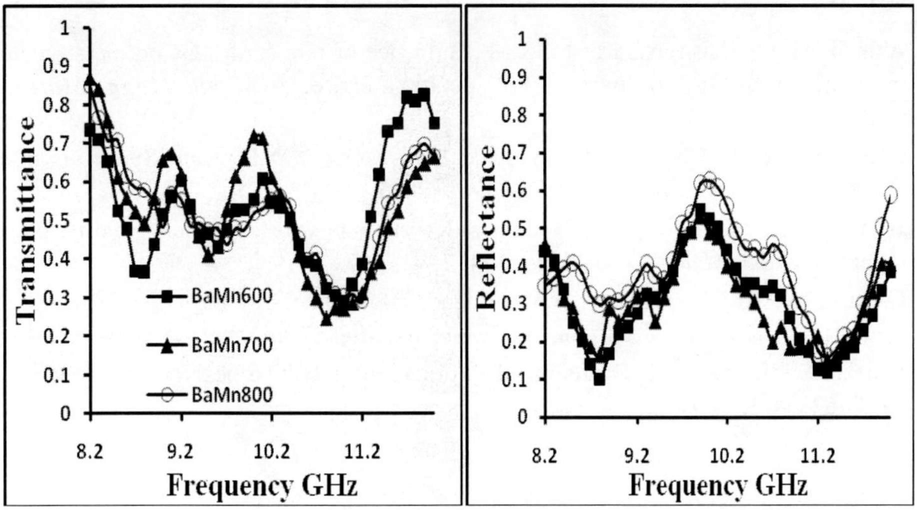

Fig. 3: Microwave Transmittance and Reflectance of $Ba_4Co_2Mn_{(0.2)}Fe_{(35.8)}O_{60}$ Ferrite Sintered at 600, 700 and 800°C

CONCLUSIONS

The $Ba_4 Co_2 Mn_{(0.2)} Fe_{(35.8)}O_{60}$ mixed hexaferrite have been synthesized successfully by a wet chemical route. The XRD patterns reveal the formation of hexagonal crystal system of ferrite. The diffractogram of $Ba_4Co_2Mn_{(0.2)}Fe_{(35.8)}O_{60}$ powders sintered at 600,700 and 800°C shows the presence of M and Y phases. Crystallite size obtained using this formula for $Ba_4Co_2Mn_{(0.2)}Fe_{(35.8)}O_{60}$ ferrite sintered at 600,700 and 800°C are 70.8 nm, 83.5 nm and 123 nm. SEM image of ferrite sintered at 600°C indicates the formation of spongy ball like grains of size 300-400 nm formed by agglomerated nanogranules of size 5-15 nm. Microwave studies reveal that the microwave transmittance increases with sintering temperature of the sample under test. Thus it can be concluded that the structural, surface morphological and microwave properties of $Ba_4Co_2Mn_{(0.2)}Fe_{(35.8)}O_{60}$ can be modified by varying sintering temperature of sample.

ACKNOWLEDGEMENTS

One of the authors Dr. Vijaya Puri is grateful to UGC for the award of Research scientist ‘C’. Authors are also thankful to DST-FIST, PURSE and UGC meritorious fellowship for the financial support.

REFERENCES

[1] Ahmed, M.A., Okash, N., Kershi, R.M., Influence of rare-earth ions on the structure and magnetic properties of barium W-type hexaferrite, *J. Magn. Magn. Mater.* 320, 1146–1150, 2008.

[2] J. Jeong a, K.W. Cho a, D.W. Hahn a, B.C. Moon b, Y.H. Han, Synthesis of Co2Z Ba-ferrites, *Mater. Lette.* 59, 3959 – 3962, 2005.

[3] Hsing-,I H., Ren-Qian, Y., Hexagonal ferrite powder synthesis using chemical co-precipitation, *Mater, Chem. Phy.*, 104, 1–4, 2007.

[4] Garcia-Casillas, P.E., Beesley, A.M., Bueno, D., Matutes-Aquino, J.A., Martinez, C.A., Remanence properties of barium hexaferrite, *J. Alloy Comp.*, 369, 185–189, 2004.

[5] Lisjak, D., Makovec, D., Drofenik, M., Formation of U-type hexaferrites, *J. Mater Res.*, 19, 2462–2471, 2004.

AMPLIFICATION AND EMERGENCE OF CHIRALITY IN NANOSTRUCTURES

SARITA ROY, SOUMEN BASAK[1] AND ANJAN K. DASGUPTA

Department of Biochemistry, University of Calcutta, Ballygunge Circular Road, Kolkata
[1]Chemical Sciences Division, Saha Institute of Nuclear Physics, Bidhannagar, Kolkata
E-mail: adgcal@gmail.com

ABSTRACT

Chirality in nanoscale has been a subject of keen recent interest. Unlike the molecular chirality the chiral nanoclusters normally synthesized on a chiral template, may assume a number of emerging properties. The importance of chirality has been underestimated in the field of nanomedicine as the modulation of nano-scale interaction as a function of chirality is hardly explored. In the present work a quantitative measure of chirality and its amplification after its embedment in a nanostructure is provided. Similarly, emergent chiral structures have been designed, that shows chiral signals away from either the plasmon or the original chiral template.

Our work thus can be divided in two parts; (i) a new representation that describes chiral structures from polarization techniques; (ii) describing the amplification and emerging properties of chiral nanostructures on the basis of such representation.

INTRODUCTION

Enantiomeric selection (selection of a given handed molecule out of its two non-superimposable mirror images) has been a challenging problem in stereochemistry because of the dearth of techniques to quantify and segment chiral molecules from their mirror image. Chirality has some tremendously useful applications in pharmaceutical and drug research. A chosen enantiomer (molecule with handedness) may have more biological activity as a drug, compared to its mirror image, whereas the mirror image molecule may have an adverse effect. Some classical

V. Rajendran, K. Saminathan, P. Paramasivam and K.E. Geckeler (eds.)
Nano Materials Synthesis and Characterisation, pp. 167-172 (2012).

examples of enantiomeric drug molecules include thalidomide (1) whose one enantiomer is useful in morning sickness during pregnancy, the other enantiomer causing birth defect (a fact that was known for a long time). Other enantiomeric drugs include ethambutol (2), naproxen (3) and the list is very long. It may be noted that under normal circumstances the probability of emergence of the two enantiomers are identical and one needs some additional selection mechanism to select one from the other.

Other than drug discovery enantiomeric selection may be important in discovery of novel catalytic materials. Loading of Pt (platinum) for example is known to enhance enantio-differentiation in a size dependent fashion (4). Again with the advent of nanomedicine, chirality in nanostructure has also been a topic of keen interest (5).

Having said this, the lacuna lies in methods to determine chirality. Existing methods, e.g. polarimetry or Circular Dichroism, are unusable for measurement of chirality of a substance which has high degree of opacity and is in its solid or powder form. A number of methods have recently been proposed in addressing the problem of chirality or molecular dissymmetry. Single molecule spectroscopy for example, probes the role of local molecular environments inducing chiro-optical response in isolated molecular systems (6). Similarly ionic liquids have been used for determination of enantiomeric composition of various drugs. The effectiveness of RTIL (chiral room temperature ionic liquid) in determination of chiral pharmaceutical product follows from such studies (7).

Apart from all the above mentioned techniques, distinguishing chiral molecule mainly chiral nanoparticle (8) by a simple lab based method to overcome the said limitations has been reported in this work. This work will also be able to provide chiral fingerprint, and compare the extent of symmetry breaking among different chiral molecules.

In this paper, we probe into the relative variation of fluorescence with variation in polarization of the excitation and emission polarizing angle to discriminate between the chiral natures of the given molecule.

EXPERIMENTAL

The amino acid mainly phenyl alanine, tyrosine, tryptophan, cysteine (both enantiomeric form, 'D'and 'L') were procured form Sigma-Aldrich of maximum purity. The stock solutions (100 mM) of the above amino acids were made in milliQ water and if necessary a slight alkaline condition is maintained.

UV/VIS/NIR steady-state Spectrofluorimeter (Model Quanta Master™ 40 Photon Technology International) was used to study anisotropy of amino acid and nanoparticle conjugated with amino acid. Typically the excitation and emission wavelengths were chosen depending on the characteristics of the fluorophore and in case of nanoparticle Mie wavelength was calculated for a given nanoparticle and the anisotropy measurement were done using that particular wavelength for the particle. Thus for tryptophan, an excitation wavelength 280 nm was used and emission wavelength was fixed at 350 nm. For a given pair of excitation and emission wavelength, the polarization angle was changed. Data was collected using PTI spectrofluorimeter. PTI Felix 32 software was programmed in such a way that for each excitation angle (i.e. polarizer in the excitation side), emission was varied from 0-360 degree at an interval of 45 degree. The experiment was continued at room temperature. Each data set was stored in a spreadsheet file which was then analyzed using MATLAB 7 software.

For chiral nanoparticle discrimination we have synthesized silver nanoparticle (SNP) by reduction of silver nitrate (Merck, Germany) with sodium borohydride (Merck, Germany) and stabilized by trisodium citrate (Merck, Germany). Chiral SNP (8) is prepared by addition of pre-synthesized SNP with different amino acid in 1:1 ratio. The final concentration of amino acid and SNP with respect to ionic silver is maintained at 1 mM. Chirality of SNP was verified by Circular Dichroism CD. The plasmonic absorption of particle is measured by UV/VIS Thermo-Scientific spectrophotometer (Evolution 300, USA).

RESULTS AND DISCUSSION

The plasmonic nature of SNP and cysteine conjugated SNP is shown in Fig. 1.

Figure 2 shows a typical polarization angle variation of fluorescent intensity pattern for asymmetric molecule (tyrosine) in which water Raman line (9) was corrected.

Anisotropy is defined as the follows:

$$A = \frac{I_{vv} - I_{vh}}{I_{vv} + 2I_{vh}}$$

To convert the above anisotropy equation into angles for measurements, denoting the vertical component V $\rightarrow$ 0° and H $\rightarrow$ 90° with respect to vertical excitation,

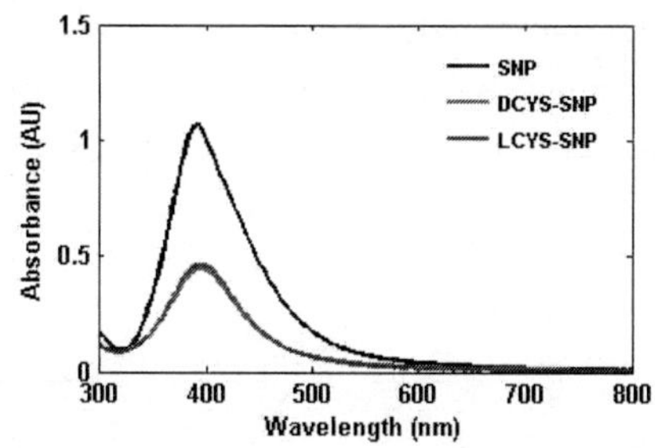

Fig. 1: SNP (black line) Shows Plasmon Maxima at 390 nm Whereas Cysteine Conjugated with Same SNP Shows Plasmon Maxima at 390 nm but with Lower Intensity Due to Conjugation

$A = [I_{0ex0em} - G\ I_{0ex90em}] / [I_{0ex0em} + 2G\ I_{0ex90em}]$; $G = I_{90ex0em} / I_{90ex90em}$, where "ex" represents the excitation angle and "em" represents the emission angle.

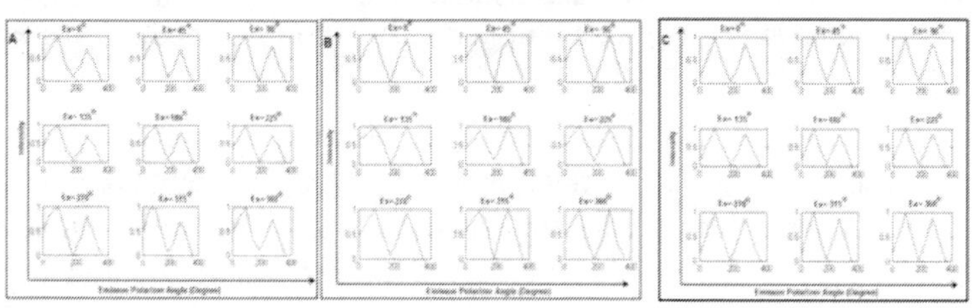

Fig. 2: The above figure shows the intensity profile of the chiral molecule tyrosine (panel A, L-tyrosine & panel B, D-tyrosine) and achiral molecule (panel C, pyrene) as a function of polarization angle, for different excitation polarization angles at their respective emission wavelength. For a highly symmetric molecule like pyrene, highly symmetric "M" like pattern arises whereas tyrosine results in asymmetric or distorted "M" like pattern. Same sets of experiments were performed in case of tryptophan and phenylalanine and distorted "M" like pattern arises.

The above mathematical expression of anisotropy involves only 0° and 90° polarizers for excitation and emission, but can be extended to an arbitrary angle θ for varying excitation and emission polarizers. Accordingly, an extended anisotropy for angle è can be defined as a function of angles between 0° and θ + 90° and is expressed as

$$A(\theta) = \frac{I_{\theta\theta} - I_{\theta,\theta+90}}{I_{\theta\theta} + 2I_{\theta,\theta+90}}.$$

In the extended anisotropy expression, wherein I is the intensity of the fluorescent emission, θ as the first subscript of I is a variable excitation polarization angle, and the second subscript of I is a emission polarization angle.

Comparison of different panel of figure 2 highlights the deviations of arms of the M. The transformations described in Fig. 3, leads to a clear distinction of the L-forms and D-forms.

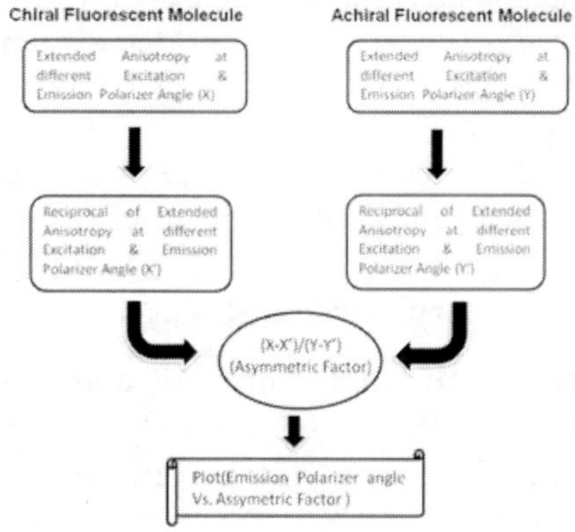

Fig. 3: The Schematic Diagram for Analysis of Chirality

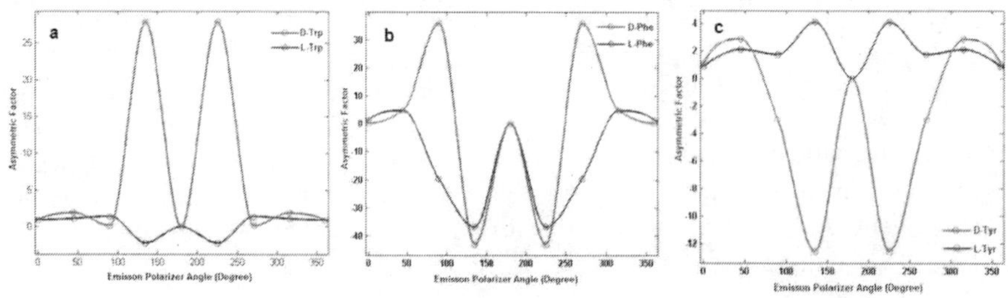

Fig. 4: Chirality of (a) Trp (b) Phe and (c) Tyr. The ordinate shows the differential anisotropy between two arms of the observed "M" like pattern by dividing them into two halves and then comparing the reflected image (mirror set at mid point (180°) perpendicular to x-axis, which represents the emission polarization angle.

An additional information, one can extract from the present approach is that, one obtains a different spectroscopic signature of the fluorophore using the differential profile. Thus, tyrosine, tryptophan and phenylalanine can be discriminated from the differential signal, the difference being taken between the emission intensity of enantiomeric molecule and its mirror image shown by Fig. 4.

To substantiate the chiral signature of a given molecule we postulate a chiral vector. The Fig. 5 demonstrates the variation of the chiral vector for the respective amino acids.

Now to verify, whether the said method is applicable in case of discriminating chiral nanoparticle we have done the same set of experiment with chiral silver nanoparticle conjugated with cysteine whose plasmonic spectra is shown in Fig. 1. The amplification and emerging properties of chiral nanostructure upon conjugation with cysteine have been illustrated in figure 6. The CD signal clearly illustrates that both amplification and emergence of new chiral signatures (~240-350 nm) have been obtained in the nanostructures (SNP-Cys) upon conjugation of silver nanoparticle with cysteine amino acid.

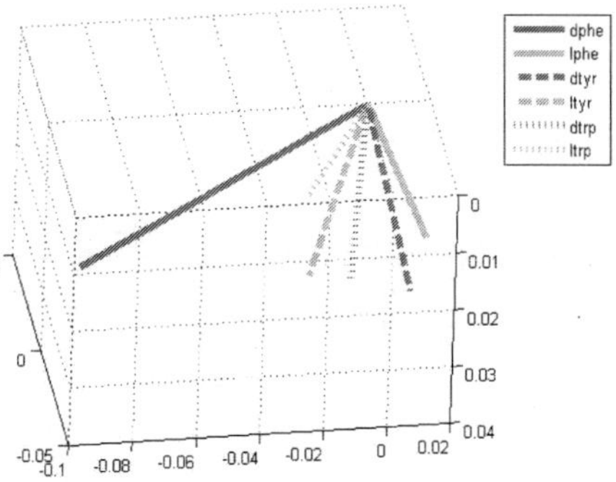

Fig. 5: Chiral Vector: The three components represent the asymmetry at three different emission polarization angles (0°, 45° and 90° respectively). The D and L types are shown by similar line type. The X, Y and Z are dimensionless quantities normalized by a typical achiral fluorescent molecule pyrene. The method can not only distinguish between L-trp and D-trp but it can also distinguish between L-trp and L-phe (as the chiral vectors will be different in each case)

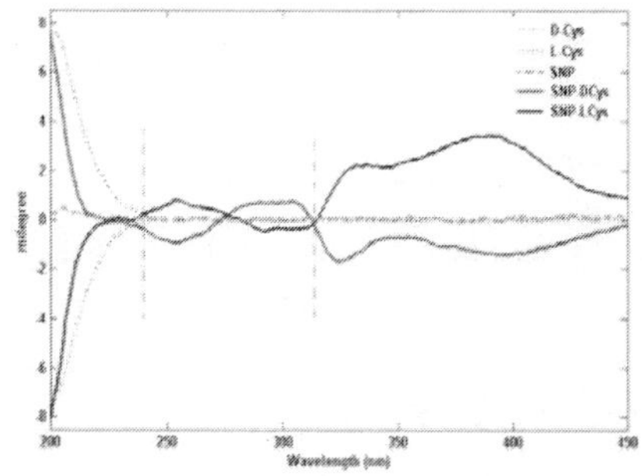

Fig. 6: CD signal of D-cysteine (red dotted line) & L- cysteine (blue dotted line) in the wavelength region mainly 200-240 nm. A yellow dashed line is the CD signal for silver nanoparticle (SNP) which passes through zero indicates no CD signal but upon conjugation with cysteine (both D and L form, represented as SNP-D Cys and SNP L Cys), show a strong Plasmonic CD signature in the range of 350-450 nm along with some new CD signal in the range of 240-350 nm

Now, this chiral nanoparticle is subjected to anisotropy measurement by taking the Mie wavelength of the nanoparticle which is characteristics of individual nanoparticle. SNP has Mie wavelength at 480 nm wheras both D- and L-conjugated cysteine nanoparticle has Mie at 490 nm. After extended anisotropy measurement both enantiomeric form of SNP (DCys-SNP, LCys-SNP) gives opposite signature as shown in Fig. 7. Hence the said method can successfully discriminate between different chiral form of nanoparticle as well as molecules.

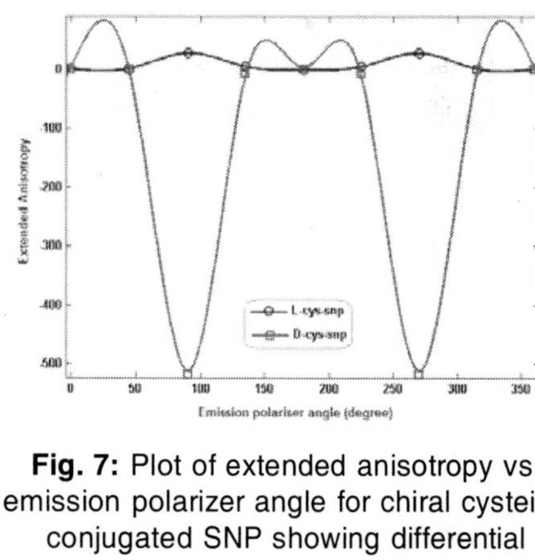

Fig. 7: Plot of extended anisotropy vs. emission polarizer angle for chiral cysteine conjugated SNP showing differential enantiomeric signature of chiral SNP

Hence, in this paper we have found a method which can lead to a miniaturized detection of chiral molecules or nanostructures. The method can quantify the extent of symmetry breaking using a simple mirror image comparison of the polarization angle profile of the emission intensity and thus can scale the breaking of symmetry in a convenient way. Here we have also designed a scale free chiral vector which can differentiate between different types of molecules from the sample based on a magnitude of the chiral vector and also With the help of chiral vector whether we could be able to address the problem of topological chirality will be subject of study.

ACKNOWLEDGEMENTS

The first author Sarita Roy thanks CSIR grant number [09/028(0817)/2010-EMR-I] for assisting senior research fellowship.

REFERENCES

[1] T. Eriksson, S. Bjorkman, P. Hoglund, Clinical pharmacology of thalidomide, *Eur J Clin Pharmacol*, 57, 365-376, 2001.

[2] Su-Ann Lim, Ethambutol-associated Optic Neuropathy, *Ann Acad Med Singapore*, 35, 274-278, 2006.

[3] N.M. Davies, M.R. Wright, A.S. Russell, and F. Jamali, Effect of the enantiomers of Flurbiprofen, Ibuprofen and Ketoprofen on Intestinal permeability, *Journal of Pharmaceutical Sciences*, 85, 1170-1173, 1996.

[4] E. Toukoniitty, P. Mäki-Arvela, A.K. Neyestanaki, T. Salmi, A. Villela, R. leino, R. Sjöholm, E. Laine, J. Väyrynen and T. Ollonqv, Catalyst selection and solvent effects in enantioselective hydrogenation of 1-phenyl-1,2-propanedione, in *Studies in Surface Science and Catalysis* 130, A. Corma, F. V. Melo, S. Mendioroz, J. L. G. Fierro (Eds), Elsevier Science B. V., 2000, 3363-3368.

[5] V. Kitaev, Chiral nanoscale building blocks-from understanding to applications, *J. Mater. Chem*, 18, 4745- 4749, 2008.

[6] R. Hassey-Paradise, A. Cyphersmith, A.M. Tilley, T. Mortsolf, D. Basak, D. Venkataraman and M.D. Barnes, Dissymmetries in fluorescence excitation and emission from single chiral molecules, *Chirality*, 21, E265-E276, 2009.

[7] C.D. Tran, D. Oliveira, Fluorescence determination of enantiomeric composition of pharmaceuticals via use of ionic liquid that serves as both solvent and chiral selector, *Analytical Biochemistry*, 356, 51-58, 2006.

[8] T. Li, H.G. Park, H.S. Lee, S.H. Choi, Circular dichroism study of chiral biomolecules conjugated with silver nanoparticles, *Nanotechnology*, 15, S660-S663, 2004.

ANNEALING EFFECT ON THE STRUCTURAL AND ELECTRICAL PROPERTIES OF n-TYPE $MoBi_2(Se_{0.5}Te_{0.5})_5$ THIN FILM

Manauti Salunkhe, Nita Pawar, Suvarta Kharade, Sanjeevani Patil, Suhas Mohite and P.N. Bhosale

Materials Research Laboratory, Department of Chemistry, Shivaji University, Kolhapur
E-mail: p_n_bhosale@rediffmail.com ; manu.11oct@gmail.com

ABSTRACT

n-type $MoBi_2(Se_{0.5}Te_{0.5})_5$ thermoelectric thin films have been deposited on glass substrate by Arrested Precipitation Technique (APT). The change in thermoelectric properties of $MoBi_2(Se_{0.5}Te_{0.5})_5$ thin films were examined after annealing at the temperature 473 K. The structural, morphological, compositional and electrical properties of thin films were characterized by X-ray Diffraction (XRD), Scanning Electron Microscopy (SEM), Energy Dispersive Spectroscopy (EDS), etc. Thermoelectric properties of the thin films have been evaluated by measurements of electrical conductivity and Seebeck coefficient in the temperature range 300-500 K. Our aim is to investigate the effect of annealing on behavior of $MoBi_2(Se_{0.5}Te_{0.5})_5$ thin films along with optostructural and optoelectronic properties.

INTRODUCTION

There has been increasing interest during the past few decades in semiconducting chalcogenide thin films, because of their wide application in various fields of science and technology, leading to reduction in the cost of production of semiconductor devices. Bi_2Te_3 based materials are one of the most classical thermoelectric materials, have attracted more attention for their potential

V. Rajendran, K. Saminathan, P. Paramasivam and K.E. Geckeler (eds.)
Nano Materials Synthesis and Characterisation, pp. 173-178 (2012).

application in microelectronics and other high technical fields [1]. The efficiency of thermoelectric conversion is evaluated by means of the thermoelectric figure of merit (ZT). It is apparent that the improvement in thermoelectric figure of merit is possible by increasing Seebeck coefficient (S) and electrical conductivity (σ) and by decreasing thermal conductivity. Annealing is critical for semiconductors, which can strongly influence electrical transport behavior. It has been reported that annealing has a positive effect on the thermoelectric properties of bismuth telluride based materials [2]. Carrier concentration is important in optimizing thermoelectric figure of merit. If carrier concentration is low, conductivity of the semiconductor is high.

To the best of our knowledge this is the first report on synthesis of quaternary nanocrystalline $MoBi_2(Se_{0.5}Te_{0.5})_5$ thin films using self organized growth process popularly known as Arrested Precipitation Technique (APT) [3-5]. The APT process is based on Ostwald ripening law [6, 7]. The purpose of present work is to establish and optimize the growth condition to produce nanocrystalline $MoBi_2(Se_{0.5}Te_{0.5})_5$ thin films. To investigate the changes in optostructural, morphological, compositional, optoelctronic properties and suitability as photoelectrode in PEC solar cell and thermocooling properties of $MoBi_2(Se_{0.5}Te_{0.5})_5$ thin films after annealing.

EXPERIMENTAL DETAILS

$MoBi_2(Se_{0.5}Te_{0.5})_5$ thin films have been deposited by using APT, it allows the Bi-TEA and Mo-TEA complex to react with Se^{2-}, Te^{2-} ions which are released slowly by the dissociation of sodium selenosulphide (Na_2SeSO_3) and sodium tellurosulphide (Na_2TeSO_3) in alkaline medium at pH 9.5 and temperature 55^{0}C. Initially high purity precursors such as ammonium molybdate and bismuth nitrate triturated separately in triethanolamine complexing agent, and after 8 hours it forms clear solution of Mo-TEA, Bi-TEA complexes respectively. Na_2SeSO_3 and Na_2TeSO_3 were prepared by refluxing selenium and tellurium metal power separately with anhydrous sodium sulphite at 90^{0}C for 8 and 12 hours respectively. The formation of $MoBi_2(Se_{0.5}Te_{0.5})_5$ thin films occur when ionic product of Bi^{3+}, Mo^{4+}, Te^{2-} and Se^{2-} exceeds the solubility product of $MoBi_2(Se_{0.5}Te_{0.5})_5$ thin film. Overall growth mechanism for the mixed metal chalcogenide thin film of Molybdenum bismuth selenotelluride is summarized as follows:

$$(NH_4)_2[Mo2N(CH_2\text{-}CH_2\text{-}O)_3] + 2(NH_4)_3[Bi2N(CH_2\text{-}CH_2\text{-}O)_3] + 5/2Na_2SeSO_3 + 5/2Na_2TeSO_3 + 13H_2O \xrightarrow{pH9.5} MoBi_2(Se_{0.5}Te_{0.5})_5 + 6[N(CH_2\text{-}CH_2\text{-}OH)_3] + 8(NH_4)OH + 5Na_2SO_4$$

The as deposited $MoBi_2(Se_{0.5}Te_{0.5})_5$ thin films were annealed at constant temperature (473 K) in muffle furnace for 3 hours and then characterized by different techniques.

RESULTS AND DISCUSSION

Optical Absorption Studies

The optical absorption study was carried out mainly to determine optical band value of material by using UV-Visible spectrometer (Model: Shimazu, UV Probe-1800 Japan) in the wavelength range 300-800 nm. The variation of photon energy (hν) vs $(\alpha h\nu)^2$ for as deposited and annealed $MoBi_2(Se_{0.5}Te_{0.5})_5$ thin films are shown in Fig. 1. It is evident that optical band gap (Eg) decreases after annealing from 1.57 to 1.40 eV. Similar results for decrease in optical band gap after annealing has been reported by other group [8]. During annealing process, bond breaking and bond arrangement can take place, which result in change in local structure of thin film. These include superior effects such as shifts in the absorption edge, more substantial atomic and molecular reconfiguration which is associated with changes in optical constants and absorption edge shift. During the process of annealing, enough vibrational energy is present to break some weaker bonds thus introducing some translational degree of freedom to the system, results an increase in the heat capacity of the system and decrease in the optical band gap.

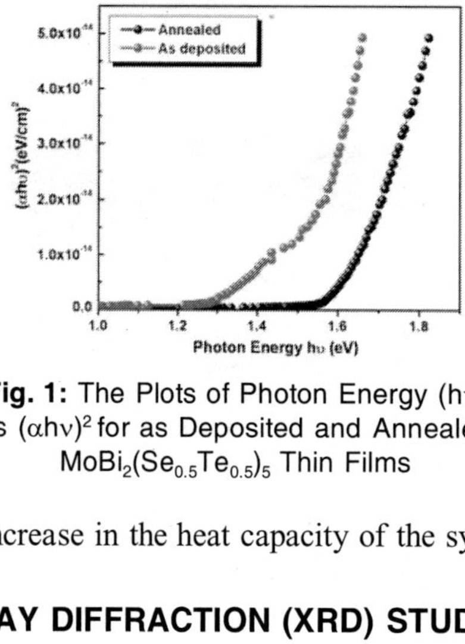

Fig. 1: The Plots of Photon Energy (hν) vs $(\alpha h\nu)^2$ for as Deposited and Annealed $MoBi_2(Se_{0.5}Te_{0.5})_5$ Thin Films

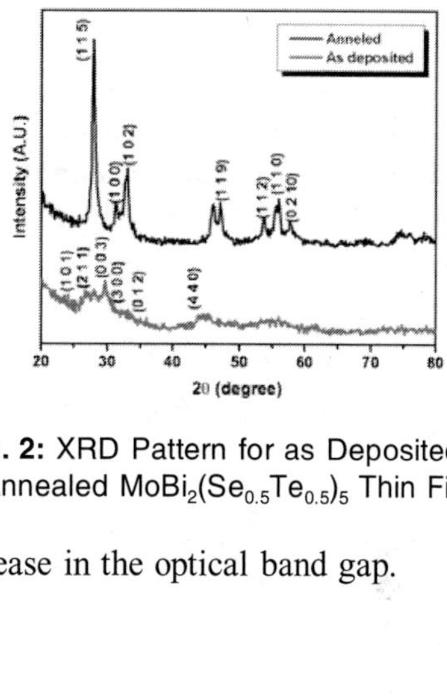

Fig. 2: XRD Pattern for as Deposited and Annealed $MoBi_2(Se_{0.5}Te_{0.5})_5$ Thin Films

X-RAY DIFFRACTION (XRD) STUDY

In the present investigation X-ray diffraction study was carried out using X-ray diffractometer [Bruker AXS Model D8 Advance X-ray Diffractometer] with Cu Kα target having wavelength 1.542 Å and the 2q angle ranging from 20-80°C. XRD patterns of as deposited and annealed $MoBi_2(Se_{0.5}Te_{0.5})_5$ thin films are shown in Fig. 2. The diffraction peaks for as deposited $MoBi_2(Se_{0.5}Te_{0.5})_5$ thin films are found at 23.20°, 27.60°, 29.36°, 30.54°, 38.20° and 43.60° corresponding to lattice planes (101), (211), (003), (300), (012) and (440) respectively. The diffraction peaks for annealed $MoBi_2(Se_{0.5}Te_{0.5})_5$ thin films are sharpened and found at 2θ values 27.96°, 31.08°, 32.74°, 47.34°, 53.58°, 55.58° and 57.71° corresponding to planes (115), (107), (102), (105), (112), (110), (0210) respectively. The XRD pattern reveals that the as deposited $MoBi_2(Se_{0.5}Te_{0.5})_5$ thin films are nanocrystalline in nature with rhombohedral structure. After annealing the $MoBi_2(Se_{0.5}Te_{0.5})_5$ thin films shows mixed phase formation, i.e. rhombohedral and hexagonal with polycrystalline in nature. Likewise the same observation was found in the chalcogenide material by Huimin Jia [9]. There are no JCPDS standard data available for different compositions of $MoBi_2(Se_{0.5}Te_{0.5})_5$ thin films. The plane indices of as deposited $MoBi_2(Se_{0.5}Te_{0.5})_5$ thin films are obtained by comparing the intensities and peak positions with these of Bi_2Te_3, Bi_2Se_3, $MoTe_2$ and $MoSe_2$ binary systems with JCPDS cards no. 08-0027, 72-2123, 23-1257 and 72-1420 respectively. For annealed $MoBi_2(Se_{0.5}Te_{0.5})_5$ thin films the peak positions were compared with Bi_2Te_3, Bi_4Se_3, $MoTe_2$ and $MoSe_2$ binary systems with JCPDS cards no. 72-2036 (rhombodral), 72-2065 (rhombodral), 72-0117 (hexagonal) and 17-0887 (hexagonal) respectively. When the $MoBi_2(Se_{0.5}Te_{0.5})_5$ thin film was annealed at 472 K, diffusion and agglomeration of atoms were accelerated and then the thin film crystallization is improved. The crystallite size was calculated by standard Debye Scherer's formula and it shows increase in crystallite size from 26.7 to 36.8 nm after annealing for $MoBi_2(Se_{0.5}Te_{0.5})_5$ thin films.

Scanning Electron Microscopy (SEM)

The surface morphology of the films was studied by using scanning electron microscope [SEM: JEOL JSM-6360 Analytical scanning electron microscope]. Figure 3 shows the surface morphology of as deposited and annealed $MoBi_2(Se_{0.5}Te_{0.5})_5$ thin films. It reveals that the as deposited $MoBi_2(Se_{0.5}Te_{0.5})_5$ thin films are fibrous reticulate framework. After annealing of $MoBi_2(Se_{0.5}Te_{0.5})_5$ thin films, the fibrous reticulate framework was not disturbed but the thickness of the fibers increases. This may occurs because of the consequent coalescing at high temperature for $MoBi_2(Se_{0.5}Te_{0.5})_5$ thin films.

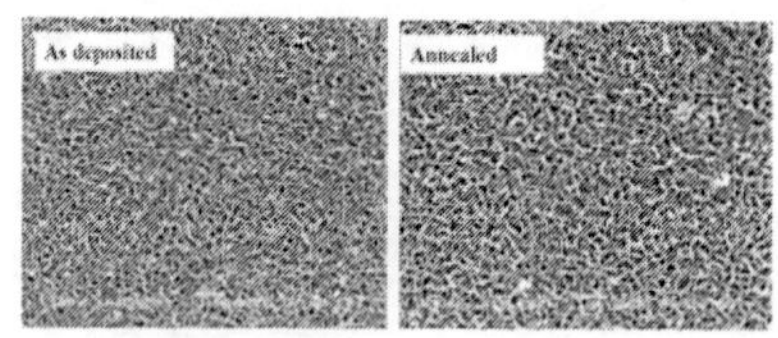

Fig. 3: SEM Images for as Deposited and Annealed $MoBi_2(Se_{0.5}Te_{0.5})_5$ Thin Films.

Energy Dispersive Spectroscopy (EDS) Studies

The compositional analysis was carried out by using energy dispersive spectroscopy [EDS: JEOL JSM-6360 Analytical Scanning Electron Microscope] for the as deposited and annealed $MoBi_2(Se_{0.5}Te_{0.5})_5$ thin films. Figure 4 shows EDS patterns of the as deposited and annealed $MoBi_2(Se_{0.5}Te_{0.5})_5$ thin films. Inset of Fig. 4 gives the atomic

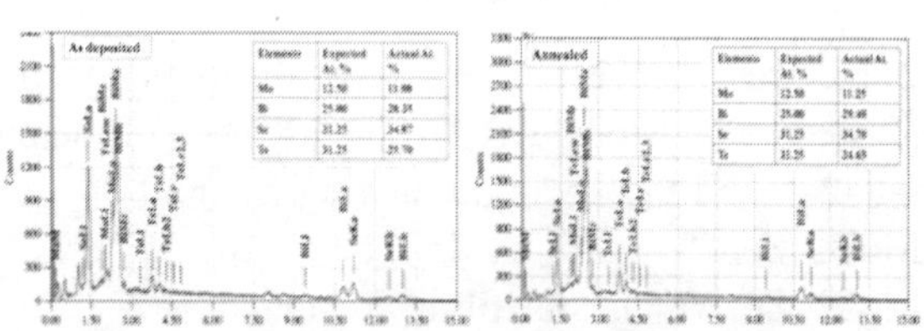

Fig. 4: EDS Patterns for as Deposited and Annealed $MoBi_2(Se_{0.5}Te_{0.5})_5$ Thin Films

percentage of Mo, Bi, Se and Te elements in the $MoBi_2(Se_{0.5}Te_{0.5})_5$ thin films which indicates that the as-deposited and annealed thin films has near stoichiometric composition and decrease in atomic percentage of selenium and tellurium after annealing. The loss of selenium and tellurium occurs in annealed thin films because atomic bonds are not equivalent in all the directions at the surface.

Electrical Conductivity (EC) Studies

Electrical conductivity measurement was done in temperature range 300 to 500 K at constant voltage (5 V); using standard DC two probe method. The variation of ln σ vs 1000/T for as deposited and annealed $MoBi_2(Se_{0.5}Te_{0.5})_5$ thin films are shown in Fig. 5. The measurement shows an increase in electrical conductivity of the materials with increase in temperature indicating the films are semiconducting in nature. From graph it reveals that electrical conductivity increases after annealing. The increase of electrical conductivity after annealing

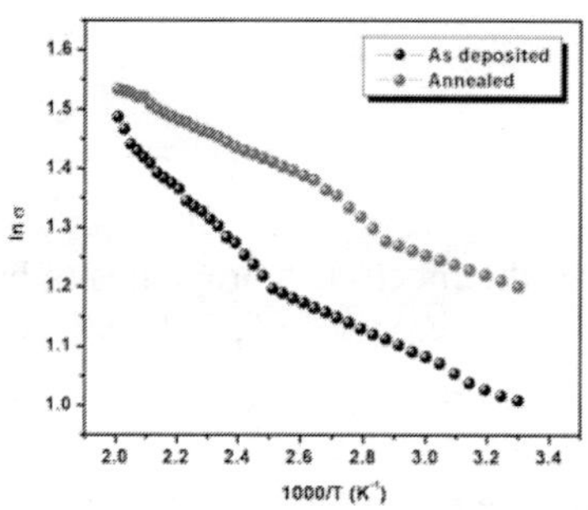

Fig. 5: Plots of ln σ vs 1000/T for as Deposited and Annealed $MoBi_2(Se_{0.5}Te_{0.5})_5$ Thin Film

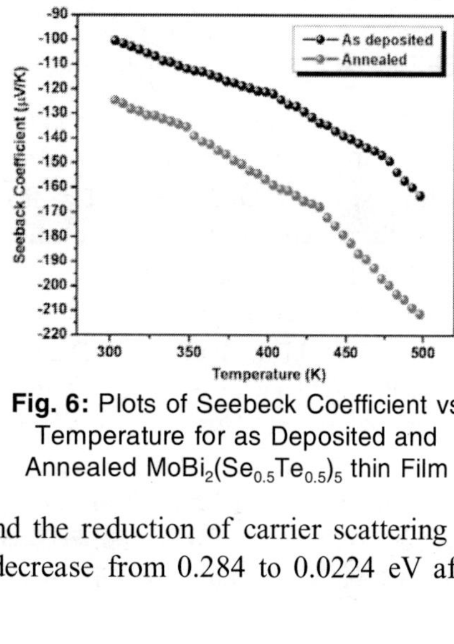

Fig. 6: Plots of Seebeck Coefficient vs Temperature for as Deposited and Annealed $MoBi_2(Se_{0.5}Te_{0.5})_5$ thin Film

is mainly due to the preferred crystal orientation and the reduction of carrier scattering by grain boundaries. Activation energy was found to decrease from 0.284 to 0.0224 eV after annealing of $MoBi_2(Se_{0.5}Te_{0.5})_5$ thin films.

THERMOELCTRIC POWER (TEP) STUDIES

The Seebeck coefficient vs temperature for as deposited and annealed $MoBi_2(Se_{0.5}Te_{0.5})_5$ thin films are shown in Fig. 6. Thermoelectric power measurements were made by using the integral method in which one end of the film was heated and the other kept at the constant low temperature and developed thermal electromotive force was measured with respect to the temperature gradient. All the samples shows –ve polarity of the generated voltage. This confirms n-type conduction of material [10]. From the figure it can be confirmed that after annealing of $MoBi_2(Se_{0.5}Te_{0.5})_5$ thin films Seebeck coefficient was decreases from –100 to – 120 μV/K.

CONCLUSIONS

Nearly stoichiometric $MoBi_2(Se_{0.5}Te_{0.5})_5$ thin films were deposited by APT. After annealing the band gap decreases from 1.57 to 1.40 eV. X-ray diffraction pattern of the as deposited $MoBi_2(Se_{0.5}Te_{0.5})_5$ thin film shows formation of nanocrystalline nature with rhombohedral structure. After annealing polycrystalline nature with mixed rhombohedral and hexagonal phase was obtained. After annealing crystallite size was increased from 26.7 to 36.7 nm. SEM images reveals the $MoBi_2(Se_{0.5}Te_{0.5})_5$ thin films are fibrous reticulate framework. After annealing of $MoBi_2(Se_{0.5}Te_{0.5})_5$ thin films, the fibrous reticulate framework was not disturbed but the thickness of the fibers increases. From electrical properties it reveals that the as deposited and annealed $MoBi_2(Se_{0.5}Te_{0.5})_5$ thin film are semiconductors with n-type conduction.

ACKNOWLEDGEMENT

One of the author (MMS) is indebted to University Grant Commission for providing the departmental facility.

REFERENCES

[1] Bell L.E., Cooling, Heating, Generating Power, and Recovering Waste Heat with Thermoelectric Systems, *Science*, 321, 1457-1461, 2008.

[2] Yamashita O. and Shoichi T., Effect of annealing on thermoelectric properties of bismuth telluride compounds doped with various additives, *J. Appl. Phys.* 95, 161-169, 2004.

[3] Patil N.S., Sargar A.M., Mane S.R., Bhosale P.N., Growth mechanism and characterisation of chemically grown Sb doped Bi_2Se_3thin films, *J. Appl. Surf. Sci.*, 254, 5261-5265, 2008.

[4] Ajalkar B.D., Mane R.K., Bhosale P.N., Optical and electrical studies on molybdenum sulphoselenide $[Mo(S_{1-x}Se_x)_2]$ thin films prepared by arrested precipitation technique (APT), *J. Sol. Energ. Mater. Sol. Cells*, 81, 101-112, 2004.

[5] R.K. Mane, B.D. Ajalkar, P.N. Bhosale, Electrical and optical properties of bismuth sulphotelluride $[Bi_2(S_{(1-x)}Te_x)_3]$ thin films prepared by arrested precipitation technique (APT), *J. Mater. Chem. Phys.*, 84, 247-250, 2004.

[6] IUPAC, Compendium of Chemical Terminology, 2nd ed. (the "Gold Book") 1997. Online corrected version: 2006.

[7] Ostwald W., 1896. Lehrbuch der Allgemeinen Chemie, vol. 2, part 1, Leipzig, Germany.

[8] Chavan S.D., Senthilarasu S., Lee J., and Lee S.H., Effect of annealing on the structural and optical properties of indium diffused $Cd_{0.7}Zn_{0.3}Se$ thin films, *J. Phys. D: Appl. Phy.* 41, 165502-165507, 2008.

[9] Huimin J., Yan H., Yiwen T., Lizhi Z., Synthesis and photoelectrochemical behavior of nanocrystalline CdS film electrodes Electrochem. *Commun.*, 8, 1381-1385, 2006.

[10] Dong L.X. and Park Y.H., Structure and transport properties of $(Bi_{(1-x)}\ Sb_x)Te_3$ Thermoelectric materials prepared by mechanical alloying and pulsed discharge sintering, *Materials Transactions*, 43, 681-687, 2002.

SYNTHESIS AND CHARACTERIZATION OF CERIUM DIOXIDE (CeO_2) NANOPARTICLES BY CHEMICAL PRECIPITATION METHOD

R. SARAVANAN, S. JOICY, V. NARAYANAN[1] AND A. STEPHEN

Department of Nuclear Physics, University of Madras, Guindy Campus, Chennai
[2]Department of Inorganic Chemistry, University of Madras, Guindy Campus, Chennai
E-mail: saravanan3.raj@gmail.com

ABSTRACT

In this study nanoparticles of CeO_2 were prepared by chemical precipitation method using cerium acetate as precursor. The structure of the resulting product was characterized by powder X-ray diffraction (XRD) which indicates that CeO_2 exhibit cubic structure and its crystallite size were found to be ~12 nm. Surface morphology and elemental analysis were analyzed by high resolution scanning electron microscopy (HR-SEM) fitted with energy dispersive X-ray spectroscopy (EDAX). Further this sample is used for photocatalytic application and the results are discussed in detailed.

INTRODUCTION

During past decades, there has been growing interest using semiconductor oxides as photocatalysts for the treatment of environmental pollution. Many efforts have been made to research on the different semiconductor catalysts such as TiO_2, ZnO and Fe_2O_3, etc. [1-2]. Besides, cerium oxide (CeO_2) is one of the semiconducting photocatalyst which is quite transparent to visible light but have an excellent UV absorption property due to its wider bandgap value to degrade the organic pollutants. Moreover, CeO_2 nanoparticles find its commercial applications in gas sensors, cosmetic materials, fuel cells, etc. [3-5].

V. Rajendran, K. Saminathan, P. Paramasivam and K.E. Geckeler (eds.)
Nano Materials Synthesis and Characterisation, pp. 179-184 (2012).

The nanoparticles are widely used as catalyst due to its high surface to volume ratio which tends to have high capacity to degrade the organic dyes. Over the years, various routes to synthesize CeO_2 nanoparticles have been investigated by many researchers. To control size and distribution of particles, chemical precipitation method emerged as most flexible and promising synthetic route in fabricating of CeO_2 nanoparticles due to advantages of mild synthesis condition, easy scale up, simple and cost effective.

In the present study, we report synthesis of CeO_2 nanoparticles via chemical precipitation method. Structural and optical properties of prepared catalyst were studied by different characterization techniques. The major aim of this work is to achieve efficient degradation of methylene blue (MB) under UV light using the prepared catalyst (CeO_2).

EXPERIMENTAL

Synthesis of CeO_2 Nanoparticles

Analytical grade cerium acetate ($Ce(CH_3COO)_3$) and sodium hydroxide (NaOH) used as starting materials in this experiment were purchased from Aldrich chemicals. Methylene Blue (MB) dye was purchased from Aldrich chemicals. All the chemicals were used as received without further purification. Throughout the experiment all aqueous solution were prepared using double distilled water. Nanoparticles of CeO_2 were synthesized via chemical precipitation method as follows: 1.583g (0.05 mol) of cerium acetate ($Ce(CH_3COO)_3$) was dissolved in 100 ml of distilled water under magnetic stirring at room temperature for 30mins to obtain clear solution. After that 0.05 mol of NaOH solution was added drop wise into the above aqueous solution with constant stirring. Resultant yellow colored precipitate was filtered and washed with double distilled water for several times. After washing, the precipitate was dried at room temperature for 24 hours. This as-prepared powder was heated at 400°C for 30 mins in muffle furnace. Finally pale-yellow color powder was obtained after the heat treatment. This synthesis procedure can be explained in equation form as follows:

$$Ce(CH_3COO)_3 + 3NaOH \rightarrow Ce(OH)_3 + 3CH_3COO^- + 3Na^+ \quad \text{... (1)}$$

$$2Ce(OH)_3 \xrightarrow{\Delta} Ce_2O_3 + 3H_2O \quad \text{... (2)}$$

$$Ce_2O_3 + H_2O \xrightarrow{\Delta} 2CeO_2 + H_2\uparrow \quad \text{... (3)}$$

Characterization Details

Crystalline structure and purity of prepared sample were studied by powder X-ray diffraction on Rich Siefert 3000 diffractometer using $CuK_{\alpha1}$ (λ = 1.5406 Å) radiation. Surface morphology and elemental analysis of catalyst (CeO_2) were carried out by using JEOL, JSM-840A scanning electron microscope. Diffuse reflectance spectra (DRS) were recorded in the wavelength range 200-800 nm using CARY 100 CON UV-Vis spectrophotometer.

Photocatalytic Degradation Experiment

To assess the photocatalytic activity of CeO_2 nanoparticles, methylene blue (MB) was preferred as a model dye. The photocatalytic procedure was reported elsewhere [6]. This photo decolorization experiment was executed in photocatalysis chamber integrated with an 8W mercury vapour lamb (365 nm) positioned at the vessel axis and magnetic stirrer arrangement within the enclosure. In order to avoid direct contact of mercury lamb with aqueous solution, it is roofed by quartz tube. 100mg of CeO_2 catalyst was suspended in 100 ml aqueous solution of MB with initial concentration of 3×10^{-5} moles/liter. The adsorption equilibrium between dye and surface of catalyst was maintained stirring in dark condition for 30mins before exposed to UV light. During irradiation, suspension was syringed at regular intervals of time and centrifuged to wipe out the catalyst. The absorption of MB was recorded on UV-Visible spectra at room temperature. Degradation efficiency can be calculated using following equation:

$$\eta(\%) = \left[\frac{C_0 - C}{C_0}\right]\% \qquad \text{... (4)}$$

where, C_0 and C are concentration of solution before illumination (t = 0) and after illumination for t minutes respectively.

RESULT AND DISCUSSION

X-ray Analysis

The XRD pattern of CeO_2 nanoparticles by chemical precipitation method is shown in fig. 1. All diffraction peaks indexed to face centered cubic structure of CeO_2 with lattice constant a = b = c = 5.411 Å is in good agreement with JCPDS card no: 43-1002 as well as data reported in reference [7]. There is no other diffraction peaks arising from $Ce(OH)_3$ and Ce_2O_3 were observed, which indicating the formation of pure-phase CeO_2.

The crystallite size of maximum diffracted peak (111) found to be ~12 nm using Scherrer formula [8].

$$D = \frac{0.9\,\lambda}{\beta\,\cos\theta} \qquad \text{... (5)}$$

where D is crystallite size; λ is wavelength of incident X-ray (1.5406 Å); θ is diffraction angle; β is full width half maximum.

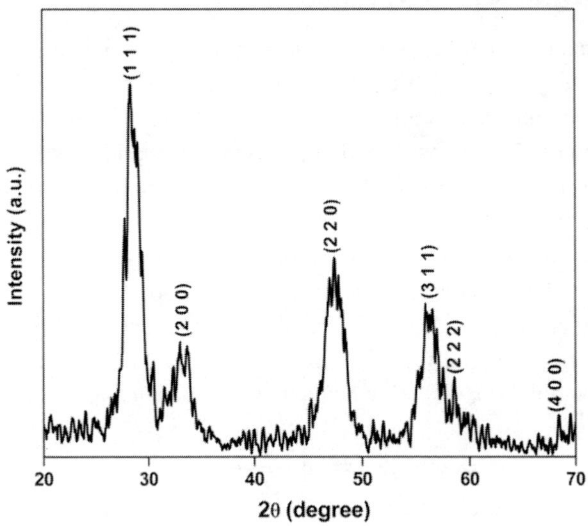

Fig. 1: XRD Pattern of CeO_2 Nanoparticles

HR-SEM and EDX Analysis

Typical HR-SEM and EDX analysis of CeO_2 nanoparticles are illustrated in Fig. 2. HR-SEM image revealed that agglomeration of CeO_2 nanoparticles are mostly spherical in shape and each spherical particle is aggregates of tiny spherical shaped particles. EDX analysis indicates that the prepared sample is only composed of Ce and O elements and no other remarkable impurities were observed.

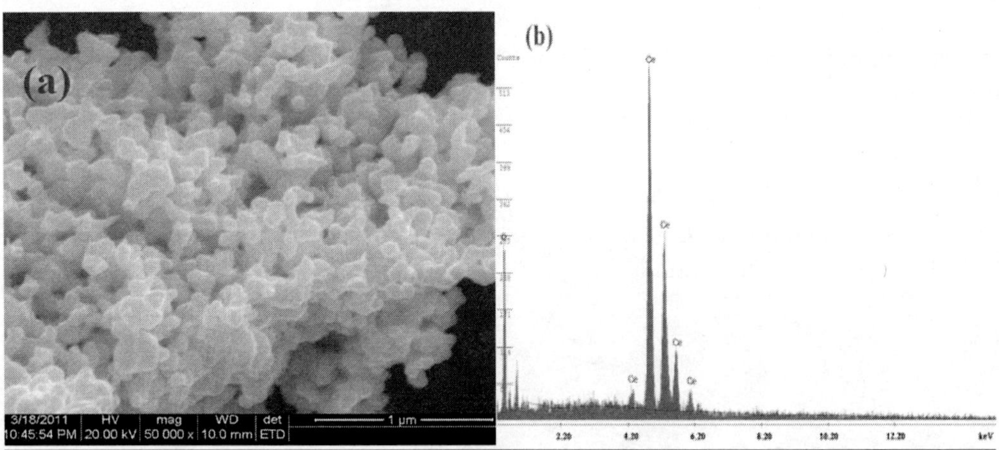

Fig. 2: (a) SEM Image and (b) EDX Spectra of CeO_2 Nanoparticles

DRS UV-Vis Analysis

Diffuse reflectance spectrum of prepared CeO_2 nanoparticles is shown in Fig. 3.

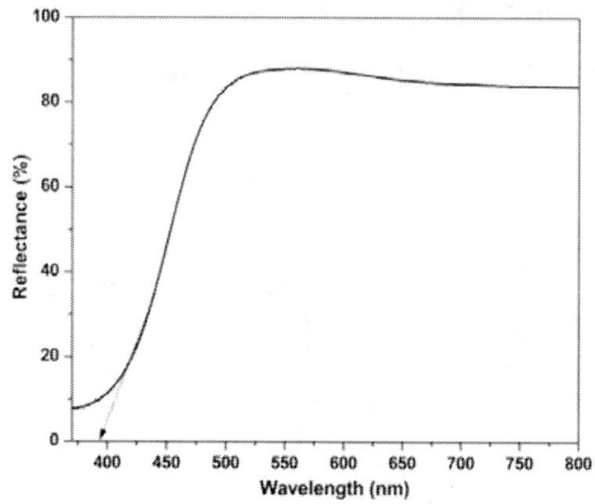

Fig. 3: Diffuse Reflectance Spectrum (DRS) of CeO_2 Nanoparticles

From this figure, it is clear that a sharp decrease in reflectance at around 397 nm due to strong absorption. Further, optical bandgap energy of prepared sample found to be 3.2 eV based on the following relation:

$$E_g = \frac{1240}{\lambda} eV \qquad \text{... (6)}$$

where, E_g represents bandgap energy, λ represents lower cutoff wavelength in nm of prepared catalyst. As a result of larger bandgap, CeO_2 nanoparticles could efficiently absorb UV light and might have potential application in photocatalyst.

Photocatalytic Degradation of MB

The absorption spectra of MB are shown in Fig. 4 using CeO_2 catalyst under UV irradiation for various time intervals. As can be seen in Fig. 4, the maximum absorption peak of MB at 664 nm gradually diminishes with an increase of UV exposure time. The photocatalytic degradation efficiency of 65% was achieved within 180 minutes. The photoctalytic mechanism is explained as following equations [9]:

$$CeO_2 + h\nu \rightarrow CeO_2(h^+_{VB} + e^-_{CB}) \qquad \text{... (7)}$$

$$h^+_{VB} + H_2O \rightarrow H^+ + OH^\bullet \qquad \text{...(8)}$$

$$e^-_{CB} + O_2 \rightarrow O_2^{-\bullet} \qquad \text{... (9)}$$

$$O_2^{-\bullet} + H^+ \rightarrow HO_2^\bullet \qquad \text{... (10)}$$

$$HO_2^\bullet + H^+ + e_{CB}^- \rightarrow H_2O_2 \qquad \text{... (11)}$$

$$H_2O_2 + e_{CB}^- \rightarrow OH^\bullet + OH^- \qquad \text{... (12)}$$

$$h_{VB}^+ + OH^- \rightarrow OH^\bullet \qquad \text{... (13)}$$

$$OH^\bullet + \text{organic pollutants} \rightarrow \text{degradation products} \qquad \text{... (14)}$$

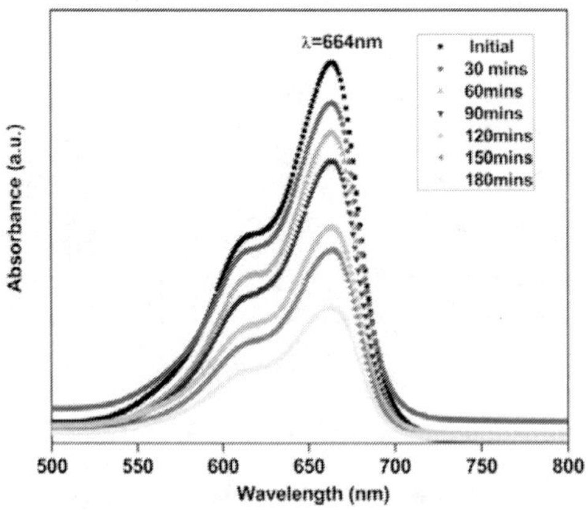

Fig. 4: UV-Visible Absorption Spectra of MB

CONCLUSION

Nanoparticles of CeO_2 were successfully synthesized by chemical precipitation method. The XRD analysis shows CeO_2 exhibits cubic in structure with crystallite size of ~12 nm. The SEM image shows the formation of spherical shaped nanoparticles and EDX analysis indicates the presence of cerium and oxygen. Bandgap value of CeO_2 was found to be 3.2 eV. The Photocatalytic efficiency of synthesized CeO_2 was attained 65% within 180 minutes under UV exposure.

REFERENCES

[1] Kavita Kabra et al, *Ind. Eng. Chem. Res.* 43, 7683-7696 (2004).

[2] Xianluo Hu et al, *Langmuir.* 26(5), 3031-3039 (2010).

[3] Noriya Izu, et al., *Sensors and Actuators* B. 87, 95-98 (2002).

[4] Bae. et al., *Mater. Lett.* 56, 610 (2002).

[5] Hung. et al., *Electrochimica Acta.* 50, 745-748 (2004).

[6] Saravanan. et al., *Int. J. Nanoscience.* 10, 253-257 (2011).

[7] Kumar. et al., *Recent Res. Sci. Tech.* 2(4), 37-41 (2010).

[8] Hung-Yi Chang et al., *J. Crystal Growth.* 283, 457-468 (2005).

[9] Ioannis et al., *Appl. Catal. B: Environ.* 49, 1–14 (2004).

STUDIES ON BIOSYNTHESIS OF IRON NANOPARTICLES AT ROOM TEMPERATURE USING AQUEOUS EXTRACT OF AQUATIC PLANT (WATER HYACINTH) AND ITS APPLICATION FOR CHROMIUM (VI) REMOVAL

Vasantha Gokilam R., Anjali Nair M., Kanakalakshmi A., Sukumar C. and Shanthi K.

Department of Environmental Science, PSG College of Arts and Science, Coimbatore
E-mail: shanthivis@gmail.com

ABSTRACT

The biosynthesis of nanoparticles by plant extracts has been an exciting discovery in the field of bionanotechnology. These nanoparticles find extensive applications in a wide spectrum of scientific, engineering and industrial areas. Iron nanoparticles were synthesized using a rapid, single step, and completely green biosynthetic method employing aqueous Water hyacinth extracts as both the reducing and capping agent. The biosynthesis of nanoparticles has been proposed as a cost effective environmental friendly alternative to chemical and physical methods. Recently, silver and iron nanoparticles of various sizes and morphologies have been synthesized using coffee and green tea extract. Iron nanoparticles of various sizes and morphologies were also formed instantaneously using Water Hyacinth extracts. The study has been carried out for biosynthesis of iron nanoparticles using aqueous aquatic plant Water Hyacinth extracts. Fe^0 nanoparticles were synthesized by the ferric chloride and were characterized by UV–Vis (UV-Visible) and FTIR (Fourier Transform Infrared spectroscopy), VSM (Vibrating sample magnetometer), XRD (X-Ray Diffraction), Scanning Electron Microscopy (SEM) and EDS. To study the

V. Rajendran, K. Saminathan, P. Paramasivam and K.E. Geckeler (eds.)
Nano Materials Synthesis and Characterisation, pp. 185-190 (2012).

efficiency of the biosynthesized Iron Nanoparticles in the removal of cr(VI) from aqueous solution. Various kinetic studies have been carried out. This environmentally friendly method of biological iron nanoparticles production provides rates of synthesis faster or comparable to those of chemical methods and can potentially be used in human contacting areas such as cosmetics, foods and medical application.

INTRODUCTION

Nanotechnology is the engineering and art of manipulating matter at the nanoscale (1–100 nm). Biosynthesis of nanoparticles by plant extracts is under exhaustive research however, an exact mechanism for the synthesis of nanoparticles using plants has not yet been elucidated although several hypothetical mechanisms have been proposed (Rai et al. 2004). The use of plant extracts for synthesis of nanoparticles is potentially advantageous over microorganisms due to the ease of scale up, the biohazards, and elaborates process of maintaining cell cultures (Sathishkumar et al., 2009). Chromium is a commonly identified contaminant in soils and groundwater. Much research has focused on the remediation of Cr (VI) and many treatment processes have been developed. The study has been carried out to biosynthesize iron nanoparticles using plant at room temperature. This study also includes the properties of iron nanoparticles and how the adsorption processes of chromium occurs by biosynthesized iron nanoparticles.

EXPERIMENTAL

Iron nanoparticles were synthesized by of ferric chloride ($FeCl_3$), with the water hyacinth plant extract. (Njagi et al., 2010). Iron nanoparticles prepared by the above method was characterized with X-Ray Diffraction (XRD), UV-Vis Spectroscopy, Scanning Electron Microscopy with energy Dispersion X-Ray Method (SEM & EDX), and Fourier Transform Infrared Spectroscopy (FTIR). Hexavalent chromium was determined spectrophotometrically by diphenyl carbazide method. Biosorption studies were carried with different chromium (VI) concentrations, pH, and Dosage. Various Kinetics and Isotherms studies were also carried out.

RESULTS AND DISCUSSION

The biosynthesized iron nanoparticles from water hyacinth were characterized for its properties and the results were shown in Figs 1, 2, 3 and 4. The Fe^0 nanoparticles showed its absorption maxima at 260 nm. This observation is similar to that obtained by Morgada et al. 2009. In the FTIR band at 3,421 cm^{-1} was ascribed to OH stretching vibration and the one at 1,641 cm^{-1} to the OH bending vibration of surface-adsorbed water which, suggests the formation of ferricoxy hydroxide (FeOOH) layer on Fe^0 nanoparticles A single peak was observed at around 3350.95 cm^{-1} and 3350.95 cm^{-1}, which is due to absorptions peak of –OH or $-NH_2$ in the iron nanoparticles reported by (Harajyoti Mazumdar *et al.,* 2011). The basal spacing of mount in XRD Fig. 2 was 1.99 nm, 3.14 nm indicates that the iron nanoparticles consisted mostly of iron (reflection at 2θ of 45°, d = 1.99 nm, 2θ of 28°, d = 3.14nm) and small amounts of crystalline iron (hydr)oxides (reflections at the 2θ of 35° and 31°, d = 2.81 nm). The SEM images and EDS spectrum of iron nanoparticles loaded with chromium are shown in Figs. 3 and 4. These images suggest that the particles agglomerate to form irregular clusters and showed peak of the iron and chromium. The spectrum also reveals the presence of chloride ions, Zirconium, Potassium, and manganese impurities with oxygen that indicates the surface of oxidation of the iron nanoparticles. (Njagi, et al., 2010).

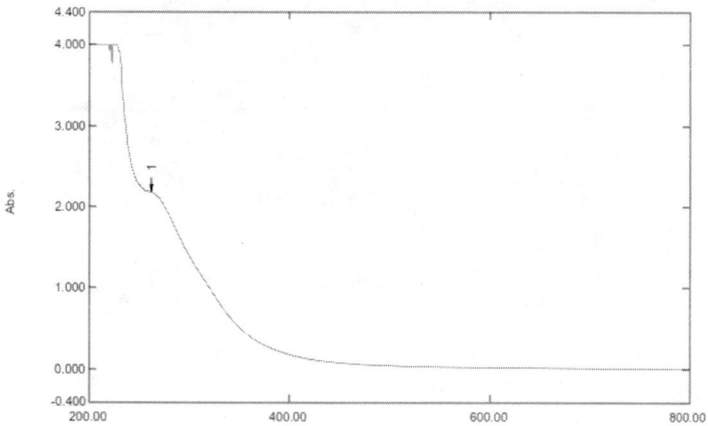

Fig. 1: UV Spectrum of BiosynthesizedIron Nanoparticle

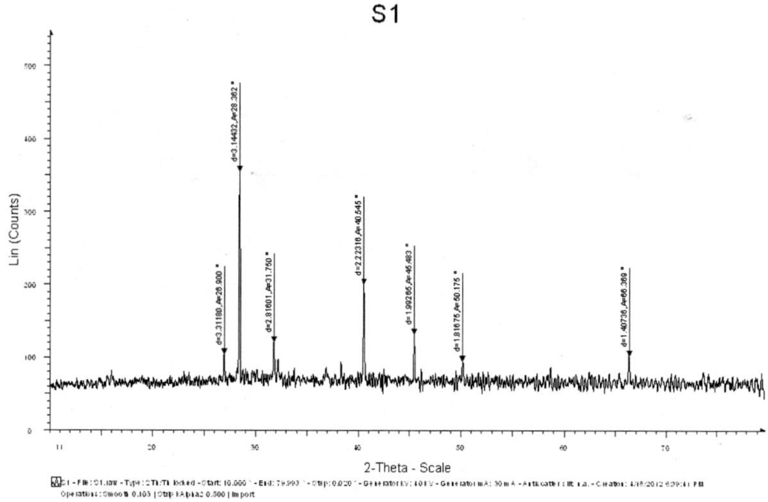

Fig. 2: XRD of Biosynthesized Iron Nanoparticle

Fig. 3: SEM of Chromium Loaded Iron Nanoparticles

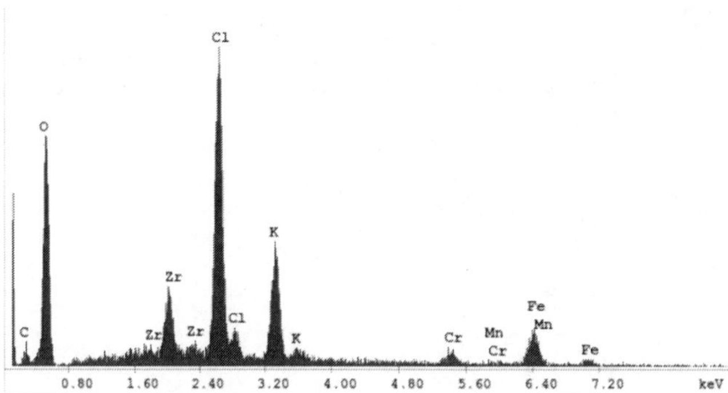

Fig. 4: EDS of Chromium Loaded Iron Nanoparticles

Sorption Studies

Studies on the effecicieny of Iron nanoparticles in the removal of chromium using different dosage, different pH, different chromium concentration and contact time were carried out and the results were presented in (Table 1). The optimum condition for the maximum removal of chromium by iron nanoparticle synthesized from plant extract was 100 mg of particle dosage, at pH 2.0 with contact time of 150 minutes from the 100 mg/L of chromium solution.

Table 1: Effect of pH, Nanoparticle Dosage and Chromium Concentration on Chromium Removal by Biosynthesized Iron Nanoparticles

pH	*% Removal*	*Iron Nanoparticle Dosage, gm/100ml*	*% Removal*	*Chromium Concentration, ppm*	*% Removal*
2	99	0.05	78	25	99.5
3	81	0.10	99	50	99.25
4	56	0.15	99.8	100	99
5	44	0.20	100	150	98
6	71			200	97
7	76				
8	31				
9	19				

Isotherm Studies

In this study, Langmuir, Freundlich and Temkin models were used to determine the adsorption equilibrium between the adsorbent and metal ions. Taking into account the correlation coefficients for Langmuir isotherm (R^2) of Iron nanoparticles was 0.995 for chromium removal which confirms the strong monolayer sorption onto specic sites. The Freundlich biosorption constants "n" values was 1.9. Same results reported by (Iram, et al., 2010). Now the R^2 value for all isotherm showed the trend of Langmuir (0.995) > Freundlich (0.979) > Harkins Jura (0.72) >Temkin (0.9773) > D-R (0.888) (Table 2). From this study it was observed that the pseudo second order kinetic model fitted very well than the pseudo first order kinetic model for chromium biosorption by Iron nanoparticles studied. The pseudo second-order equation tted the data verywell in a large quantity of literature for biosorption (Ho, 2006) (Table 3).

Table 2: Isotherm Studies

Freundlich			*Langmuir*			*Tempkin*		
K_f	n	R^2	Q_{max}, mg/gm	b, L/mg	R^2	A_t	b_t, l/gm	R^2
81.7	1.9	0.98	226	0.8	0.995	10.9	43	0.98
Harkins Jura			*DR-Isotherm*					
A	B	R^2	Q_{DR}, mg/gm	β	E, kJ/mole	R^2		
1193	0.52	0.72	141	5.86	0.29	0.89		

Table 3: Kinetic Studies

Kinetics	*Pseudo second order*			*Pseudo second order*			
Cr (VI), mg/L	*K, L/min*	*Q_e(cal), mg/gm*	R^2	*K^2, gm/mg/ min*	*Q_e(cal), mg/gm*	*h*	R^2
25	10.4	24.5	0.885	0.03	0.76	0.025	0.991
50	13	49	0.839	0.005	2.00	0.011	0.941
100	13	98	0.904	0.002	4.16	0.010	0.941
150	11	145	0.851	0.003	5.00	0.019	0.974
200	11	191	0.849	0.003	6.25	0.024	0.981

CONCLUSIONS

A "green" one-step approach has been developed for the synthesis of nanoscale iron particles at room temperature.The Biosynthesized Iron nanoparticles samples were assessed with Chromium removal from aqueous solution. This simple method offers agglomerate and irregular clusters of iron nanoparticles with unique size and shape when compared with control samples. The study reveals that Fe^0 nanoparticles play a key role in Cr (VI) removal through reduction. Fe^0 nanoparticles and factors like concentration, pH and time of treatment facilitate the process, which may lead to an innovative remediation technology that is likely more cost effective and less environmentally disruptive. This biological method is potentially attractive for large scale synthesis of metallic and metal oxide nanomaterials.

REFERENCES

[1] Harajyoti Mazumdar and Nabanita Haloi, A study on biosynthesis of iron nanoparticles by *Pleurotus* sp. *J. Microbiol. Biotech. Res.*, 2011, 1 (3): 39-49.

[2] Ho, Y.S., 2006, Review of second-order models for adsorption systems, *J Hazard Mater.*, 136:103-111.

[3] Iram M. Guo C., Guan Y.P. Ishfaq, A. Liu, H.Z. Adsorption and magnetic removal of neutral red dye from aqueous solution using Fe_3O_4 hollow nanospheres, *J. Hazard. Mater.* 181 (2010) 1039–1050.

[4] Morgada M.E., Levy I.K., Salomone V., Farias S.S., Lopez G., Litter M.I. (2009) Arsenic (V) removal with nanoparticulate zerovalent iron: Effect of UV light and humic acids. *Catal Today* 143 : 261–268.

[5] Njagi, Eric C. Hui Huang, Lisa Stafford, Homer Genuino and Hugo M. Galindo, Biosynthesis of Iron and Silver Nanoparticles at Room Temperature Using Aqueous Sorghum Bran Langmuir 2011, 27(1), 264–271.

[6] Rai, D. D.A. Moore, N.J. Hess, L. Rao, S.B. Clark, Chromium (III) hydroxide solubility in the aqueous system: a thermodynamic model, *J. Solution Chem.* 33 (10) (2004) 1213–1242.

[7] Sathishkumar, M; Sneha, K; Kwak, I.S. Mao, J.; Tripathy, S.J.; Yun, Y.-S., Phyto-crystallization of palladium through reduction process using *Cinnamom zeylanicum* bark extract, *J. Hazard. Mater.* 2009, 171, 400–404.

EFFECT OF SINTERING TEMPERATURE ON THE STRUCTURAL AND MAGNETIC PROPERTIES OF COBALT FERRITE NANOPARTICLES

Sheena Xavier, Smitha Thankachan, Binu P. Jacob and E.M. Mohammed

Research Department of Physics, Maharaja's College, Ernakulam
E-mail: emmohammed_2005@yahoo.com

ABSTRACT

Cobalt ferrite nanoparticles have been synthesized by sol-gel method. The prepared sample was sintered at four different temperatures (300°C, 400°C, 500°C and 700°C) for four hours. The structural characterizations of all the prepared samples were done using XRD, TEM and FTIR. Crystallite size is found to be increasing with sintering temperature and it can be attributed to the grain growth of the particles. The particle size of the sample was determined using TEM. The FTIR spectra show two strong absorption bands in the range of 1000-400 cm^{-1}, characteristic of spinel ferrites. The room temperature magnetic measurements showed a strong influence of sintering temperature on saturation magnetization and coercivity.

INTRODUCTION

Nanosized spinel ferrite particles have attracted much attention in recent years because of their potential applications in high density magnetic recording, magnetic fluids, spintronics, data storage and gas sensors [1, 2]. Among the ferrite nanoparticles, cobalt ferrite has been widely studied due to its excellent chemical stability, mechanical hardness, reasonable saturation magnetization and high magneto-crystalline anisotropy [3]. These properties make it a promising candidate for many applications, namely magnetic data storage, magnetic drug targeting, biosensors and magnetic refrigeration [4].

V. Rajendran, K. Saminathan, P. Paramasivam and K.E. Geckeler (eds.)
Nano Materials Synthesis and Characterisation, pp. 191-196 (2012).

Nano ferrites are simultaneously good magnetic and dielectric materials. These properties of the nano ferrites are affected by the preparation conditions, chemical composition, sintering temperature and the method of preparation [5].

Several chemical and physical methods such as spray pyrolysis, sol-gel, co-precipitation, combustion technique, high energy milling, etc., have been used for the fabrication of stoichiometric and chemically pure nano ferrite materials [6]. Among the available chemical methods, the sol-gel technique is an excellent method to synthesize nanoparticles with maximum purity [7]. This method has the advantage of good stoichiometric control and the production of ultrafine particles with a narrow size distribution. In the present paper structural and magnetic properties of nanocrystalline cobalt ferrite in relation to sintering temperature were investigated.

EXPERIMENTAL

Synthesis

Nano particles cobalt ferrite was synthesized by the sol-gel combustion method. Stoichiometric ratio of cobalt nitrate and ferric nitrate (AR grade MERCK) were dissolved in ethylene glycol using a magnetic stirrer. The solution was then heated at 60°C for 3 hours until a wet gel of the metal nitrates was obtained. The gel was then dried at about 120°C self ignites to a highly voluminous and fluffy product which was ground to obtain fine powder. The combustion can be considered as a thermally induced redox reaction of the gel wherein ethylene glycol acts as the reducing agent and the nitrate ion acts as an oxidant. The nitrate ion provides an oxidizing environment for the decomposition of the organic component. The obtained powder was ground well and sintered for 4h in a muffle furnace at four different temperatures 300, 400, 500 and 700°C.

Characterization

The cobalt ferrite samples were characterized by using X-ray powder diffractometer (XRD, Bruker AXS D8 Advance) using Cu-Kα radiation (λ = 1.5406A°) at 40 kV and 35 mA. Lattice parameter was calculated assuming cubic symmetry. The crystal structure, crystallite size and X-ray density were determined. The particle size was determined by subjecting the sample to Transmission electron microscopy (Philips, CM200). Wavelength Dispersive X-ray Fluorescence (WD-XRF) Spectrometer (Bruker S4-Pioneer) was used for elemental analysis. The Fourier transform infrared (FTIR) absorption spectra of the samples were recorded using FTIR spectrometer (Thermo Nicolet, Avatar 370) in the wave number range 4000-400 cm^{-1} with Potassium bromide (KBr) as solvent. The Magnetic characterization was carried out using a vibrating sample magnetometer (VSM; Lakeshore 7410) at room temperature up to a maximum field of 20kOe.

RESULTS AND DISCUSSION

Structural Analysis

The XRD patterns of $CoFe_2O_4$ nanoparticles sintered at 300, 400, 500 and 700°C are depicted in Fig. 1 and are typical of spinel structure.

Comparing the XRD pattern with the standard data (JCPDS PDF card No. 22-1086), the formation of single phase cobalt ferrite nanoparticles was confirmed. The diffraction peaks are broad because of the nanometer size of the crystallite. The crystallite size 'D' of the samples has been estimated from the broadening of XRD peaks using the Scherrer equation [7]. The lattice parameter 'a' is calculated for prominent peak (311) using Bragg's equation. The actual X-ray density, ρ_x was calculated using the formula

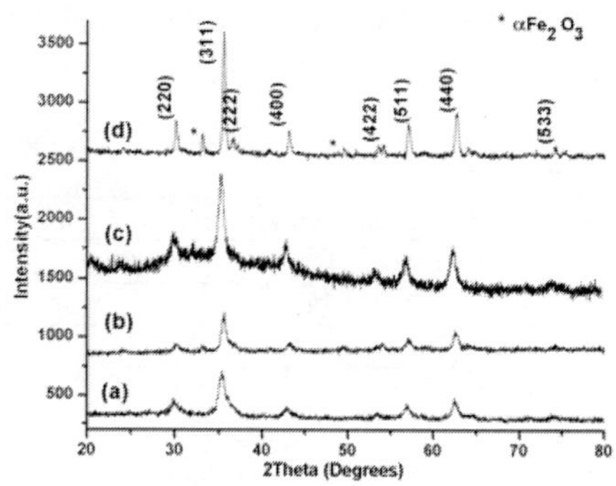

Fig. 1: XRD Patterns of $CoFe_2O_4$ Sintered at (a) 300° C (b) 400° C (c) 500° C (d) 700° C

$$\rho_x = \frac{8M}{Na^3} \qquad \text{... (1)}$$

where, 'M' is the molecular weight, 'N' is Avogadro's number and 'a' is the lattice parameter. Calculated values of lattice parameter, crystallite size and X-ray density of all the samples are listed in Table 1.

Table 1: Lattice Parameter, Crystallite Size and X-ray Density of $CoFe_2O_4$

Sintering Temperature (°C)	*a (nm)*	*D (nm)*	*X.D (g/cc)*
300	0.8408	12.815	5.243
400	0.8362	18.467	5.330
500	0.8422	19.718	5.217
700	0.8375	52.488	5.306

TEM and WD-XRF Analysis

Figure 2 shows the TEM image of $CoFe_2O_4$ sample sintered at 400°C. In the figure most of the nanoparticles appear with an almost spherical shape and are agglomerated. The average particle sizes determined from TEM analysis were found to be in reasonable agreement with that obtained from XRD analysis.

The stoichiometry of the powder samples were checked by WD-XRF analysis. The composition of the elements present in the sample sintered at 400°C is shown in Table 2. From the table it is obvious that the samples show the expected stoichiometry. No trace of any impurity was found, which indicates the purity of the samples.

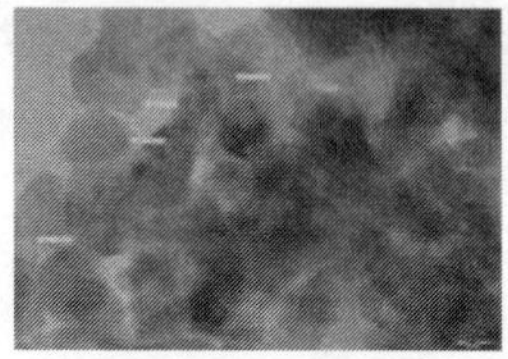

Fig.2: TEM Image of $CoFe_2O_4$ Sintered at 400°C

Table 2: Elemental Analysis of $CoFe_2O_4$

Elements Present	*Expected (wt. %)*	*WD - XRF (wt. %)*
Co	25.11	25.25
Fe	47.608	47.3
O	27.261	27.3

FTIR Analysis

FT-IR spectra of the investigated samples are shown in Fig. 3. In the wave number range 1000–300 cm^{-1}, the infra red bands of solids are usually assigned to vibration of ions in the crystal lattice.

Two main broad metal-oxygen bands are seen in the IR spectra of all spinels, and ferrites in particular. The highest one V1 generally observed in the range 600 – 550 cm^{-1}, corresponds to intrinsic stretching vibrations of the metal at the tetrahedral site, whereas the V2, the lowest band, usually observed in the range 450-385 cm^{-1}, is assigned octahedral-metal stretching[8].

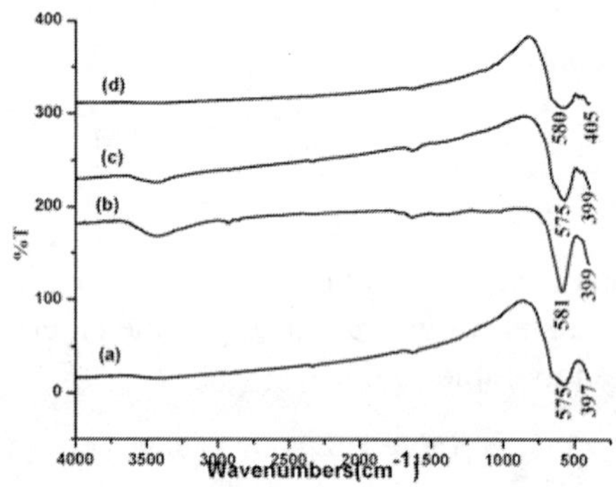

Fig. 3: FTIR Spectra of $CoFe_2O_4$ Sintered at (a) 300° C (b) 400° C (c) 500° C (d) 700° C

Magnetic Properties

The hysteresis loops were measured to determine magnetic parameters such as the saturation magnetization M_s, coercivity H_c and remanent ratio R for all the samples. Figure. 4 show the typical magnetic hysteresis loops of $CoFe_2O_4$ sintered at 300, 400, 500 and 700°C.

The changes in the magnetic properties of cobalt ferrite can be attributed to the modification of the particle sizes dependent on the sintering temperature. The coercivity first increases as the particle size increases, reaches a maximum value of 1893 Oe, and then decreases for any further increase in particle size. This can be understood as due to the enhanced role of the surface and its strong anisotropy, as opposed to the weaker bulk anisotropy [9]. The saturation magnetization seems to increase linearly with the sintering temperature. This behaviour can be attributed to the surface effects that are the result of finite–size scaling of nanocrystallites, which in turn leads to a non-collinearity of magnetic moments on their surface [10].The results of the magnetization measurements of all the samples are shown in Table 3.

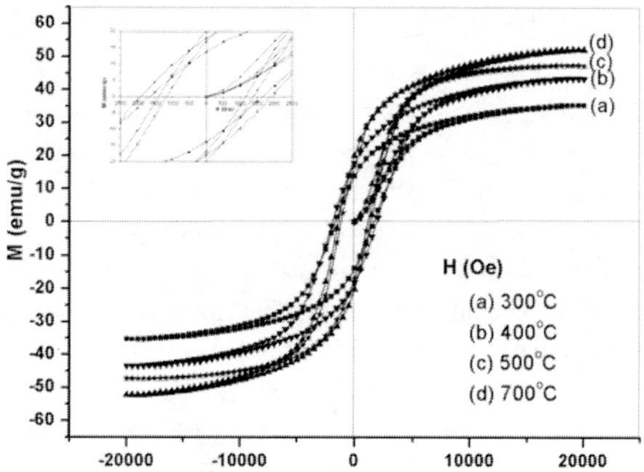

Fig. 4: Room Temperature Magnetic Hysteresis of $CoFe_2O_4$

Table 3: Magnetic Parameters of $CoFe_2O_4$ Nanoparticles

Sintering Temperature	M_S *(emu/g)*	M_R *(emu/g)*	H_C *(Oe)*	*R*
300	35.38	13.85	1640.3	0.3915
400	43.5	17.95	1893.4	0.4126
500	47.37	21.84	1652.9	0.4611
700	52.24	20.07	1379.2	0.3842

CONCLUSIONS

Nanocrystalline cobalt ferrite has been successfully synthesized by sol-gel method. XRD results confirmed the formation of single phase cubic structure of the nanoparticles. The crystallite size of the samples increases with the increase in the sintering temperature. The crystallite size varies from 13 to 52 nm and is in reasonable agreement with the results obtained from TEM. The FTIR analysis confirmed the spinel structure of the samples. The saturation magnetization increased gradually with the increasing sintering temperature, while coercivity initially increases and then decreases. These results have been explained based on the size and surface effects of the ultrafine materials.

ACKNOWLEDGEMENTS

Authors thank SAIF, CUSAT Kochi, SAIF, IIT Madras and SAIF IIT Mumbai for providing XRD, FTIR, VSM and TEM measurement facilities.

REFERENCES

[1] Veena Gopalan E., Joy P.A., I.A Al-Omari, Sakthi Kumar D. and and Yasuhiko Yoshida, Anantharaman M.R, On the Structural, Magnetic and Electrical Properties of Sol–gel derived Nanosized Cobalt Ferrite, *J. Alloys compd* 485, 711-717, 2009.

[2] Yue Zhang, Zhi Yang, Di Yin, Yong Liu and Chun Long Fei, Composition and magnetic properties of cobalt ferrite nanoparticles prepared by the co-precipitation method, J. Magn. Magn. Mater., 322, 3470-3475, 2010.

[3] Zhenfa Zi, Yuping Sun, Xuebin Zhu and Wenhai Song, Synthesis and magnetic properties of $CoFe_2O_4$ ferrite nanoparticles, *J. Magn. Magn. Mater.*, 321, 1251-1255, 2009.

[4] Veena Gopalan E., I.A Al-Omari, Sakthi Kumar D., Yasuhiko Yoshida and Joy P.A. et al. Inverse magnetocaloric effect in sol-gel derived nanosized cobalt ferrite, *Appl. Phys. A: Materials Science & Processing*, 99(2) 497-503, 2010.

[5] Gul I.H. and Masqood A., Structural, magnetic and electrical properties of cobalt ferrites prepared by the sol–gel route., *J. Alloys compd* 465, 227-231, 2008.

[6] Pawan Kumar, Sharma S.K., Knobel M., Singh M. and Effect of La^{3+} doping on electric, dielectric and magnetic properties of cobalt ferrite processed by co-precipitation technique, *J. Alloys compd*, 508, 115-118, 2010.

[7] Binu P. Jacob, Smitha Thankachan, Sheena Xavier and E M Mohammed, Effect of Gd^{3+} doping on the Structural and Magnetic Properties of nanocrystalline Ni-Cd mixed ferrite, *Physica Scripta*, 84, 045702-045708, 2011.

[8] Ahmed M.A., Ateia E., Salah L.M. and El_Gamal, Structural and electrical studies on La^{3+} substituted Ni-Zn ferrites, *Mater. Chem. Phys.* 92 ,310-321, 2005.

[9] Toksha B.G., Sagar E. Shirsath, S.M. Patange and K.M. Jadhav, Strutural investigations and magnetic properties of cobalt ferrite nanoparticles prepared by sol-gel auto combustion method, *Solid State Commun.* 147, 479-483, 2008.

[10] Lunhong Ai and Jing Jiang, Influence of annealing temperature on the formation, microstructure and magnetic properties of spinel nanocrystalline cobalt ferrites, *Curr. Appl. Phys.* 10, 284-288, 2010.

SYNTHESIS AND CHARACTERIZATION OF NANO $La_{0.6}Pb_{0.4-x}Ca_xMnO_3$ ($0.005 \leq x \leq 0.015$) PEROVSKITE MAGNATITE MATERIALS

G.K. SIVASANKARAYADAV, K. SAKTHIPANDI[1], K. THYAGARAJAN AND V. RAJENDRAN[1]

Department of Physics, JNTUA College of Engineering, Pulivendula, Andhra Pradesh
[1]Centre for Nano Science and Technology, K.S. Rangasamy College of Technology, Tiruchengode, Tamil Nadu
E-mail: ktrjntu@gmail.com

ABSTRACT

In recent years, the interest on the elevations of the structural, magnetic and electrical properties of two divalent doped perovskite magnetite materials are increasing demand due to its potential applications like fuel cells and sensors. In the present paper, an effort has been made towards the synthesis and characterization of nano $La_{0.6}Pb_{0.4-x}Ca_xMnO_3$ ($0.005 \leq x \leq 0.015$) perovskite magnetic materials. The above perovskite have been prepared in bulk and nanoscale using solid state reaction and sonochemical reactor methods. The structural properties of the prepared samples were investigated using FTIR, XRD, SEM, EDAX, and TEM studies. The observed XRD results indicate that the prepared bulk and nano samples are in the rhombohedral crystal structure. The estimated particle size of bulk and nano perovskites is 350-400 nm and 80-50 nm respectively. The structural properties were studied using SEM and TEM.

V. Rajendran, K. Saminathan, P. Paramasivam and K.E. Geckeler (eds.)
Nano Materials Synthesis and Characterisation, pp. 197-202 (2012).

INTRODUCTION

The research on $Ln_{1-x}A_xMnO_3$ perovskites (Ln = rare earth, A = alkaline earth element) have strongly influenced by many researchers [1, 2]. Young et al. [3] have studied the crystal structure and magnetic properties of $La_{0.7}Pb_{0.3}Mn_{1-x}Co_xO_3$ ($0 \leq x \leq 1$) perovskites. For $x = 0$ these authors found a rhombohedral structure and also a rather large magneto resistance effect, about 50% below 200 K and about 25% around 300 K. Troyanchuk et al[4] have observed that the $La_{1-x}Pb_xMnO_3$ perovskites (x = 0.4-0.6) have a rhombohedral (slightly distorted) cubic structure. Huang et al [5] have studied the crystal structure and the magnetic scaling behavior of $La_{1-x}Pb_xMnO_3$ perovskites (x = 0.0-0.5) and have shown that all the samples crystallize in the rhombohedral structure. The present work focus on the synthesis and the structural studies of $La_{1-x}Pb_xMnO_3$ manganite.

EXPERIMENTAL PROCEDURE

Sample Preparation

By the using of solid state reaction technique LPCM perovskite Manganite materials with different molar ratio values of x = 0.005, 0.01, 0.015 were prepared with $La_2(NO_3)_3.6H_2O$ (MERCK 99%), $Pb(NO_3)_2$ (MERCK 99%), $CaCO_3$ (SRL 99.5%) and $MnCO_3$ (Aldrich 99.9%) to obtain the nominal compositions of $La_{0.6}(Pb_{0.4"x}Ca_x)MnO_3$. On the other hand for the making of Nano phase material, we used sono chemical reactor method. The calcinated power at 1000°C for 12h in an atmospheric air was used for different characterization techniques.

RESULTS AND DISCUSSION

X-ray Diffraction Analysis

The XRD patterns were obtained using Cu Kas a radiation source by applying voltage and current ratings of 40 kV and 30 m. The recorded XRD patterns of as-prepared and sintered samples are shown in Figs. 1 & 2. It is inferred that the observed XRD patterns of the as-prepared sample reveals a crystalline nature as reported elsewhere [5, 6]. Further, XRD peaks reveal the existence of mixed crystalline nature namely rhombohedral and tetragonal crystalline phases in the as-prepared sample. The peaks observed at 27.616% and 48.365% correspond to tetragonal phase and other peaks correspond to rhombohedral phase. On the other hand, rhombohedral phase with the space group of R3c along with the hexagonal structure is observed in the samples sintered at 1273 K. Nevertheless, the diffraction peaks observed in the as-prepared sample is quite broad in nature, indicating the crystalline nature. On the other hand, in sintered sample, the peaks are very narrow and slightly shifted towards its lower 2θ values which indicate the nanocrystalline nature of the sintered sample. The decrease in diffraction peak width (FWHM) for (1 1 0) planes in the sintered sample conrms the decrease in nanocrystalline size upon sintering. The observed results indicate a decrease in the crystalline size from 50 to 15 nm while comparing the sintered sample with the as-prepared sample.

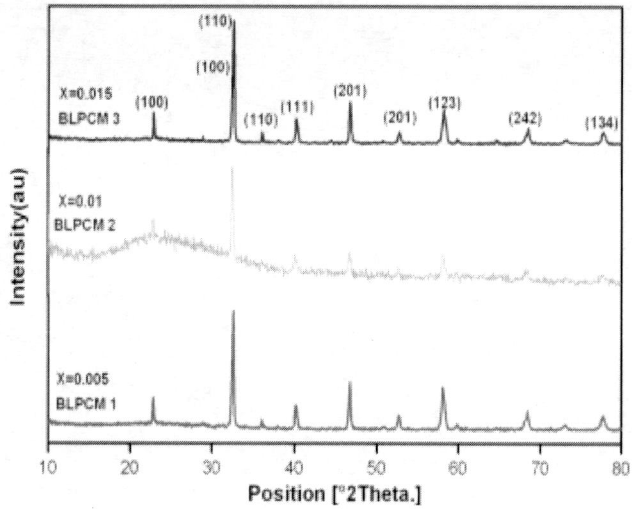

Fig. 1

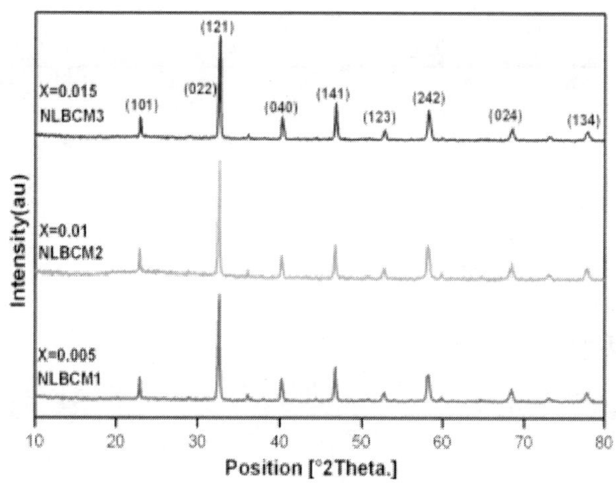

Fig. 2

Fourier Transformation Infrared Spectroscopy Analysis

FTIR studies were done in the as-prepared LPCM samples. The FTIR spectrum of the as-prepared sample is given in Figs. 3 & 4. The observed band at 600 cm^{-1} corresponds to s stretching mode which involves the internal motion of change in Mn–O–Mn bond length [6]. One peak observed at 1630 to the deformation and the O–H stretching mode of absorbed water molecules [7].

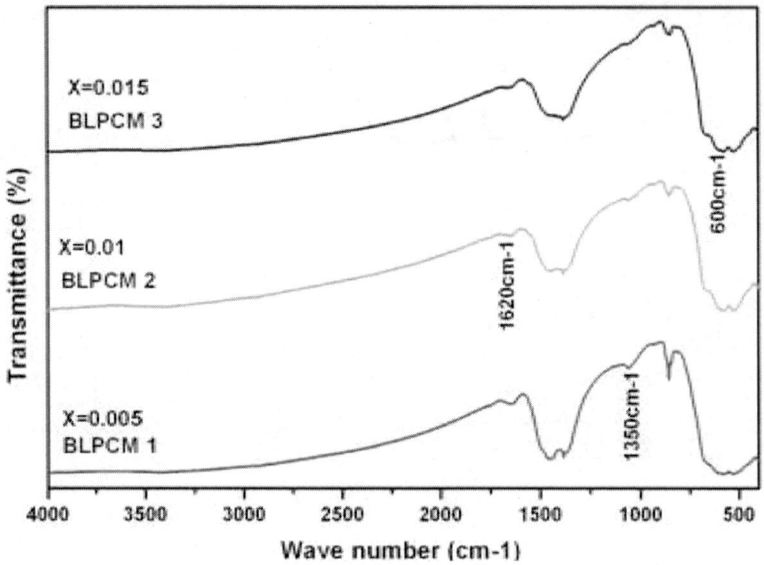

Fig. 3

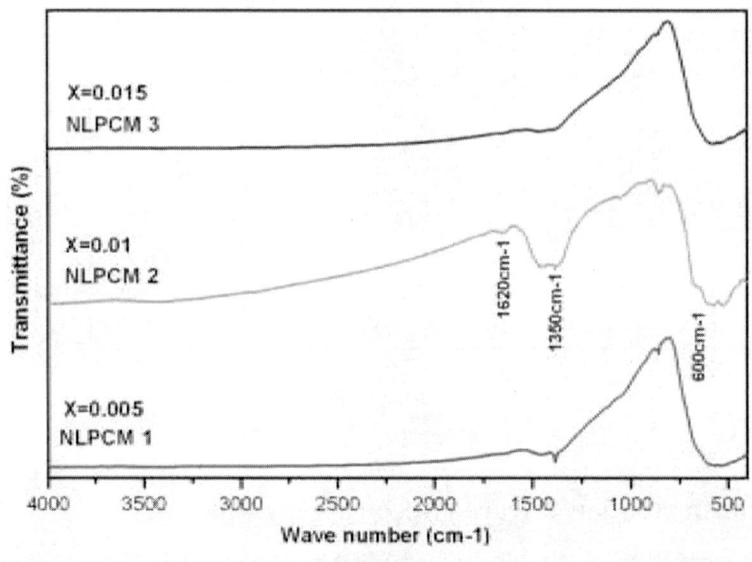

Fig. 4

Microscopic Studies

The SEM image of the as prepared and sintered samples is shown, respectively, in Fig. 5. SEM micrograph of as-prepared sample reveals that the obtained LPCM samples are voluminous and porous in view of the large amount of gases evolved during the sonication and preheat treatments. It is evident from that the existence of irregular shapes of agglomerates along with the cage-structure in the as-prepared sample. Due to the elongated aggregates and poorly dened boundaries, the grain size determination of particles is difficult from the observed SEM images. The approximate estimate of the grain size is 60 and 25 nm, respectively. It is evident from the observed XRD and FTIR results, the as-prepared sample contains a mixture of crystalline structures of rhombohedral and tetragonal [8, 9] and O–H stretching mode of absorbed water molecules [10] Thus, it reveals that an agglomerate structure with the approximate grain size of 60 nm. Generally, during the sintering, the crystalline size increases while the surface area of the particles decreases. However, in the sintered sample, the particle size reduced to 25 nm due to agglomeration effects of the existence of rhombohedral phase instead of mixed structure as observed in the as-prepared sample [8, 9]. Therefore, a decrease in the particle size is observed in the sintered samples leads to a decrease in surface area when compared to as-prepared sample. The TEM images of the as-prepared and sintered samples are shown in Fig.6. The TEM studies also conrm a decrease in particle size upon sintering, as discussed earlier from the results of XRD and SEM results.

BLPCM

NLPCM

Fig. 5

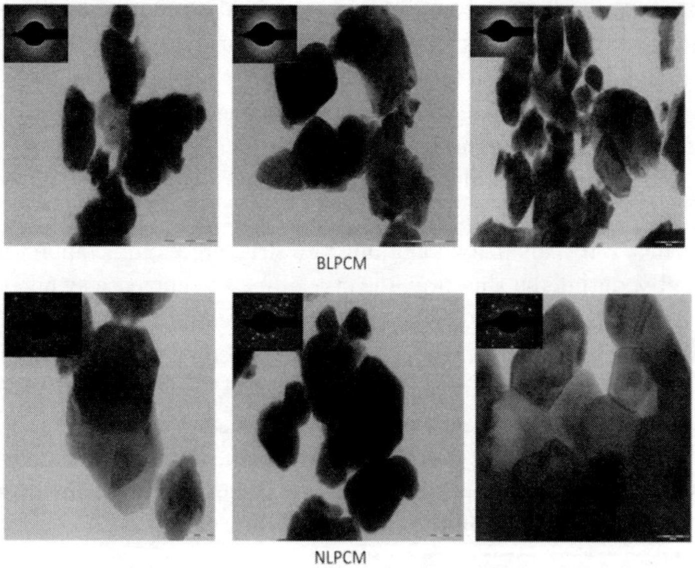

Fig. 6

REFERENCES

[1] H. Kong and C. Zhu, Ultrasonic anomalies in $La_{1/3}Sr_{2/3}MO_3$ (M = Fe, Mn), *Appl. Phys. Lett.* 88 (2006), pp. 41920–41926.

[2] X. Wu, L.Xu, B. Yang and E. Weng, Surface Characterization and catalytic performance of $La_{0.7}Sr_{0.3}MnO_{3+ë}$ coating deposited by plasma Spraying, *Surf. Coat. Technol.* 184 (2004), pp. 40–46.

[3] S.L. Young, Y.C. Chen, Lauce Horng, T.C. Wu, H.Z. Chen and J.B. Shi, *J. Magn. Magn. Mater.* 289 (2000) 5576.

[4] I.O. Troyanchuk, D.D. Khalyavin and H. Szymczak, *Mater. Sci. Bull.* 32 (1997) 1637.

[5] T. S. Huang, C. H. Chen, M. F. Tai, *Mater. Res. Soc. Symp. Proc.* 674 (2001) U 3.4.1.

[6] P. Palanichamy, P. Kalayanasundram and *B.Raj, Br. J. Non-destruct. Test.* 31 (2) (1989) 78.

[7] S. Roy and S. Bandyopadhyay, *J. Mater. Sci. Lett.* 15 (1996) 1872.

[8] Z.F. Zi, Y.P. Sun, X.B. Zhu, Z.R. Yang, J.M. Dai and W.H. Song, *J. Magn. Magn. Matter* 321 (2009) 2378.

[9] A. Gedanken, *Mater. Res. Bull.* 38 (2003) 11.

[10] M. Quijada, J. Cerne, J.R. Simpson, H.D. Drew, K.H. Ahn, A.J. Millis, R. Shreekala, R. Ramesh, M. Rajeswari and T. Venkatesan, *Phys. Rev. B* 58 (1998) 16093.

EFFECT OF GROWTH TIME ON THE MORPHOLOGY OF TiO_2 MICROSPHERES ASSEMBLED BY NANOFIBERS AND THEIR CHARACTRIZATION

G. Arthi, J. Archana[1], M. Navaneethan[1], S. Ponnusamy, Y. Hayakawa[1] and C. Muthamizhchelvan

Department of Physics, SRM University, Kattankulathur, Tamil Nadu, India
[1]Research Institute of Electronics, Shizuoka University, 3-5-1 Johoku, Naka-ku, Hamamatsu, Shizuoka 432-8011, Japan
E-mail: selvancm@gmail.com

ABSTRACT

Hierarchical TiO_2 microspheres were synthesized by a simple hydrothermal method at a low temperature of 100°C. The as-prepared TiO_2 nanoparticles and hierarchical microspheres were analyzed by X-ray diffraction (XRD), UV-VIS absorption spectroscopy (UV) and Scanning electron microscope (SEM). On the basis of XRD results, TiO_2 nanoparticles and TiO_2 microspheres both exhibited rutile phase. The increased XRD peak intensity indicates the enhanced crystallinity of the TiO_2 microspheres. The reduction in particle size is observed from UV results. The SEM image reveals that, the diameter of the prepared TiO_2 nanoparticles is 100 nm.

INTRODUCTION

Nanomaterials and nanostructures are attracting tremendous attention in both fundamental studies and applications due to their unique physical and chemical properties [1-3]. Many efforts have been concentrated on a wide variety of metal oxides, such as TiO_2, SnO_2, VO_2, and ZnO. Among them TiO_2 and TiO–derived materials have been extensively used for various

V. Rajendran, K. Saminathan, P. Paramasivam and K.E. Geckeler (eds.)
Nano Materials Synthesis and Characterisation, pp. 203-208 (2012).

applications such as dye sensitized solar cell, gas sensors, catalysts, water treatment. In particular, TiO_2 have been used as photoanodes in dye sensitized solar cells. [4–6], TiO_2 have been used in variety of applications since it has high stability, low cost and environmental friendly [7,8]. TiO_2 nanomaterials have been prepared by many methods like sol-gel, sonochemical method and hydrothermal method. Hydrothermal method is the suitable method for the preparation of different types of nanostructures [9,10]. TiO_2 nanostructures with complex morphology are relatively very hard to synthesize owing to the rapid hydrolysis rate of Ti-containing precursors in aqueous solution especially when they are compared to the well developed SiO_2, ZnO, SnO_2.Therefore the synthesis of TiO_2 nanostructures with tunable size and morphology still remains a major challenge, particularly in designed complex configuration. Thus, it still remains a significant challenge to develop facile and effective surfactant-free methods for the synthesis and architecture control of TiO_2 hierarchical nanostructures at low temperature [11-13]. Here, we have reported the preparation of rutile TiO_2 microspheres which consists of nanofibers and also studied the effect of the growth time on the morphology of TiO_2 microspheres. The morphology of Microspheres assembled by nanofibers which is very siutabler for dye sensitized solar cell applications.

EXPERIMENTAL

All chemicals were of analytical grade and used without further purification. In a typical process, 1 ml of titanium butoxide was added drop wise into the mixture of hydrochloric acid and distilled water (1 : 6). The above solution was stirred well for an hour. Then the solution was poured into autoclave and the temperature was maintained at 100°C for 6 hr. When the autoclave was cooled to room temperature, the reaction products were filtered, washed with deionized water, and then dried in air at 80°C. To find the effect of growth time on the morphology of TiO_2 the above solution was maintained at same temperature but different growth time 12 hr.

XRD patterns were recorded using an X'per PRO (PANalytical) diffractometer using CuKα (λ = 1.5406) radiation at a scan rate of 0.2°/sec. A UV-visible (UV-Cary 5E) spectrophotometer was used to obtain optical absorption spectra of the samples. Surface morphology was characterized by SEM (JEM 3010-JEOL) with an operating voltage of 200 keV.

RESULTS AND DISCUSSION

Figure 1 shows the XRD patterns of prepared hierarchical TiO_2 superstructures. Figure 1(a, b) presents the time-dependent phase transformation of the product obtained at 100 °C for two different growth time 6 hr and 12 hr. From Fig.1(a) All observed diffraction peaks were indexed to the mixture of anatase and rutile crystal structure of TiO_2, matched to standard JCPDS Card: 89-0552. When the growth time prolongs from 6 h to longer reaction time 12 h there is existence of phase transformation from mixture of two phases to pure rutile phase. The reaction time would help to improve the crystallinity of rutile phase. The obtained peak from Fig. 1. (b) corresponds to JCPDS card no: 89-0552.

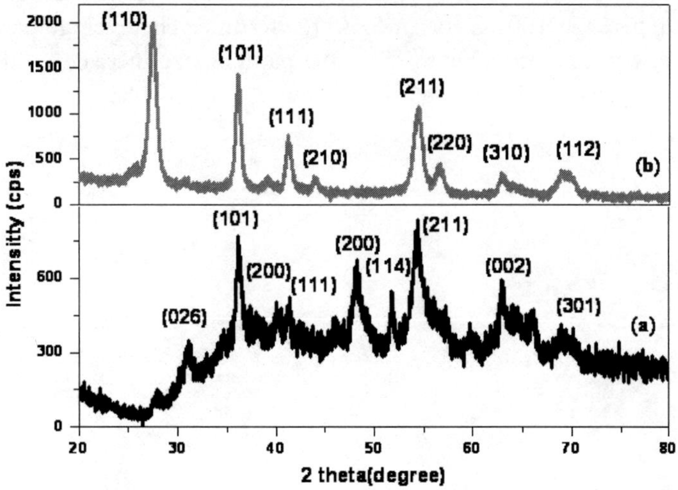

Fig. 1. XRD Patterns of Samples (a) TiO_2 Nanoparticles (b) Microsphers

Figure 2 shows the UV-vis absorption spectra of the typical rutile TiO_2 samples. It clearly exhibits the absorption edges around 409.6 nm for sample a, and 436.12 nm for sample b, corresponding to their optical bandgaps of 3.02 and 2.8 eV, respectively. This red shift was due to increasing in the size.

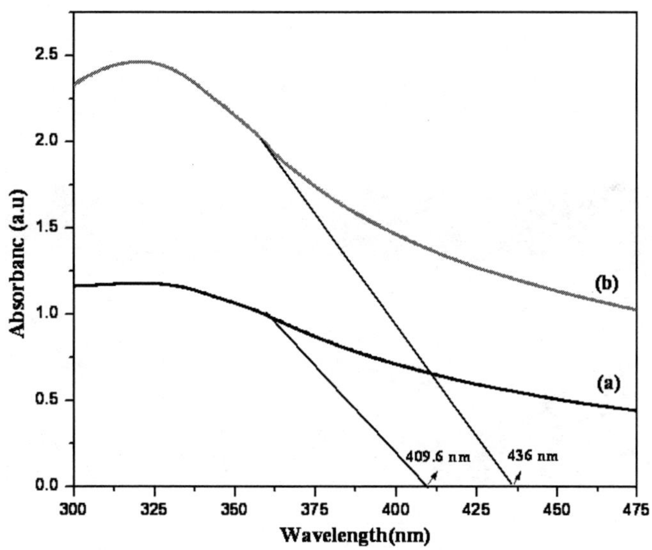

Fig. 2. UV-vis Absorption Spectra of Samples (a) TiO_2 Nanoparticles (b) Microsphers

Figure 3 shows the SEM result of prepared TiO_2 hierarchical structures. Figure 3 (a, b) shows samples prepared at 100 °C for an 6 hr growth time. The average particle size is 100 nm. When the growth time prolongs to 12 hr the particle size increases. Since the Ostwald ripening was taking place the smaller particles grew like microspheres.

Fig. 3. FESEM Images of Prepared TiO_2 Nanoparticles Synthesized at 100 °C-6 hr. (a) High Magnifications (b) Low Magnification

Figures 4 (a, b, c) shows the different magnifications of the obtained microspheres for 12 hr growth time. It is seen that most micro-spheres on the surfaces are cracked as shown in Fig. 4a, which might be due to the shrinkage of the microspheres. The microspheres were attached continuously as a result of oriented attachment process. Figure 4(b) shows the separate microsphere which was formed by nanofibers. Furthermore, the image of the cross section of a broken sphere is shown in Fig. 4 (c) which suggests that the microspheres are composed of number of nanofibers radially aligned towards their surfaces.

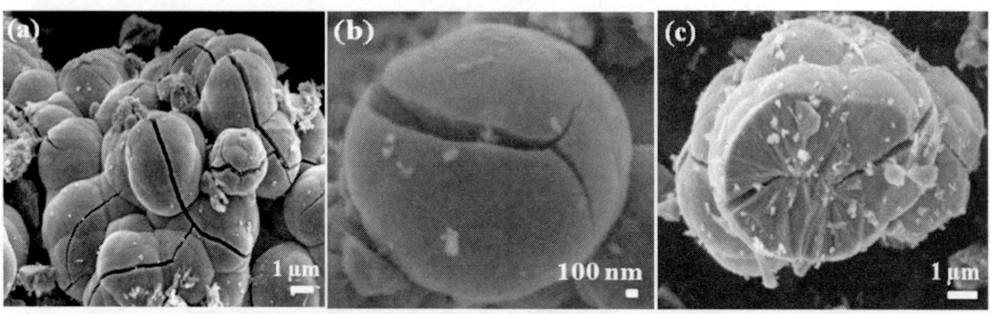

Fig. 4. FESEM Images of TiO_2 Microspheres Synthesized at 100°C-12 hr. (a) Low Magnification, (b) High Magnification, (c) Cross-section

CONCLUSIONS

Crystalline phase, particle size, and morphology of final products were tunable with the synthetic reaction time. By changing growth time the microsphere which is constructed by enormous nanofibers were prepared by hydrothermal method. The microsphere which has 1-dimensional TiO_2 nanostructure may be beneficial in dye sensitized solar cell applications for better electron transport.

ACKNOWLEDGEMENTS

One of the authors G.Arthi is thankful to the Shizuoka University, Research Institute of Electronics for providing necessary infra structure to carry out this research.

REFERENCES

[1] Astruc, D., Nanoparticles and Catalysis., Ed.,Wiley-VCH: Weinheim, Germany, 2008.

[2] Ferrando, R., Jellinek, J. Johnston, R.L., Nanoalloys: From Theory to Applications of Alloy Clusters and Nanoparticles, *Chem. Rev.*, 108, 845,2008.

[3] Burda, C., Chen, X., Narayanan, R. and El-Sayed., M. A., Chemistry and Properties of Nanocrystals of Different Shapes, *Chem. Rev.*, 105, 1025, 2005.

[4] Qiu Y., Chen W. and Yang S., Angew Chem Int Ed Engl, Double-Layered Photoanodes from Variable-Size Anatase TiO_2 Nanospindles: A Candidate for High-Efficiency Dye-Sensitized Solar Cells, 49,575–679,2010.

[5] Luo Y., Li D. and Meng Q. Towards optimization of materials for dye-sensitized solar cells, Adv Mater,21,4647–51, 2009.

[6] Kim Y.J., Lee M.H., Kim H.J., Lim G., Choi Y.S. and Park N.G. et al., Formation of highly efficient Dye-sensitized solar cells by hierarchical pore generation with nanoporous TiO_2 spheres, *Adv Mater*, 21, 3668–73, 2009.

[7] Linsebigle, A.L., Lu, G.Q. and Yates, J.T., Photocatalysis on TiO_2 Surfaces: Principles, Mechanisms, and Selected Results, *Chem. Rev.*, 95, 735,1995.

[8] Fujishima, A., Rao, T. N. and Tryk, D., A. Titanium Dioxide Photocatalysis, *J. Photochem. Photobiol*, C, 1, 1, 2000.

[9] Strohm, H. and Loebmann, P. Porous TiO_2 hollow spheres by liquid phase deposition on polystyrene latex-stabilised Pickering emulsions, *J. Mater. Chem.*, 14, 2667, 2004.

[10] Yin, J.,Chen, H., Li, Z., Qian, X., Yin, J., Shi, M. and Zhou, G., Preparation of PS/TiO_2 core-shell microspheres and TiO_2 hollow shells, *J. Mater. Sci. 38*, 4911, 2003.

[11] Wan, Y. and Zhao, D.Y., On the Controllable Soft-Templating Approach to Mesoporous Silicates, *Chem. Rev.*, 107, 2821, 2007.

[12] Zhang, Q., Dandeneau, C. S., Zhou, X. and Cao, G., ZnO Nanostructures for Dye-Sensitized Solar Cells, *Adv. Mater.*, 21, 4087, 2009.

[13] Wang, L., Liu, X., Lu, G.. and Wang, Y., and Evolution of SnO_2 nanoparticles into 3D nanoflowers through crystal growth in aqueous solution and its optical properties, *Mater. Chem. Phys.*, 127, 114, 2011.

GROWTH OF ZnO NANOWIRES FROM VAPOUR PHASE

J.K. Radhakrishnan, S. Neethu and Neha Srivastava[1]

Defence Bioengineering and Electro medical Laboratory (DEBEL), DRDO, C.V. Raman Nagar, Bangalore
[1]National Institute of Technology, Kurukshetra
E-mail: jkrkrish@yahoo.co.in

ABSTRACT

ZnO nanostructures are technologically important material systems in view of its semiconducting properties and excellent optoelectronic characteristics. In this paper we will discuss the results of Growth of ZnO nanowires from vapour phase through VLS/VS mechanism. The resulting nanostructures were studied primarily with scanning electron microscopy, aided further by EDS and X-ray diffraction measurements. Our preliminary growth experiments at growth temperatures of 500C, show predominantly crawling growth. The typical dimension of the grown ZnO nanowires are; 50 nm diameter and 2-5 μm length. Though the growth of these wires are in random directions, some of these wires exhibit a preference to join back at small sized gold dots. Attempts will be made to explore this further, and exploit it to achieve directed growth between two chosen gold regions, in order to realize device structures.

INTRODUCTION

ZnO nanostructures are an exciting material system to study, because of the ease of nanostructure growth and the exotic diverse nanostructures that can be grown. It is a technologically important material in view of its semiconducting properties and excellent optoelectronic characteristics.

V. Rajendran, K. Saminathan, P. Paramasivam and K.E. Geckeler (eds.)
Nano Materials Synthesis and Characterisation, pp. 209-214 (2012).

ZnO nanostructures have been prepared through a large number of routes, ranging from wet-chemical synthesis to growth from vapor phase. An advantage that accompanies the growth from vapor phase are; it can result better purity of the grown materials, and besides by using suitable templates, the device structures can be fabricated in-situ, during growth. In this paper we will discuss the results of Growth of ZnO nanowires from vapor phase through Vapor Liquid Solid (VLS)/Vapor Solid (VS) mechanism. Si-wafers with its native oxide, were sputtered with Gold film of different thicknesses, and annealed at 880C, in Argon atmosphere, to break them into nano islands. ZnO nano wires were grown over these Gold islands, at growth temperatures of 500C, by Carbothermal reduction of ZnO. The resulting nanostructures were studied primarily with scanning electron microscopy, aided further by EDS and X-ray diffraction measurements.

EXPERIMENTAL DETAILS

(100) oriented n-type Si-wafers (carrier conc. 10^{+18}) with its native oxide, were sputtered with Gold films of different thicknesses. These films were annealed at 880C for 60minutes in a single zone furnace (shown in Fig 1), under Argon flow, to break them into nano islands. These annealed samples were used for ZnO wire growth, by Carbothermal reduction of ZnO.

A horizontal, tubular, single Zone furnace was used for the growth experiments (Fig. 1). Equal amounts of ZnO powder (Alfa Aesor), and Graphite powder (Alfa Aesor), were ground together in Mortar and Pestle, and approximately 1.5 gram of this mixture is placed in an alumina boat, inside the quartz reactor, at a position corresponding to temperature 950-960C. Down stream of the tube, the substrate with gold film (already annealed), is kept over a flat quartz boat, at a location corresponding to temperature 500-520C. After an initial flushing with Argon, the furnace is heated to the required temperature in 60minutes of under Argon flow. After leaving for a period of 40min for the reaction at this temperature, the furnace is switched off, and cooled to RT under argon flow, before removal of the samples.

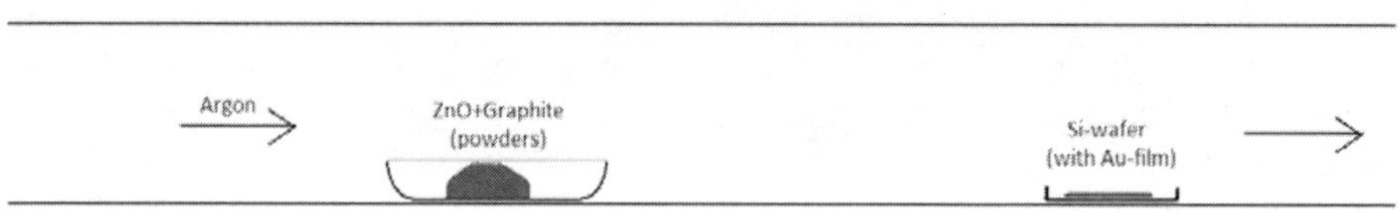

Fig. 1: Schematic of the Reactor Used for the Growth of ZnO Nanowires from Vapor Phase

RESULTS AND DISCUSSION

Figure 2 (a,b,c) show the scanning electron micrographs of the god islands that form upon annealing the Silicon wafer sputtered with Gold film. Figure 2(d) shows the Energy Dispersive Spectra (EDS) for scan over 50 nm × 50 nm area gold in Fig. 2(c).

It can be seen that, the annealing at 900C, breaks the continuous gold film into faceted gold particles of size in the range 500 nm to 100 nm. The hexagonal faceting of gold is a known phenomena in annealed gold films. The EDS spectra taken over the islands indicates mainly gold, and a small amount of Silicon (indicating probably the formation of some amount of Gold-silicide).

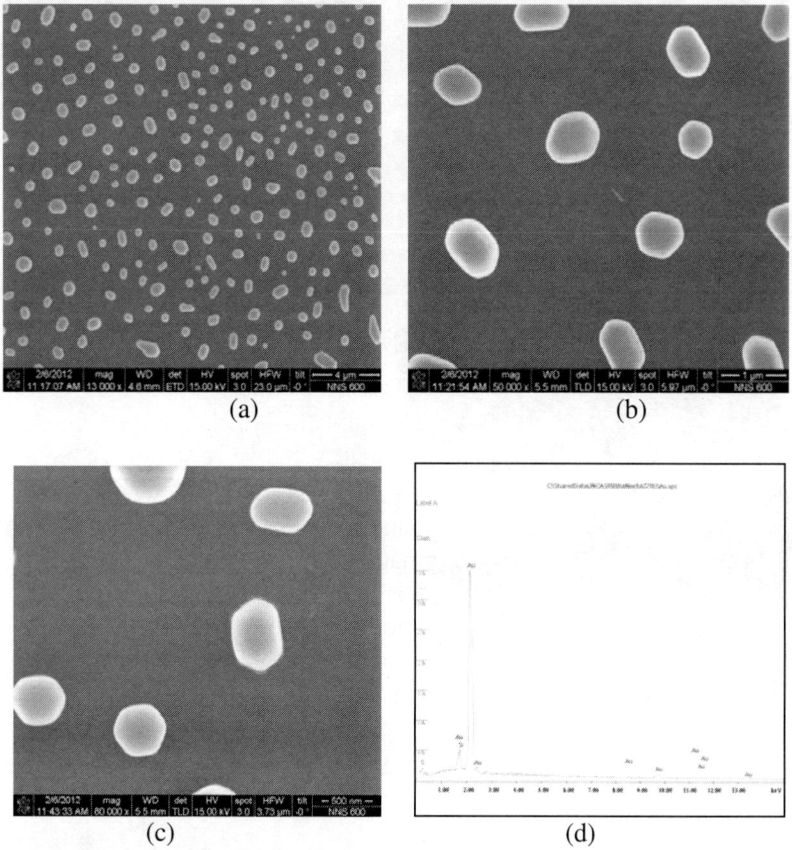

Fig. 2 (*a, b, c*): SEM Images of Gold Particles Formed Upon Annealing the Silicon Wafer Sputtered with Gold Film (at Different Magnifications). Energy Dispersive Spectra (EDS) for Scan over 50 nm x 50 nm area of Gold Particle in (c).

Figure 3(a,b,c) show the scanning electron micrographs of ZnO Nanowires grown by Carbothermal reduction of ZnO. Figure 3(d, e) show the Energy Dispersive Spectra scanned over 50 nm × 50 nm central areas of the Figure 3(b,c).

It can be seen that nanowires of approximate dia 50nm, grow from faceted gold particles and grow predominantly horizontally in the plane of the wafer, growing randomly in all directions, some of them terminating at very tiny particles of diameter few nm. Most of the nanowires seem to start from the edge of the faceted gold particles, grow in random directions and even change direction during growth. The length of the nanowires for current reaction times used in our experiments is in the range 2 to 5 microns.

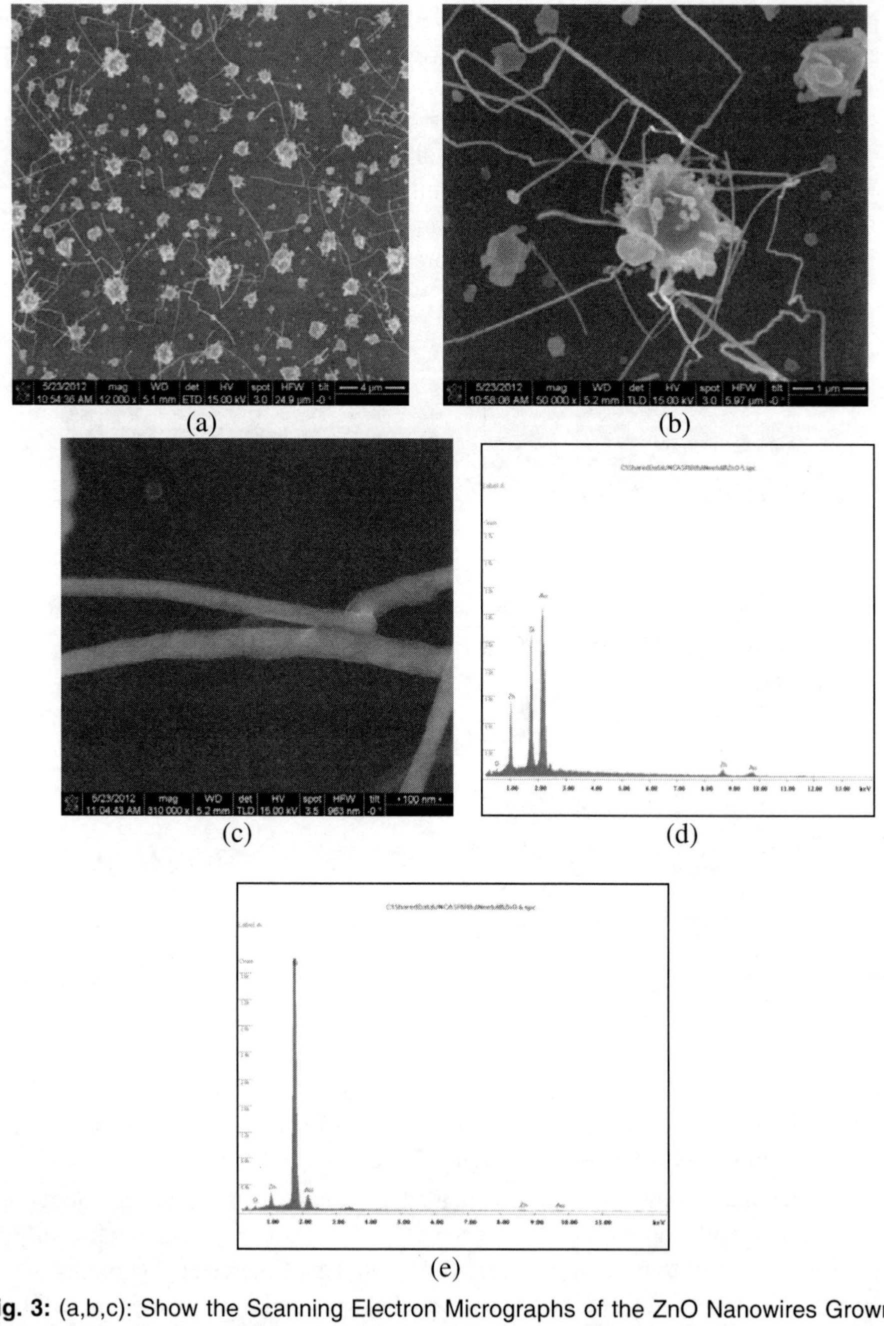

Fig. 3: (a,b,c): Show the Scanning Electron Micrographs of the ZnO Nanowires Grown by Carbothermal Reduction of ZnO. 3(d, e): Show the Energy Dispersive Spectra Scanned over 50 nm x 50 nm Central Areas of the (b,c).

The EDS spectra on the nanowires (Fig: 3(e)), shows the presence of both Zn and O, though not in good stoichiometry, which seems to be the normal case for all ZnO grown from vapor phase (this can be improved further by providing Oxygen along with the Argon in the carrier gas, during the growth). The presence of the gold in the EDS spectra of the nanowires (Fig 3 (e)) indicate that some amount of gold is dissolved in the ZnO nanowires. The presence of Silicon is from the Si-wafer in the background behind the nanowire. However the EDS spectra (Fig 3(d)) over the gold particle from which the nanowires grow, shows predominantly gold, besides Silicon and ZnO too (which are present there).

Figure 4 (a,b,c), show the X-ray diffraction spectra taken over the samples. The dominant peak is the (004) Si peak (Fig 4(a)). An expansion of the weaker peaks show (Fig 4(b,c), (111) Au reflection and (100) ZnO reflection.

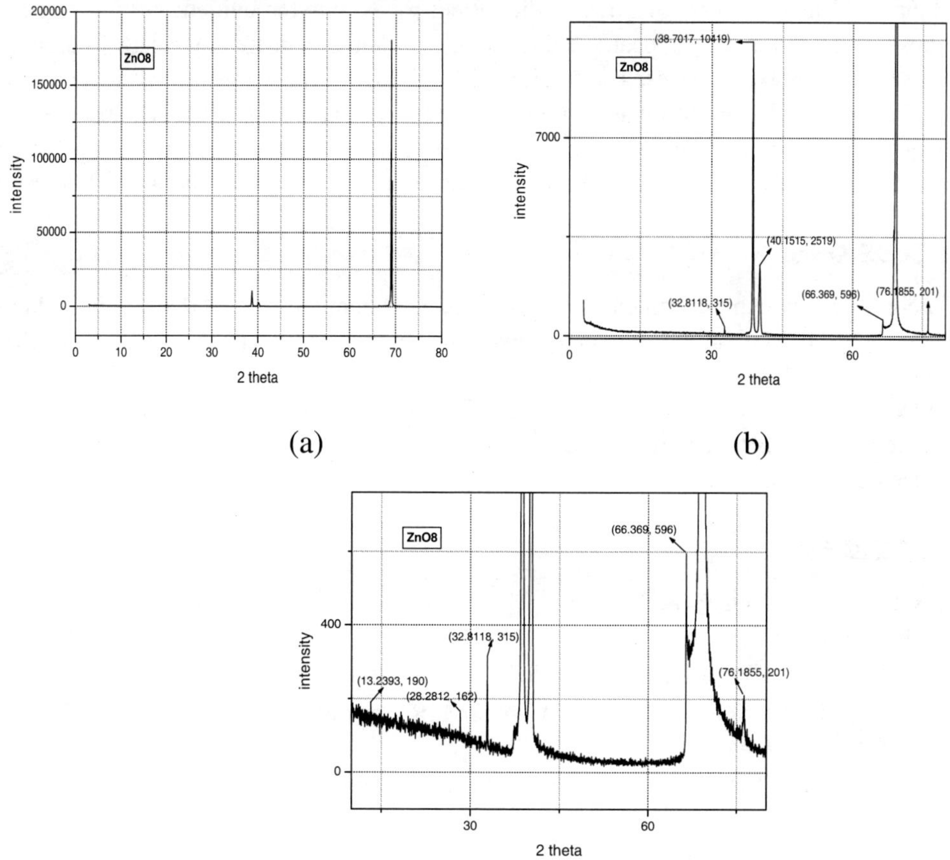

(c)

Fig. 4: (a) X-ray Diffraction Spectra taken over the Samples, (b,c): An Expansion of the weaker Peaks

In a comprehensive study on the growth of patterned ZnO wires using gold templates, it has been found that the main factors affecting the crawling growth are the gold dot size and the temperature [2-4]. While 3 nm thick gold dots resulted in single vertical wires, 5 nm thick dots show additional crawling growth at the base of the vertical ones [2], the reason being thicker gold dots break into multiple dots at elevated temperature, and the dots one the edge catalyse the crawling growth [2,3]. The effect of temperature was that, the crawling growth was preferred at higher growth temperatures of 660C (growth was completely vertical at 640C).

In our case, the presence of purely the horizontal crawling growth, without the presence of vertical wires, and at the low growth temperatures of 500C used in our experiments, indicate the interplay of some other growth mechanisms in these processes. Besides, A good examination of all the SEM images of our nanowires didnot not reveal any the evidence of Au particle in their tips (which will usually appear as shiny particles). Neither is there any faceted morphology to the grown Zinc Oxides wires. The traditional VLS growth mechanism for ZnO reported in literatures requires that Au droplets be found in the tips of nanorods/wires. The absence of the shiny gold tips in the our ZnO nanowires indicate the growth is via VS growth mechanism, and the growth of ZnO from vapor phase at lower temperatures is usually due to VS growth.

CONCLUSIONS

Our preliminary growth experiments on Growth of ZnO nanowires by carbothermal reduction of ZnO, show predominantly crawling growth, at the chosen growth temperatures of 500C. Though the growth of these wires are in random directions, some of these wires exhibit a preference to join back at some very small gold dots. Attempts will be made to explore it further, to exploit this to achieve a directed growth between two chosen gold regions in order to realize device structures.

REFERENCES

[1] R.T. Rajendra Kumar, E. McGlynn, M. Biswas, R. Saunders, G. Trolliard, B. Soulsetin, J.R. Declere, J.P. Mosnier and M.O. Henrym, *Jl. Appl Physics*, 104 (2008) 084309.

[2] H.J.Fan and M.Zacharias, J. Mater. Sci. Technol, 24 (2008) 589.

[3] P. Yang, H. Yan, S. Mao, R. Russo, J. Johnson, R. Saykally. N. Morris, J. Pham, R. He and H. Choi, *Adv. Funct. Mater.*, 12 (2002) 323.

[4] J. Song and S. Lim, *J. Phys. Chem.* C 111 (2007) 596.

SYNTHESIS AND CHARACTERIZATION OF PURE NANOCRYSTALLINE $LiCoO_2$ PARTICLES

D. Narsimulu, K. Hari Prasad, M. Venkateswarlu[1] and N. Satyanarayana

Department of Physics, Pondicherry University, Pondicherry
[1]R&D, Amaraja Batteries, Thirupathi
E-mail: nallanis2011@gmail.com

ABSTRACT

Rechargeable Li-ion batteries have become the most promising power sources for variety of portable electronic devices like cellular phones laptop computers, iPod, etc. $LiCoO_2$ is the most commercialized cathode material, because of its easy preparation, simple structure, high energy density, low cost, etc. Recently, nano particles show better electrical and electrochemical properties and hence, could develop improved capacity, energy density and better cycle life Li ion batteries. Synthesis processes play an important role in developing various size and shape of the nanoparticles. In the present work, simple and low cost combustion process was employed and synthesized the lithium cobalt oxide nanoparticles. Thermal behavior, crystalline phase, structure formation and electrical conductivity of the $LiCoO_2$ particles were investigated through the TG/DTA, XRD, FTIR and impedance spectroscopy respectively.

INTRODUCTION

Rechargeable Li-ion batteries with high energy density have become the most promising power sources for variety of portable electronic devices. Lithium transition metal oxides $LiCoO_2$, $LiNiO_{-2}$, $Limn_2O_4$, have been studied extensively studied has cathodes of the rechargeable lithium batteries high output voltage and high energy density with long cycle life [1]. Among

V. Rajendran, K. Saminathan, P. Paramasivam and K.E. Geckeler (eds.)
Nano Materials Synthesis and Characterisation, pp. 215-220 (2012).

these oxides, $LiCoO_2$ has been used as cathodes since 1990, due to its ease of manufacture, good capacity retention, favorable rate capability and safety.

$LiCoO_2$ is well known metal oxide for Li-ion batteries because it is easily prepared compared with other oxides. The most common used methods for preparation of metal oxides is solid state reactions, however, it is involve high temperature and long heat treatment time. Other methods such as hydrothermal, sol-gel, Co-precipitation process have been used to synthesize $LiCoO_2$ [2]. The combustion synthesis is another method for oxide preparation and it is reaction between the metal nitrates and appropriate fuels. In this process, exothermic reaction is enough to prepare the several materials. In other words, heat required to drive the chemical reaction is obtained from the reaction itself and not for the external source [3, 4]. During the combustion reaction, several gases based on the oxygen, hydrogen, nitrogen are produce. Other advantages of the combustion synthesis are the formation of fine particles powder with large surface area and small particle size [5].

In this paper, we present the synthesis of $LiCoO_2$ by combustion process, crystallite size and electrical conductivity of $LiCoO_2$ was measured.

EXPERIENTAL

For the combustion synthesis, $LiNO_3$ (99.8%, S.d. Fine-chem.ltd.) and Co $(NO_3)_2.6H_2O$ (GR grade, Merck, India) were used as metal nitrates and citric acid anhydrous (GR grade, Merck, India), was used as fuel.

For the synthesis of $LiCoO_2$ by combustion process, the required stoichiometric quantities of metal nitrate solutions were mixed with citric acid solution, by keeping meal nitrates to citric acid ratio 1:1, under constant stirring condition the resultant pink color solution was evaporated at 80°C for 6 hours and it turned into pink color solution. Further, the resin was dried at 150°C, for 24 hours and polymeric intermediate was obtained. The obtained intermediate was grounded and calcined at various temperatures to obtain nanocrystalline $LiCoO_2$ powder.

RESULT AND DISCUSSION

TG/DTA

The TG/DTA thermogram for polymeric intermediate of lithium cobalt oxide are shown in Fig. 1. The weight loss about 6% between the 55°C and 147°C with endothermic (Heat absorption during the reaction) peak observed in DTA curve is due to the removal of adsorbed moisture. The sharp exothermic (Heat emission during the reaction) peak observed at 167^0C with major weight loss about 16% is due to the combustion of polymeric intermediate and it was completed 197°C. The weight loss 13% between 355°C and 377°C is due to the decomposition of organic compound accompanied by exothermic peak in DTA curve. No more weight loss is observed after 450°C, it confirms the complete decomposition of Citric acid derivative as well as formation of the $LiCoO_2$. Hence, 500°C is optimized as the calcining temperature for the polymeric intermediate of the $LiCoO_2$ sample for the complete removal of organic derivatives.

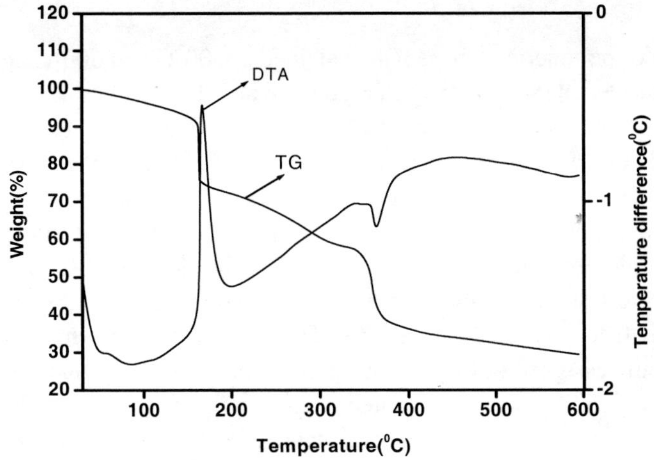

Fig. 1: TG/DTA Thermograms of Polymeric Intermediate of $LiCoO_2$ Sample

XRD

Figure 2 shows the XRD patterns of $LiCoO_2$ sample calcined at different temperatures. X-ray diffraction pattern of polymeric intermediate of $LiCoO_2$ does not shows any peaks, it indicate that polymeric intermediate may be in amorphous nature. From Fig. 2, the crystalline phase began to form at 300°C and pure crystalline phase of the $LiCoO_2$ was obtained at 500°C and these results supported by TG/DTA results. The observed peaks at 19°, 37° and 45° respectively for (003), (101) and (104) planes confirm the formation of crystalline $LiCoO_2$ phase. The crystallite size of the $LiCoO_2$ was calculated by using Scherer's formula and it is found to be 51.98 nm. Hence, XRD results confirm the formation of nanocrystalline $LiCoO_2$ phase.

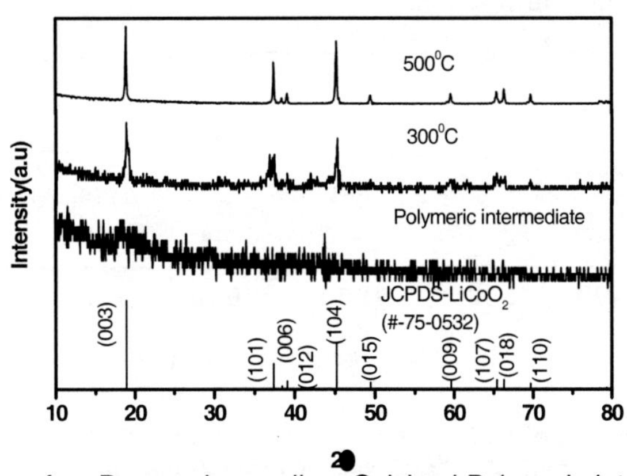

Fig. 2: XRD Patterns of as Prepared as well as Calcined Polymeric Intermediate of $LiCoO_2$ Sample at Different Temperatures along with JCPDS Data

FTIR

The FTIR spectra of polymeric intermediate, calcined at 300°C and 500°C are shown in Fig.3. For polymeric intermediate, the observed broad peak at 3400 cm^{-1} is assigned to the structural OH group (due to the adsorbed moisture) vibrations. The sample was thermally treated at 80^0C and also the band at 1681 cm^{-1} assigned to the bending mode of the water molecule [6]. The IR peak at 1725 cm^{-1} is corresponds to the asymmetric vibrations of bridging COO^- group and symmetric vibrations bridging COO^- at 1389 cm^{-1}. The IR peak observed at 1621cm^{-1} symmetric stretching vibration of COO^- groups [6, 7]. The IR peak observed at 1207 cm^{-1} and 1058 cm^{-1} respectively are due to the asymmetric and symmetric stretching vibrations of C-O-C group, which confirm the esterification. The FTIR spectra at 300^0C, the observed low intense IR peak at 3463 cm^{-1} assigned to the stretching frequencies of OH groups. groups [6]. The IR peak at 603 cm^{-1} and 507 cm^{-1} are attributed to the asymmetric stretching modes of (CoO_6) octahedral in $LiCoO_2$ structure. So, it confirm that $LiCoO_2$ phase formation and it begans to form at 300^0C and it is consisted with XRD and TG/DTA results. The intensity of the FTIR peak related to the organic derivative are decrease, increases in calcining temperature, indicate the removal of large portion of residual molecule and disappeared completely, when the polymeric intermediate was calcined at 500°C, which confirm the formation of pure phase of nanocrystalline $LiCoO_2$ at 500°C and it is consistant with XRD and TG/ DTA results.

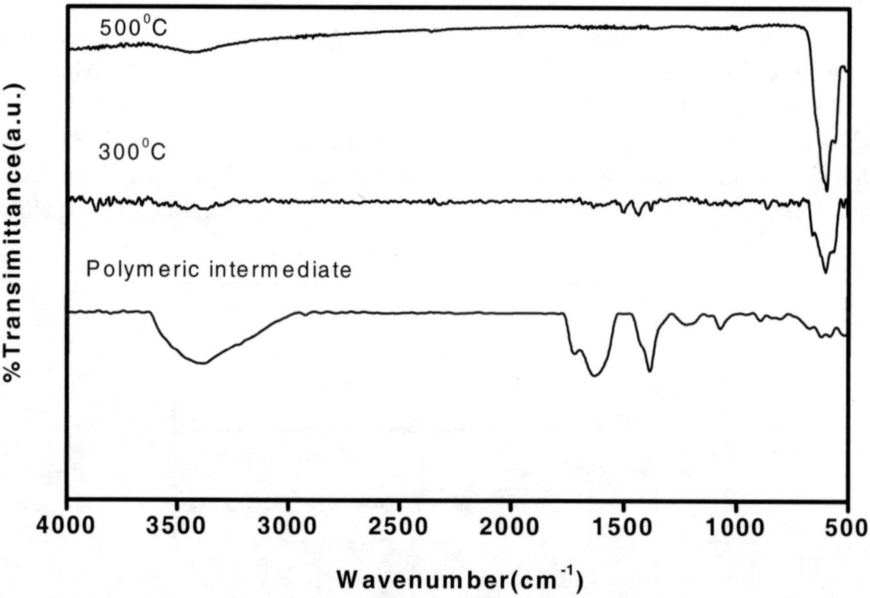

Fig. 3: FTIR Spectra of as Prepared as well as Calcined Polymeric Intermediate of $LiCoO_2$ Sample at Different Temperature

ELECTRICAL CONDUCTIVITY

Fig 4, Shows the impedance plot obtained at 300°K for the $LiCoO_2$ sample. The impedance plot is fitted using win fit software by adjusting the resistance and capacitance values. From the fit, best grain boundary resistance (R_{gb}), bulk resistance (R_b), grain boundary capacitance (Cgb) and bulk capacitance (C_b) values as well as the electrical behavior of the sample in the form equalent circuit are obtained. The conductivity of $LiCoO_2$ sample was calculated using bulk and grain boundary resistance (R_b and Rgb) and dimension of the pellets. The grain interior and grain boundary conductivities of nanocrystalline $LiCoO_2$ are found to be 8.185×10^{-6} ohm^{-1} cm^{-1} and 5.325×10^{-6} $ohm^{-1}cm^{-1}$ respectively. The total conductivity of the $LiCoO_2$ is found to be 1.3509×10^{-5} ohm^{-1} cm^{-1}.

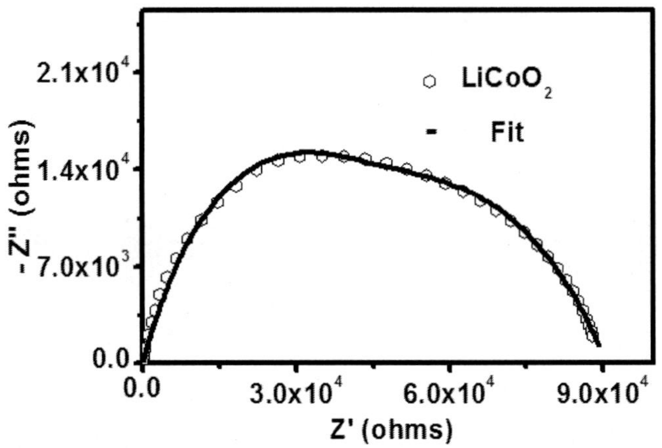

Fig. 4: Impedance Plots (Z' and Z") of $LiCoO_2$ Sample.

CONCLUSION

$LiCoO_2$ powder was synthesized by combustion process. The crystallite size of the $LiCoO_2$ is calculated by using Scherer's formula and it is found to be 51.98 nm. The electrical conductivity of the $LiCoO_2$ is found to be 1.3509×10^{-5} ohm^{-1} cm^{-1}

ACKNOWLEDGEMENT

Dr. N. Satyanarayana gratefully acknowledges, DST-NANOMISSION, Government of India for providing financial support in the form of major research project DST-NANO sanction letter No: SR/S5/NS-49/2005, dated on 30-07-2010.

REFERENCES

[1] M.M. Thackeray, J.O. Thomas and M.S. Whittingham, *MRS Bull.* (2000) 39.

[2] K. Brandt, *Solid State Ion.* 69 (1994) 173.

[3] A. Varma and J.P. Lebrat, *Chem. Eng. Sci.* 47 (1992) 2179.

[4] Y. Zhang and G.C. Stangle, *J. Mater. Res. 9* (1994) *1997*.[5] P. Ravindranathan, S. Komarneni and R. Roy, *J. Mater. Sci.*12 (1993) 369.

[6] G. Socrates, Infrared and Raman Characteristic Group Frequencies, John Wiley and Sons, New York (2001).

[7] G.V. Rama Rao, D.S. Surya Narayana, U.V. Varadaraju, G.V. Rao, S.Venkadesan, J. *Alloys Compd*, 217 (1995) 200-208.

STUDIES ON THE SYNTHESIS AND CHARACTERIZATION OF CO-PRECIPITATED NANOCRYSTALLINE Zn-Pb-O BULK SYSTEM

S.I. Inamdar and K.Y. Rajpure

Electrochemical Materials Laboratory, Department of Physics, Shivaji University, Kolhapur
E-mail: chb_phy@unishivaji.ac.inrajpure@yahoo.com

ABSTRACT

Zinc lead oxide samples are synthesized by a simple chemical precipitation method using zinc nitrate & lead nitrate precursors in 2:1 proportion. The mixed orthorhombic and tetragonal phases of Zn-Pb-O system have been confirmed from structural analysis. Scanning electron microscopy reveals the surface morphology of prepared Zinc lead oxide powder. The composition and structure of the Zn-Pb-O compound have been investigated by Raman spectroscopy. Frequency dependent dielectric constant (ε') shows the usual dispersion behavior, while linear variation of AC conductivity (σ_{ac}) with frequency confirms the conduction due to small polaron.

INTRODUCTION

The oxides of Zn [1] and Pb [2] are among the most studied and extensively used semiconducting materials. ZnO and PbO have widespread applications in many fields. Lead oxide decomposes well below the 400°C at atmospheric pressure. For this reason, zinc lead oxide compounds have not previously been studied [3]. In this present work we used co-precipitation method for the synthesis of Zn-Pb-O sample. In co-precipitation, physical properties of synthesized material depend on the pH of solution, annealing temperature and time. Keester et al. [3] studied the phase relations in the system PbO-ZnO-O at 4 kbars oxygen partial pressure. And discovered

V. Rajendran, K. Saminathan, P. Paramasivam and K.E. Geckeler (eds.)
Nano Materials Synthesis and Characterisation, pp. 221-226 (2012).

ternary phases, viz., Zn_2PbO_4, and $Zn_xPb_{1-x}O$ and characterized by using X-ray powder diffraction data. The physical properties of Zn-Pb-O nanoparticles are useful due to its high dielectric constant and co-existence of various phases may be applied in various applications. In the present investigation, we have made efforts to prepare well crystallized Zn-Pb-O sample with co-precipitation method.

EXPERIMENTAL DETAILS

The zinc lead oxide samples were prepared by Co-precipitation method using AR grade Zinc and lead nitrates (1:2 proportions) in double distilled water. The pH of the solution was maintained to neutral by adding aqueous ammonia to preserve hydroxide phases of Zn and Pb. White precipitate formed was washed with double distilled and dried to get white powder. The white powder formed was sintered at 400, 600, 800 and 1000°C (rate of 5°C/min) for 4 hours followed by natural cooling down to room temperature inside the furnace. These compositions were further mixed with polyvinyl alcohol as a binder and pressed into pellets of 15 mm diameter and 2–4 mm thickness using a hydraulic press.

The structural characterization of samples were analyzed by X-ray diffractometer (Philips, Model PW-3710) using Cu K_α radiation (λ = 1.54056 Å). The surface morphology was investigated using a JEOL JSM- 6360 scanning electron microscope (SEM). The Raman analysis was carried using Jobine Yvon HR 800 UV system. The frequency dependence of dielectric permittivity (ε'), dissipation factor (tanδ) and AC conductivity (σ_{AC}) were studied using a precision LCR meter bridge (HP 4284 A) in the range from 20 Hz to 1 MHz.

RESULTS AND DISCUSSION

X-ray Diffraction Studies

X-ray diffraction patterns of zinc lead oxide samples 1 are analyzed by JCPDS data card nos. 23-1496 (orthorhombic Zn_2PbO_4) and 23-1497 (Tetragonal $Zn_xPb_{1-x}O$). Diffraction patterns reveal that samples contain mixed ternary phases, viz, Zn_2PbO_4 and $Zn_xPb_{1-x}O$. Characteristic peaks, viz., (130), and (221) of Zn_2PbO_4 have been detected. For $Zn_xPb_{1-x}O$ phase reflections are (004), (104), (114), (212), (204), (105) and (202). This is due to crystal reorientation effect occurred due to sintering. So the stable phases Zn-Pb-O system prepared by co-precipitation in the present work could be the mixture of polycrystalline Zn_2PbO_4, $Zn_xPb_{1-x}O$ and traces of oxides of Zn and Pb in amorphous form.

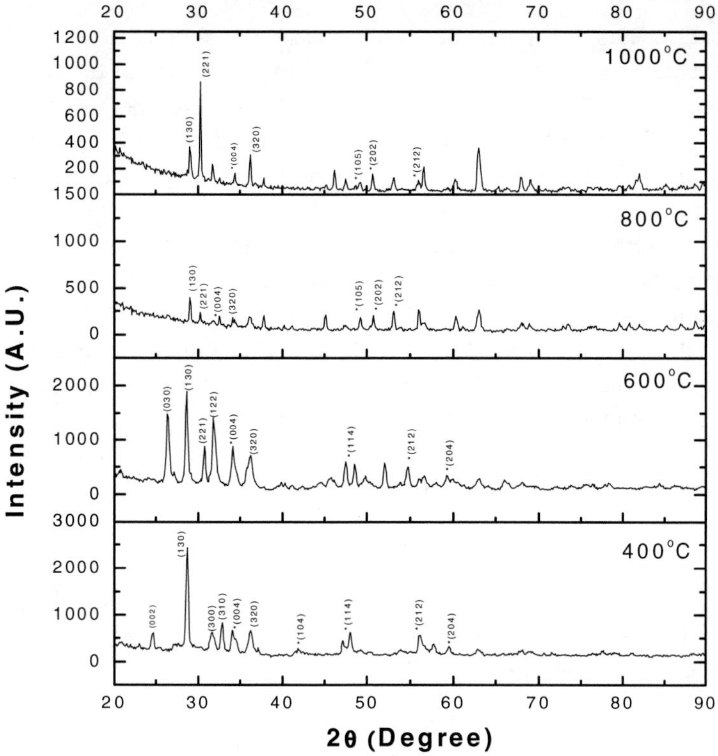

Fig. 1: X-ray Diffraction Patterns of Zinc Lead Oxide Samples Sintered at Temperature 400-1000 °C

Surface Morphology

The Fig. 2 shows SEM images for the samples sintered at 400°C-1000°C. The samples show porous surface nature due to evolution of lead for sintering temperatures 400, 600 and 800°C. Elongated grains are seen clearly in phase of Zn-Pb-O sample sintered at higher temperature 1000°C.

Fig. 2: Scanning Electron Micrographs of Zn-Pb-O Sample Sintered at Temperature 400-1000°C

Raman Spectroscopy

The intense peaks occurs in Raman spectra of the samples sintered at 400-1000°C shown in Fig. 3. The most intense peaks observed at 78cm^{-1} in the spectra of the samples sintered at temperature 400, 800 and 1000°C shown in Fig 4 reveals the presence of PbO in the material. Peaks observed at 123 and 143 cm^{-1} for the sample sintered at temperature 600°C shows the orthorhombic phase of lead oxide. From fig it is seen that the sample sintered at 600°C and 1000°C exhibit the 290 cm^{-1} position which corresponds to the B_1g phonon mode of orthorhombic lead monoxide [4]. Appearance of peak near 386 cm^{-1} is attributed to A_1 (TO) mode of ZnO. In aman spectra there is one more peak observed at position 408 cm^{-1} for Zn-O bond. The aman spectra of the samples sintered at 400, 600 and 800°C shows sharp peaks at 1041, 1039 and 1040 cm^{-1} respectively. D.U. Wiechert et al. reported the compound Hydro cerussite (white lead) at around 1050 cm^{-1}. [5] The theoretically predicted phonon frequencies [4-5] and the corresponding phonon mode assignments are listed in Table 1.

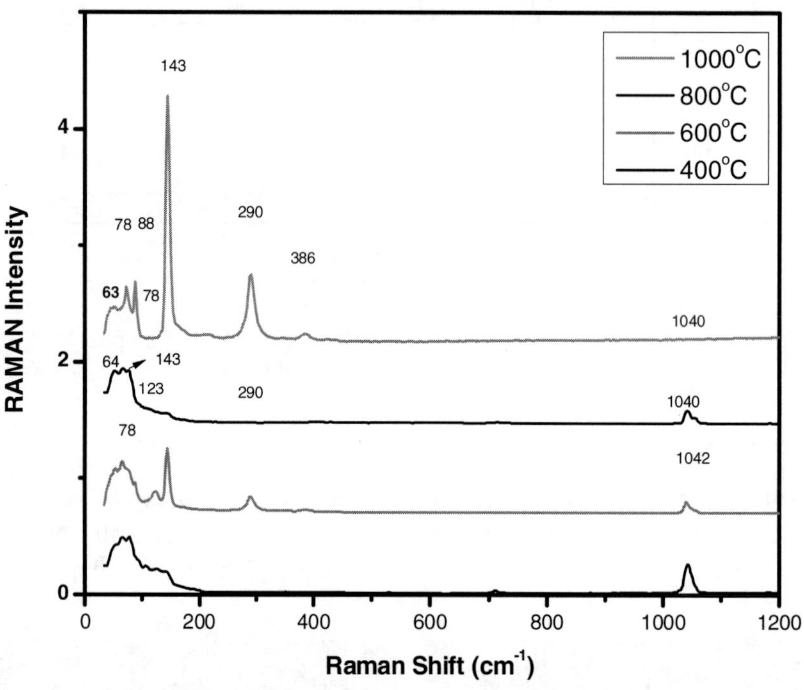

Fig. 3: RAMAN Spectra for Zn-Pb-O Sample Sintered at Temperature 400-1000°C.

Mode assignment Theoretically predictedfrequencies (cm^{-1}) Experimentally obtained frequencies (cm^{-1})

Table 1: Theoretically Predicted and Experimentally Obtained Phonon Frequencies in Raman Spectrum of the Samples Sintered at Temperature 400-1000°C.

Mode Assignment	*Theoretically Predicted Frequencies (cm^{-1})*	*Experimentally Obtained Frequencies (cm^{-1})*
E_g	81.1	78.47
B_{1g}	290	290
A_1(TO)	379	386
E_1(TO)	410	403
-	1050	1040, 1041, 1042

Dielectric Properties

Figure 4 (a) shows that dielectric constant decreases abruptly at lower frequencies and remains constant at higher for samples sintered at 400, 600, 800 and 1000°C respectively. The pattern at lower frequencies may be attributed to different types of polarization (electronic, atomic, interfacial and ionic, etc.). At higher frequencies it arises due to the contribution from electronic polarization. The frequency dependence of the ac electrical conductivity σ_{AC} measured at room temperature is illustrated in Fig. 4 (b) shows an increasing trend with the increase in frequency for all the samples [6-9]. The low-frequency ac conductivity is related to the resistive grain boundaries, whereas the high-frequency ac conductivity is due to the conductive grains.

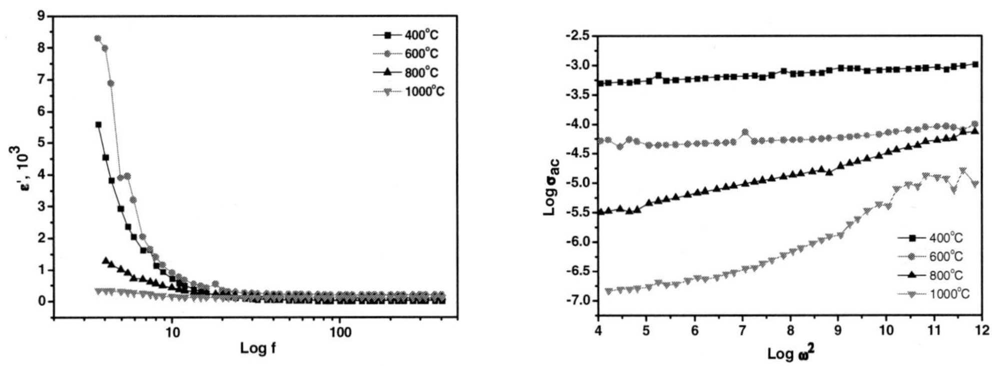

Fig. 4: Variation of Room Temperature (a) Dielectric Constant with Frequency (b) AC Conductivity with Frequency

CONCLUSIONS

Crystalline Zinc Lead oxide was successfully prepared by a chemical co-precipitation method for the first time. The formation of mixed orthorhombic and tetrahedral phases of Zn-Pb-O system has been confirmed from X-ray diffraction study. The two polymorphic forms of Zn-Pb-O bulk system, i.e. the orthorhombic and tetragonal phase can be well distinguished by their Raman spectra. Observed dielectric dispersion with frequency has been explained on basis of electron hole hopping mechanism.

REFERENCES

[1] Sun, J., Liu, F.J., Huang, H.Q., Zhao, J.W., Hu, Z.F., Zhang, X.Q. and Wang , Y.S., Fast response ultraviolet photoconductive detectors based on Ga-doped ZnO films grown by radio-frequency magnetron sputtering, *Appl. Surf. Sci.* 257 (2010) 921-924.

[2] Clarke, J.R. and Greene, J.E., Reactively evaporated photoconductive PbO: Phase transformations induced by water vapor, *Thin Solid Films*, 66 (1980) 339.

[3] Keester, K.L. and White, W.B., Phase relations in the system PbO -ZnO-Oat high oxygen pressures, *J. Inorg, Nucl. Chem.* 33 (1971) 3269-3284.

[4] Wiechert, D.U., Grabowski, S. P. and Simon, M., Raman spectroscopic investigation of evaporated PbO layers, *Thin Solid Films* 484 (2005), 73.

[5] Alim, K.A., Fonoberov, V.A., Shamsa, M. and Balandina, A.A., Micro-Raman investigation of optical phonons in ZnO nanocrystals, *J. Appl. Phys.* 97 (2005) 124313.

[6] Ahmed, M.A., El Hiti, M., A., *J. Physique III* 5 (1995) 775.

[7] Koops, C.G., On the Dispersion of Resistivity and Dielectric Constant of Some Semiconductors at Audio frequencies, *Phys. Rev.* 83 (1951) 121.

[8] K.W. Wagner, *Ann. Phys.* 40 (1973) 317.

SYNTHESIS AND CHARACTERIZATION OF MnO_2 NANO PARTICLES FOR HYDROGEN PEROXIDE SENSING

C. REVATHI, S. BAGYALAKSHMI AND R.T. RAJENDRAKUMAR

Advanced Materials and Devices Laboratory, Department of physics, Bharathiar University, Coimbatore
E-mail: rtrkumar@buc.edu.in

ABSTRACT

Manganese dioxide nanoparticles were prepared by wet chemical route at low temperature 40°C, 80°C, 100°C respectively. MnO_2 particles were characterized by XRD, UV-Vis absorption, and cyclic voltammetry. From UV-Vis absorption spectra analysis it was observed that pure MnO_2 nanoparticles shows an absorption peak edge in the range 555-618 nm. The electrochemical property of MnO_2 particles was investigated by Cyclic Voltammetry analysis. Cyclic voltammetry detection of H_2O_2 with different concentration (0.01 ml, 0.1 ml) was carried out by using Pt electrode as the working electrode. MnO_2 nanoparticles prepared at 40°C, 80°C, 100°C show H_2O_2 sensing capability with anodic peak current value 28.1μ A , 38.1μA, 43.9ìA, respectively (for 0.01 ml H_2O_2). This indicates that sensing capability of nanoparticles was increased by increasing the growth temperature.

INTRODUCTION

Metal oxides are the most common and richest class of materials which possess unique physical properties due to their nanoscale size and a high density of surface sites. Although various techniques such as chemical vapor deposition, hydro thermal method, Electrodeposition have been employed to synthesize nanostructures, many of them demand stringent reaction conditions

V. Rajendran, K. Saminathan, P. Paramasivam and K.E. Geckeler (eds.)
Nano Materials Synthesis and Characterisation, pp. 227-232 (2012).

such as high/low pressure either in vapour phase, temperature, etc. Hydrogen peroxide is very important intermediate in environmental and biological reactions. The monitoring of H_2O_2 with reliable, rapid and economic method is of great significance for numerous processes. Metal oxide particles such as iron oxide, lead oxide, ruthenium oxide, cobalt oxide have been successfully used for immobilization of enzymes, and proteins and in fabrication of H_2O_2 sensor. Among the various materials, Manganese di oxide has attracted special attention due to their distinctive properties and wide applications in catalysis, ion exchange, biosensors, energy storage [1]. In recent years, researchers proved that MnO_2 materials are interesting for H_2O_2 oxidation [2] and various methods like cyclic voltammetry and amperometric method has been used for H_2O_2 sensing. In this work MnO_2 nano particles were synthesized by varying growth temperature using wet chemical approach. MnO_2 modified Pt electrode without any surfactant or ionic liquids is prepared. Tests show that MnO_2 nano particles exhibit electrocatalytic activity towards the oxidation of H_2O_2.

EXPERIMENTAL SECTION

Synthesis of MnO_2 Nanoparticles

All the chemicals used in experiment were of analytical purity. 0.2g of potassium permanganate ($KMnO_4$) was dissolved in 320 mL double distilled water. And then 10 mL of formaldehyde ($HCONH_2$) was added into above solution under stirring for several minutes and kept in static oil bath at 40°C for 6 hours. The product was collected by centrifugation and washed with doubled distilled water and ethanol to remove any possible residual reactants. Finally, the product was dried in air at 60°C for 8 hours. The same procedure was repeated at 800°C, and 100°C.

All the Electro chemical measurements were done using (Metrohm (μ Autolab-type III)) three electrodes setup Pt, Ag/AgCl, Pt wire were served as working, reference, counter electrode respectively at room temperature. For preparing MnO_2 modified Pt electrode, 0.2 mg of MnO_2 powder were dissolved in 5mL ethanol and sonicated 30 min. 2μL MnO_2- solution will be dropped on Pt electrode and dried in air at room temperature. The modified electrode was used for H_2O_2 sensing.

RESULTS AND DISCUSSIONS

Figure. 1(a) shows XRD pattern of as prepared MnO_2 powders at 40°C, 80°C, and 100°C. There were no diffraction peaks observed for the sample grown at 40°C, and 80°C shows poor crystalline quality. Whereas the sample grown at 100°C shows diffraction peak at $2\theta = 37.05°$ which is matching with JCPDS 44-099 and the crystalline size found to be 29 nm using scherrer formula.

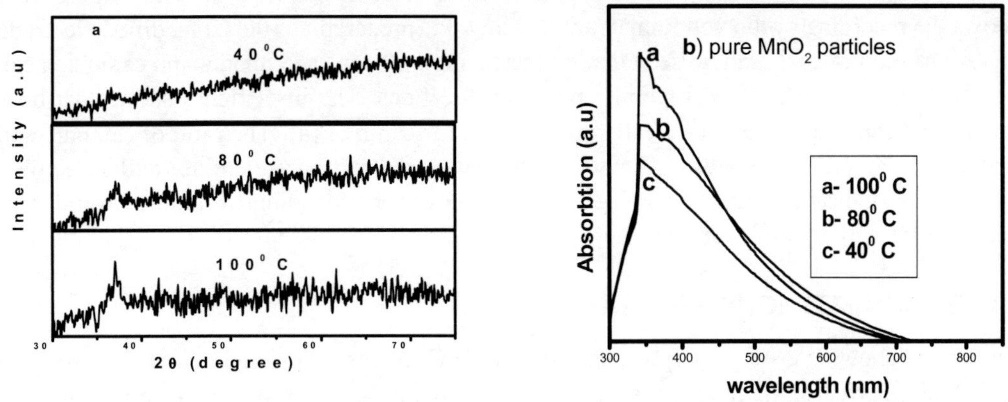

Fig. 1: (a) XRD Pattern of MnO_2 Nanoparticles (b) UV-Vis Spectrum for Pure MnO_2 Particles Prepared at Growth Temperatures 40°C, 80°C, and 100°C

Figure 1(b) shows UV-Vis spectrum for pure MnO_2 particles with the absorbance band edge range 555-618 nm [3]. The band edge absorption becomes sharper by increasing the growth temperature. This shows improvement of crystalline quality with growth temperature which is in consistent with XRD results.

Figure 2(a) shows Cyclic Voltammogram for pure MnO_2 nanoparticles measured in range of -0.4 V to 0.6 V at scan rate of 0.09 mV/s. The spectrum shows anodic peak (oxidation) potential at 0.04V, 0.07V, 0.1V, for the sample grown at 40°C, 80°C and 100°C respectively.

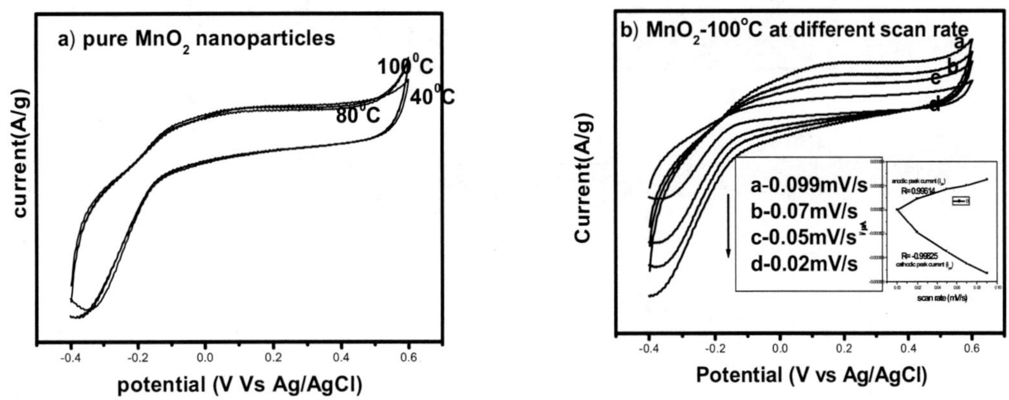

Fig. 2: Cyclic Voltammogram of (a) MnO_2/Pt Electrode which is Prepared at Different Temperature (40°C, 80°C and 100°C). (b) MnO_2/Pt Electrode in 0.1M NaOH Solution at Different Scan Rates. Inset Plots of Anodic Peak (i_{pa}) and Cathodic Peak (i_{pc}) Current vs Scan Rate (ν)

MnO_2 particles prepared at 100°C shows higher electro catalytic activity. Figure 2(b) shows the recording cyclic voltammogram of MnO_2/Pt (prepared at 100°C) modified electrode at various scan rates. It can be seen that on increasing the scan rate, anodic and cathodic peak potentials shifted towards positive and negative directions. The inset figure shows that both anodic and cathodic peaks are directly proportional to scan rate [4]. The ratio of cathodic and anodic peak current is about unity. These results over all redox process confined at the electrode surface also indicate a surface confined redox process corresponding to rapid conversion of surface film without diffusion [5].

DETECTION METHODS FOR H_2O_2 SENSING

Figure 3(a) shows UV- Vis absorption spectrum for H_2O_2 detection. On addition of H_2O_2 solution with MnO_2 particles, the intensity of absorption band edge decreased as compared with pure MnO_2 nanoparticles. This may be due to reduction of nanoparticles from the solution [6]. By increasing the growth temperature, the intensity of absorbance band edge becomes sharper and has high catalytic ability for the detection H_2O_2 solution.

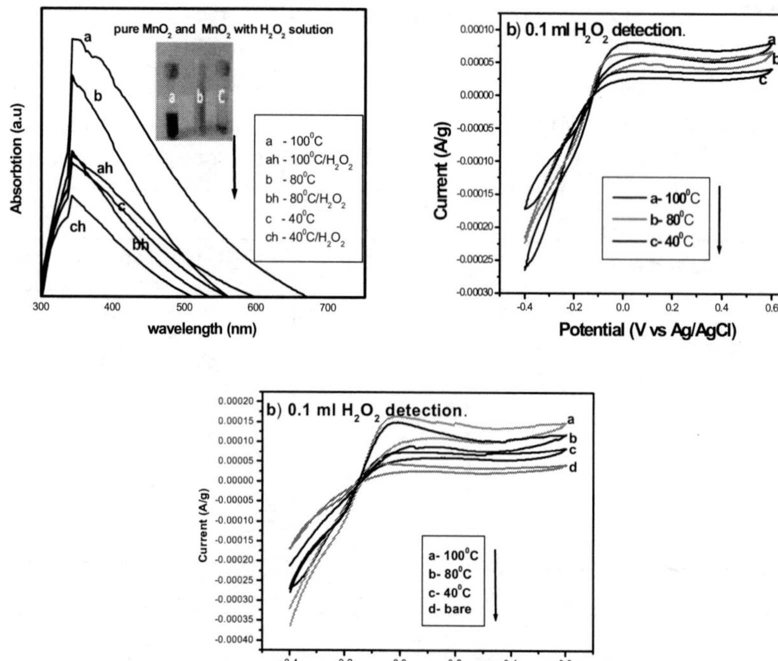

Fig. 3: (a) UV-Vis Spectrums for H_2O_2 Detection and Inset Figure Show (a) Pure H_2O_2, (b) when H_2O_2 Added (c) After Addition of H_2O_2 Solution. **(b)** C-V for MnO_2/Pt Electrode with 0.0 1 ml H_2O_2 and 0.1 ml H_2O_2 in 0.01M NaOH Solution at Scan Rate 0.09 mV/s

Figure 3(b) shows Cyclic Voltammogram for H_2O_2 detection. To verify the electro catalytic activity of MnO_2 modified Pt electrode for H_2O_2 detection, the electrochemical experiment was carried out in the presence of 0.1 ml and 0.01 ml H_2O_2 solution From this, bare electrode shows no oxidation response for H_2O_2 detection between -0.4V to 0.6V potential range. In the presence of H_2O_2, oxidative peak current of MnO_2 increased. The increased peak current value of H_2O_2 oxidation confirms that MnO_2 particles have catalytic ability for H_2O_2 detection. This result confirms manganese oxide will be suitable for hydrogen peroxide detection. It is interesting to note that rate of H_2O_2 detection increased with increase in growth temperature. Also on increasing the concentration of H_2O_2 solution, the oxidative current increased.

CONCLUSION

In conclusion, MnO_2 nanostructures were prepared by varying growth temperature using wet chemical approach. Based on our results by increasing the growth temperature the crystalline quality of the nanoparticles was improved. Electrochemical studies indicating the electrochemical behavior of MnO_2 particles increases with increasing growth temperature. The particles prepared at 100°C shows high H_2O_2 detection capability as compared with the particles prepared at 40°C, 80°C. Further, these MnO_2 nanostructures may find potential applications in non enzymatic glucose sensors.

REFERENCES

[1] Jipeng Ni, Wencong Lu, Liangmiao Zhang, Boahua Yue, Xingfu Shang and Yong Lv, "Low-Temperature Synthesis of Monodisperse 3D Manganese Oxide Nanoflowers and Their Psudocapacitance Properties", *J. Phys. Chem. C.*, 113, 54-60, 2009.

[2] Li Zhang, Zhen Fang, Yonghong Ni and Guangchao Zhao, "Direct Electrocatalytic Oxidation of Hydrogen Peroxide Based on Nafion and Microspheres MnO_2 Modified Glass Carbon Electrode", *Int. J. Electrochem. Sci.*, 4, 407-413, 2007.

[3] Taku Fujimoto, Yoshiteru Miznkoshi, Yoshio Nagata, Yasuaki Maeda and Ryuichiro Oshima, "Sonolytical Preparation of Various Types of Metal Nanoparticles in Aqueous Solution", *J. scripta mater.*, 44, 2183 – 2186, 2001.

[4] Zhi-Yuan Wu, Soundappan Thiagarajan, Shen-Ming Chen, and Kuo-Chiang Lin, "Electrochemical Preparation and Characterization of MnO2-PEDOT Hybrid Film and its Application in Electrocatalytic Oxidation of Memantine Hydrochloride", *Int. J. Electrochem. Sci.*, 7, 1230 – 1241, 2012.

[5] Soundappan Thiagarajan, Tsung hsuan Tsai and Shen – Ming Chen, "Electrochemical Fabrication of Nano MnO_2 Modified Electrode for the detection of H_2O_2", *Int. J. Eelectrochem.*, 6, 2235 – 2245, 2011.

[6] Xi – Liang Luo, Jing – Juan Xu, Wei Zhao and Hong – Yuan Chen, "A novel glucose ENFET Based on the special reactivity of MnO_2 nanoaprticles", *Biosensors and Bioelectronics.*, 19, 1295 – 1300, 2004.

PREPARATION OF COBALT–COPPER FERRITE ($Co_{0.8}Cu_{0.2}Fe_2O_4$) NANOPOWDER VIA CO-PRECIPITATION METHOD AND INVESTIGATION OF ITS MAGNETIC PROPERTIES

J. BALAVIJAYALAKSHMI, N. SURIYA NARAYANAN[1] AND R. JAYAPRAKASH[2]

Department of Physics, PSGR Krishnammal College for Women, Coimbatore, Tamil Nadu
[1]Department of Physics, Govt. College of Technology, Coimbatore, Tamil Nadu
[2]Nanotechnology Laboratory, Department of Physics, Sri Ramakrishna Mission Vidyalaya College of Arts and Science, Coimbatore, Tamil Nadu
E-mail: balavijayalakshmiroopa@gmail.com

ABSTRACT

Copper substituted Cobalt ferrite nano particles ($Co_{0.8}Cu_{0.2}Fe_2O_4$) are synthesized using co-precipitation method. Cobaltous Chloride [$CoCl_2.6H_2O$], Cupric Chloride [$CuCl_2.2H_2O$] and anhydrous Ferric Chloride [$FeCl_3$] along with sodium hydroxide [NaOH] are used as the raw materials. Cobalt-Copper ferrite sample sintered at 900ºC is subjected to X-ray diffraction to calculate the average nano-crystallite size using Debye – Scherrer formula. The FT-IR spectra of the samples are recorded to ensure the presence of metallic compounds. The magnetic properties of the copper doped cobalt ferrite nano particles are studied using Vibrating Sample Magnetometer (VSM). The morphological analysis of the sample is done using SEM.

INTRODUCTION

Nanocrystalline magnetic materials have gained considerable importance, because of their large novel physical aspects like spin canting, superparamagnetism and surface anisotropy due to their large surface to volume ratio as compared to the bulk material. Due to spin anisotropy

V. Rajendran, K. Saminathan, P. Paramasivam and K.E. Geckeler (eds.)
Nano Materials Synthesis and Characterisation, pp. 233-238 (2012).

and high coercivity at room temperature, these nano magnetic particles have wide ranging applications in high density information storage devices, transformer cores, microwave devices, magnetic fluid, etc. Among the spinel ferrites, $CoFe_2O_4$ is interesting because of its cubic magnetocrystalline anisotropy, high coercivity, moderate saturation magnetization, high chemical stability, magnetic and electrical conductivity. Many reports prove that the property of spinel ferrites can be further improved by doping MFe_2O_4 with metal elements, leading to enhanced crystal anisotropy [1]. Also the structural and magnetic properties of nano crystalline Copper ferrite $CuFe_2O_4$ are quite interesting. So a novel idea of using Cu as an additive element to Cobalt ferrite, Cobalt Copper ferrite $Co_{0.8}Cu_{0.2}Fe_2O_4$ is chosen in the present work.

To obtain nanosized spinel ferrite particles, various preparation techniques have been developed, such as hydrothermal synthesis [2–4], co-precipitation [5–6], sol–gel [7-8] and microemulsion processes [9-10]. Among the various preparation methods, co-precipitation method is chosen in the present work for preparing Co-Cu mixed ferrites.

EXPERIMENTAL

Nano crystalline Cobalt ferrites $Co_{0.8}Cu_{0.2}Fe_2O_4$ are prepared by co-precipitation method. The desired composition is obtained by taking stoichiometric amounts of Cobaltous Chloride [$CoCl_2.6H_2O$], Cupric Chloride [$CuCl_2.2H_2O$] and anhydrous Ferric Chloride [$FeCl_3$] dissolved in distilled water. The neutralization is carried out with sodium hydroxide solution. The pH of the solution is maintained at 10 and it is stirred for 2hrs. The precipitate is thoroughly washed with distilled water until they are free from impurities. The product is dried to remove water contents. The dried powder is mixed homogeneously and is sintered at 900°C.

Cobalt-Copper ferrite samples sintered at 900°C are subjected to X-ray diffraction to calculate the average particle size using Debye–Scherrer formula. The FT-IR spectra of these samples are recorded to ensure the presence of the metallic compounds. The magnetic properties of the samples are studied using VSM. The morphological analysis of the sample is done using SEM.

RESULTS AND DISCUSSION

FT-IR Spectral Analysis

Figure 1(a & b) shows the FT-IR spectra of $CoFe_2O_4$ and $Co_{0.8}Cu_{0.2}Fe_2O_4$ nano particles recorded between 4000cm^{-1} to 400cm^{-1}. The main absorption bands ν_1 and ν_2 found around 600 cm^{-1} and 400 cm^{-1} respectively, are attributed to tetrahedral and octahedral site clusters in spinel structure [11]. By adding the copper as dopant, the absorption band at ν_1 is shifted to lower frequencies, while the position of ν_2 slightly changed [12,13]. The absorption band observed around 1100 cm^{-1} in $CoFe_2O_4$ corresponds to the stretching and bending modes of –OH group. From the spectrum it appears that hydroxyl groups are retained in the samples during preparation of the spinel ferrites by precipitation method and are not completely removed even after sintering at 900°C [14].

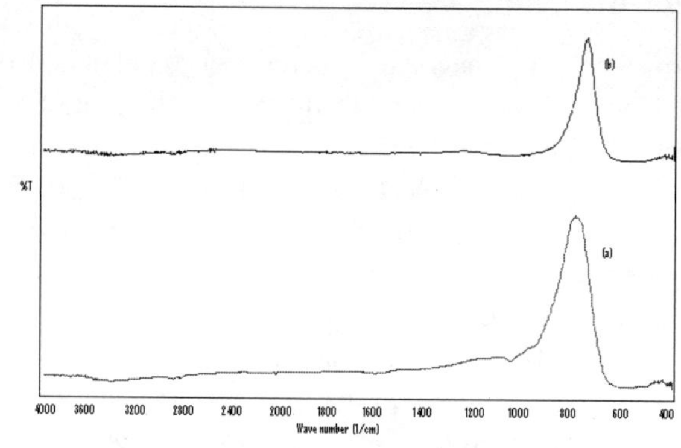

Fig. 1: FT-IR Spectrum of (a) $CoFe_2O_4$ (b)$Co_{0.8}Cu_{0.2}Fe_2O_4$ Nano Particles Obtained by Co-Precipitation Method

Structural Analysis

Figure 2 (a & b) represents the powder X-ray diffraction patterns of $CoFe_2O_4$ and $Co_{0.8}Cu_{0.2}Fe_2O_4$ sintered at 900°C. The XRD results show that the crystallinity of the samples is increased as the samples are sintered at 900°C. The peaks are well indexed with the standard pattern as reported in JCPDS card to the crystal plane of spinel ferrite (220), (311), (400), (511) and (440). The average crystallite size (D) of the samples is calculated using the Debye– Scherrer formula [15].

$$D = 0.9\lambda/\beta \cos \theta$$

where λ is the X-ray wavelength, θ the angle of Bragg diffraction and β the FWHM in radians. The average crystallite size is found to be in the range 21-24 nm. It is also observed that as the sintering temperature increases, the crystallite size also increases.

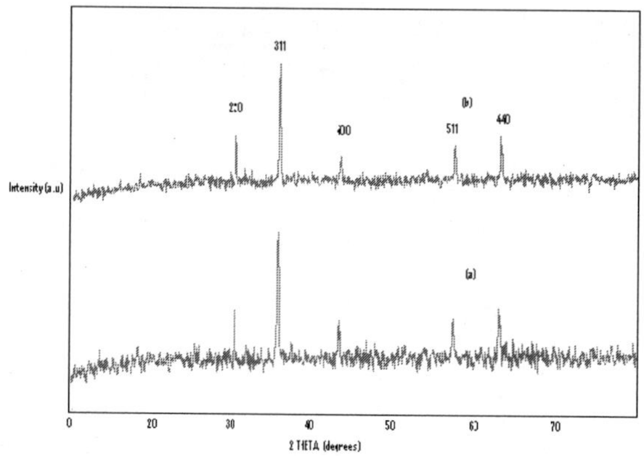

Fig. 2: XRD Pattern of (a) $CoFe_2O_4$ (b) $Co_{0.8}Cu_{0.2}Fe_2O_4$ Nano Particles Obtained by Co-Precipitation Method

Magnetic Measurements Using VSM

Figures 3 (a) & 3(b) show the typical magnetic hysteresis loops obtained from room temperature VSM measurement for $CoFe_2O_4$ and copper substituted cobalt ferrite $Co_{0.8}Cu_{0.2}Fe_2O_4$ sintered at 900°C. The Saturation Magnetization (M_s), remnant magnetization (M_r), coercivity (Hc) and squareness ratio of $CoFe_2O_4$ are found to be 74.01 emu/g, 37.50 emu/g, 926.82 Oe and 0.507 respectively. The saturation magnetization (M_S) and remnant magnetization (M_r) are found to be in good agreement with those obtained by Maaz et al. [16]. By the inclusion of copper as the dopant to cobalt ferrite $Co_{0.8}Cu_{0.2}Fe_2O_4$, the saturation Magnetization (M_s), remnant magnetization (M_r), coercivity (H_c) and squareness ratio are found to decrease. This may be attributed to the presence of the Cu^{2+} ions in the A-sites displacing the same amount of Fe^{3+} ions towards the B-sites as suggested by Kenfack and Langbein [17]. This influence can compensate the decrease of the magnetization due to the substitution of Co^{2+} by Cu^{2+}. This replacement of Co^{2+} ions by Cu^{2+} ions, which have preferential octahedral site occupancy, results in the reduction of the super-exchange interaction between the A and B sites. In other words, the inclusion of copper results in decrease in the magnetization of the B-site and so the net magnetization decreases [13].

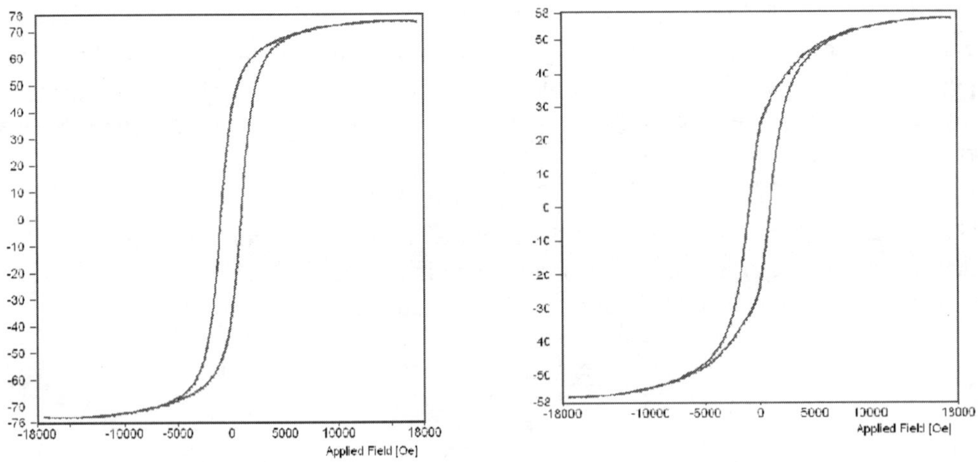

(a) (b)

Fig. 3: Vibrating Sample Magnetometer Measurements for (a) $CoFe_2O_4$ (b) $Co_{0.8}Cu_{0.2}Fe_2O_4$

Scanning Electron Microscopy

The SEM micrograph of the sample is shown in Fig. 4. The SEM picture gives the appearance of spongy structure with voids and pores present in them. It shows that the particle size and crystallinity of the samples increase with increasing sintering temperature.

Fig. 4: SEM micrographs of $Co_{0.8}Cu_{0.2}Fe_2O_4$ Nano Particles

CONCLUSION

$Co_{0.8}Cu_{0.2}Fe_2O_4$ nanoparticles are successfully prepared at room temperature by co-precipitation method. The FT-IR spectra show main absorption bands at 600 cm^{-1} and 400 cm^{-1} respectively which are attributed to tetrahedral and octahedral site clusters in spinel structure.XRD pattern reveals that synthesized ferrites consist of nano crystalline particles in the range of 21-24 nm and the crystalline nature of the sample is increased as the sintering temperature increases. It is also confirmed by the SEM micrographs. The saturation magnetization (*Ms*), remnant magnetization (*Mr*) and coercivity (*Hc*) of the cobalt ferrite decrease on doping with copper.

REFERENCES

[1] H.M. Xiao, X.M. Liu and S.Y. Fu, Synthesis, magnetic and microwave absorbing properties of core-shell structured $MnFe_2O_4/TiO_2$ nanocomposites, *Compos. Sci.Technol.* 66 (2006), 2003–2008.

[2] X. Wang, J. Zhuang, Q. Peng, and Y. Li, A general strategy for nanocrystal synthesis, *Nature* 437 (2005), 121–124.

[3] D. Zhao, X. Wu, H. Guan and E. Han, Study on supercritical hydrothermal synthesis of $CoFe_2O_4$ nanoparticles, *J. Supercrit. Fluid* 42 (2007) 226–233.

[4] Y. Xuan, Q. Li, and G. Yang, Synthesis and magnetic properties of Mn–Zn ferrite nanoparticles, *J. Magn. Magn. Mater.* 312 (2007), 464–469.

[5] B.L. Cushing, V.L. Kolesnichenko and C.J. O'Connor, Recent Advances in the liquid-phase syntheses of inorganic nanoparticles, *Chem. Rev.* 104 (2004), 3893–3946.

[6] A.L. Willis, N.J. Turro and S. O'Brien, Spectroscopic characterization of the surface of iron oxide nanocrystals, *Chem. Mater.* 17 (2005), 5970–5975.

[7] S.G. Christoskova, M. Stoyanova and M. Georgieva, Low-temperature iron-modified cobalt oxide system part I. preparation and characterization, *Appl. Catal. A-Gen.* 208 (2001), 235–242.

[8] C.O. Arean, M.P. Mentruit, E.E. Platero, F.X.L. Xamena and J.B. Parra, Sol–gel method for preparing high surface area $CoAl_2O_4$ and Al_2O_3–$CoAl_2O_4$ spinels, *Mater. Lett.* 39 (1999), 22–27.

[9] C. Liu, B. Zou, A.J. Rondinone and J.Z. Zhang, Reverse micelle synthesis and characterization of superparamagnetic $MnFe_2O_4$ spinel ferrite nanocrystal-lites, *J. Phys. Chem. B* 104 (2000), 1141–1145.

[10] J.P. Ge, W. Chen, L.P. Liu, and Y.D. Li, Formation of disperse nanoparticles at the oil/water interface in normal microemulsions, *Chem. Eur. J.* 12 (2006), 6552–6558.

[11] P. Priyadharsinia, A. Pradeepa, P.S. Rao and G. Chandrasekarana., *Mater. Chem. Phys.* 116 (2009), 207.

[12] X. Tan, G. Li, Y. Zhao, and C. Hu, *Mater. Res. Bull.* 44 (2009) 2160.

[13] Gabal M.A., Al Angari Y.M. and Kadi M and W. Structural and magnetic properties of nano crystalline $Ni_{(1-x)}Cu_xFe_2O_4$ prepared through oxalates precursors. *Polyhedron 2011*; Volume 30:1185–90.

[14] S. Dey and J. Ghose, Synthesis, characterisation and magnetic studies on nanocrystalline $Co_{0.2}Zn_{0.8}Fe_2O_4$, *Materials Research Bulletin* 38 (2003), 1653–1660.

[15] K.Mandal, S.P. Mandal, P. Agudo and M. Pal,A study of nanocrystalline (Mn–Zn) ferrite in SiO_2 matrix, Appl. Surf. Sci.182(2002) 386–389.

[16] K.Maaz, Arif Mumtaz, S.K. Hasanain and Abdullah Ceylan, Synthesis and magnetic properties of cobalt ferrite ($CoFe_2O_4$) nanoparticles prepared by wet chemical route, *Journal of Magnetism and Magnetic Materials*, Volume 308, Issue 2, 289-295.

[17] F. Kenfack and H. Langbein, *J. Mater. Sci. 41* (2006), 3683.

SYNTHESIS OF CARBON NANOTUBES FROM ETHANOL USING RF-CCVD AND Fe-Mo CATALYST

RAMAKRISHNAN S, JELMY E.J., DHAKSHNAMOORTHY M.[1], MURALI RANGARAJAN, VASANTHAKUMARI R.[1] AND NIKHIL KOTHURKAR

Department of Chemical Engineering and Materials Science,
Amrita Vishwa Vidyapeetham, Coimbatore
[2] Polymer Nanotechnology Centre, B S Abdur Rahman University, Chennai, Tamil Nadu
E-mail: nikhil.kothurkar@gmail.com; k_nikhil@cb.amrita.edu

ABSTRACT

Multiwalled carbon nanotubes (MWCNTs) were synthesized by RF heating at 850°C using an MgO-supported bimetallic Fe-Mo catalyst. Ethanol was used as the carbon source. The MWCNTs were tested using Raman spectroscopy, transmission electron microscopy (TEM) and thermo gravimetric analysis (TGA). TEM studies show nanotubes with an average diameter of 11–14nm. In Raman spectroscopy, the ratios of intensities of the G and D band and the 2D and G bands were 0.90 and 0.54 respectively. These ratios are generally indicative of high crystallinity and purity of the MWCNTs. The MWCNT purity as calculated from TGA studies was 90%.

INTRODUCTION

Since their discovery in 1991, multiwalled carbon nanotubes[1] (MWCNTs) have been at the forefront of the nanomaterials research. These nanometer size materials can think to be arising from rolling of graphene sheets into a cylindrical form. Soon after discovery, extensive research efforts on these tubular graphite structures revealed that CNT are the materials of great interest from fundamental point of view as well as for future applications in the areas of hydrogen storage, electronics, catalysis and polymer composites [5, 6]. These carbon nanotubes were

V. Rajendran, K. Saminathan, P. Paramasivam and K.E. Geckeler (eds.)
Nano Materials Synthesis and Characterisation, pp. 239-244 (2012).

synthesized by arc discharge method, laser ablation method and catalytic chemical vapor deposition (CCVD) method. Of these, CCVD is the most widely used because of the advantages that it presents in controlling the growth morphology of these nanostructures. The catalyst system and the reaction parameters like temperature and the flow rates of the carrier and hydrocarbon gases can be directly controlled [11, 12]. Moreover, by using RF-CCVD [15-17], high quality CNTs (high crystallinity and defect free) are possible. Inductive heating in the CVD process can significantly reduce the energy consumption and enable better control on the heating rate compared to external furnace methods. Alcohols are particularly attractive as the hydrocarbon source in CVD synthesis of CNTs because they are cheap, easily available and safe to handle. They also produce high purity CNTs due to the etching effect by oxygen-containing radicals generated during alcohol decomposition, which oxidize the amorphous carbon [13, 14]. We propose to characterize in detail the morphologic and structural properties of the nanotubes synthesized at 850°C by RF-CCVD from ethanol on MgO supported catalysts.

EXPERIMENTAL

Materials

The catalyst used was Fe $(NO_3)_3$ $9H_2O$, $(NH_4)_6Mo_7O_{24}.4H_2O$ supported on MgO (Sigma Aldrich, 99.99%, purity). Ethanol (Laboratory Grade) was used as the carbon source.

Catalyst Preparation and Carbon Nanotube Synthesis

The Fe-Mo/MgO catalyst system was prepared in the weight ratio of 1:0.1:12. The melt salts Fe $(NO_3)_3.9H_2O$ and $(NH4)_6Mo_7O_{24}$. $4H_2O$ were dissolved separately in distilled water. These metal solutions were mixed with prepared MgO suspension and sonicated for 1 hr. The solvent was removed using a rotary evaporator at 90°C. The powder was further dried overnight at 150°C for 15 hours and calcinated in air at 700°C for 7 hours [2]. About 100 mg of the Fe-Mo/MgO catalyst was uniformly spread into a thin layer in a graphite boat and placed in the 2" quartz tube of the RF-CCVD setup. Heat was applied by adjusting the RF power of the generator (350 kHz) to maintain a constant temperature of 850°C for the duration of the reaction. The tube was purged with 800 mL/min flow of Ar for 10 min to remove the ambient air.

Ethanol was introduced using a syringe pump at 4 mL/h for about 30 min. The tube was rinsed with Ar and allowed to cool to room temperature. The reaction product (powder) was collected from the graphite boat. It was sonicated in aqueous hydrochloric acid (v/v ratio: 1:1) for 30 min to remove the catalyst and support, followed by washing in water to neutral pH. The obtained CNTs were filtered and dried overnight at 120°C.

Characterization

High resolution transmission electron microscopy (HRTEM) was carried out on a JEOL JEM 2100 instrument. HRTEM samples were prepared by ultrasonicating a small amount of the CNTs in 2-propanol for 10 minutes, putting a few drops of the suspension on the TEM grid and drying. Thermogravimetric analysis (TGA) was conducted on a EXSTAR6200 (TG/DTA) thermal analyzer at 10 °C/min from room temperature to 900°C in air. Raman analysis carried out using a Renishaw (RM 1000) instrument with an Argon Ion Laser source of wavelength 514.5 nm. The exposure time for the each measurement is 100 second and 2 accumulations for the each measurement.

RESULT AND DISCUSSION

The Raman spectrum presented in the Fig. 1 shows the characteristic bands corresponding to multiwall carbon nanotubes. The D band (between 1305 and 1345 cm^{-1}) is associated with the vacancies, defects, and carbonic impurities (amorphous carbon, glassy carbon, etc. that destroy the symmetry of the graphene sheet. The G band (between 1500 and 1600 cm^{-1}) corresponds to the splitting of the E_{2g} stretching mode for graphite, and the 2D band (between 2450–2650 cm^{-1}) corresponds to the second harmonic of the D band and is associated to the degree of crystallinity of the graphitic layers. The calculated ratios of the intensities of the G to D and 2D to G bands in the Raman spectra of carbon nanotubes were 0.89 and 0.47 respectively [7,8,9,10].

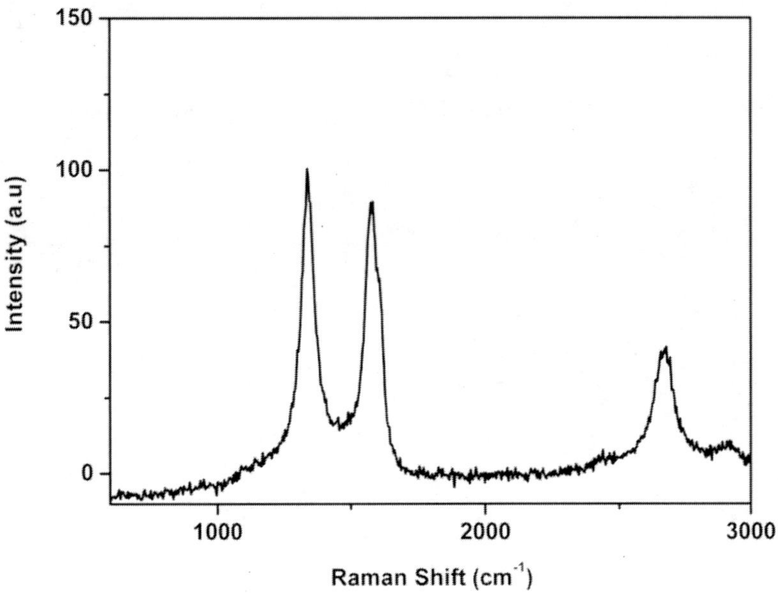

Fig. 1: Raman Spectra of MWCNTs

Morphology of the formed MWCNTs was analyzed by using HRTEM. HRTEM analyses of 15 nanotubes show an average diameter of 12.2 ± 3.23nm (95% confidence interval) for the MWCNTs. Figures 2A and 2B show the presence of the agglormorated MWCNT while figures 2D and 2C show the formation of parallely arranged MWCNTs. Small amounts of amorphous carbon could also be observed.

TGA analysis was performed on the formed MWCNT in order to find out the quality and purity of the formed MWCNTs. Three distinct weight losses were observed in the TGA analysis. The first weight loss (3%) below 150°C was due to the removal of moisture present in the MWCNTs. The second weight loss (4.46 %) in the interval of 200–400°C was due to the removal of the amorphous carbon present in the material[3]. A large weight loss (66%) in the interval of 400°C–670°C was due to the decomposition of crystalline carbon present in MWCNTs[4]. The TGA analysis thus shows the formation of over 66% of MWCNTs.

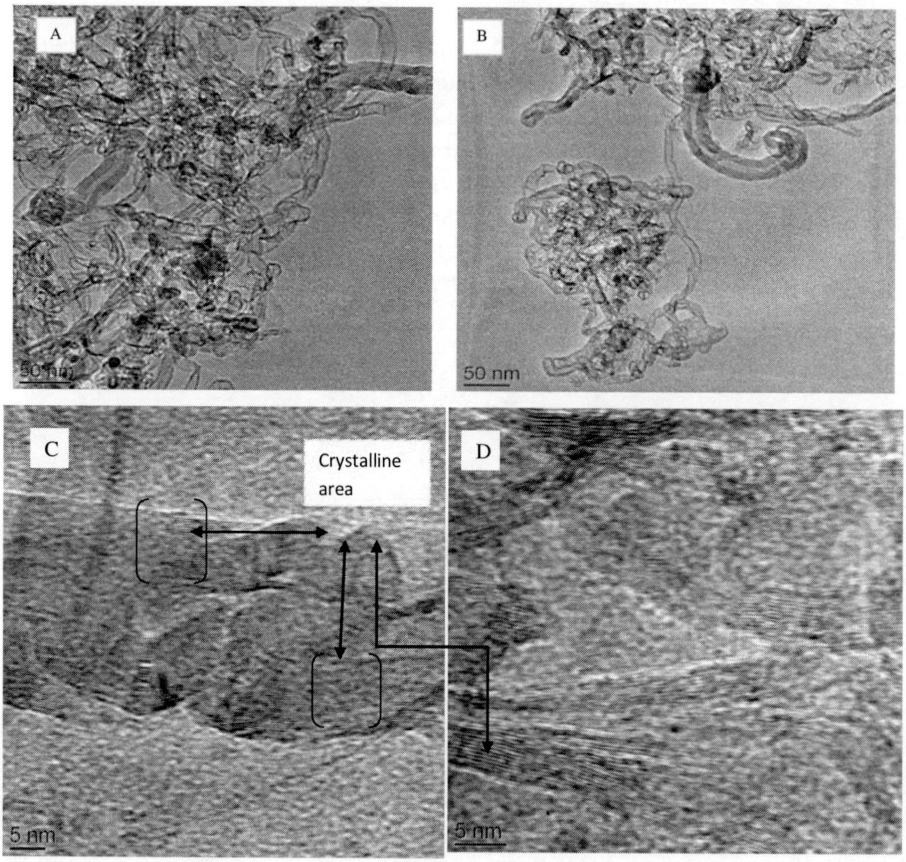

Fig. 2: (A,B,C and D) HRTEM image of the MWCNTs

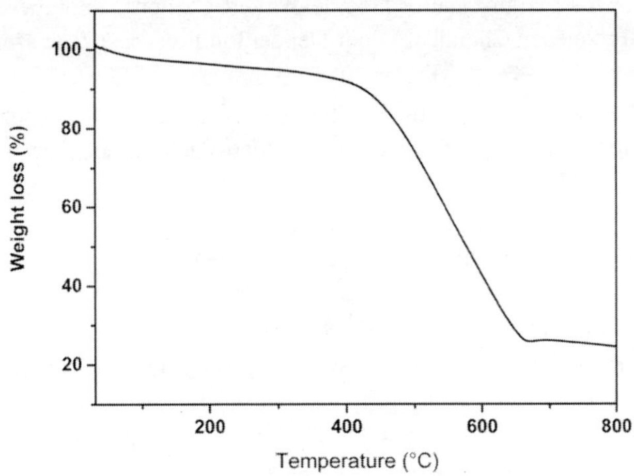

Fig. 3: TGA Curve of the MWCNTs

CONCLUSIONS

Multiwalled carbon nanotubes were synthesized successfully using RF induction heating method and using ethanol source and a bimetallic catalyst of Fe-Mo on MgO support material. The main advantage of using ethanol as the precursor for the preparation of highly crystalline MWCNTs, is its ready availability and low cost. Raman spectroscopy revealed a G to D band intensity ratio of 0.86. The product contained over 66% MWCNTs with average outer diameters of 12.2 ± 3.23nm (95 % confidence interval) out of a population of 15 nanotubes.

ACKNOWLEDGEMENTS

We would like to express our gratitude for the valuable inputs of Dr. Alexandru Biris in doing this research. We are also grateful to DRDO (Grant No. ERIP/ER /0703668/M/01/2796/D (R&D)/1071) for their generous funding.

REFERENCES

[1] Iijima S. and Ichihashi T., Single-shell carbon nanotubes of 1-nm diameter, *Nature* 1993; 363 (6430): 603-5.

[2] Biris A.R., Li Z., Dervishi E., Lupu D., Xu Y., Saini V. et al., Effect of hydrogen on the growth and morphology of single wall carbon nanotubes synthesized on a FeMo/MgO catalytic system, *Physics Letters A* 2008; 372(17): 3051-7.

[3] Biris A R, Biris A S, Lupu D., Trigwell S., Dervishi E., Rahman Z et al., Catalyst excitation by radio frequency for improved carbon nanotubes synthesis, *Chemical Physics Letters* 2006; 429(1-3): 204-8.

[4] Alexandru R. Biris., High-Quality Double-Walled Carbon Nanotubes Grown by a Cold-Walled Radio Frequency Chemical Vapor Deposition Process, *Chem Mater* 2008; 20: 3466–72.

[5] Derycke V., Auvray S., Borghetti J., Chung C-L, Lefèvre R., Lopez-Bezanilla A et al. Carbon nanotube chemistry and assembly for electronic devices, *Comptes Rendus Physique* 2009; 10(4): 330-47.

[6] Faris Y inodot, lmaz, Zuhal K. Conducting polymer composites of multiwalled carbon nanotube filled doped polyaniline, *Journal of Applied Polymer Science* 2009; 111(2): 680-4.

[7] Santangelo S., Messina G., Faggio G., Lanza M., Milone C. Evaluation of crystalline perfection degree of multi-walled carbon nanotubes: correlations between thermal kinetic analysis and micro-Raman spectroscopy, *Journal of Raman Spectroscopy* 2011; 42(4): 593–602.

[8] Tzong-Ming W and Shih-Hsiang L. Synthesis, characterization, and electrical properties of polypyrrole/multiwalled carbon nanotube composites. *Journal of Polymer Science Part A: Polymer Chemistry* 2006; 44(21): 6449-57.

[9] Dresselhaus M.S, Jorio A., Hofmann M., Dresselhaus G and Saito R. Perspectives on Carbon Nanotubes and Graphene Raman Spectroscopy, *Nano Lett* 10(3): 751-8.

[10] Yoshida A., Kaburagi Y. and Hishiyama Y., Full width at half maximum intensity of the G band in the first order Raman spectrum of carbon material as a parameter for graphitization, *Carbon* 2006.44(11).2333-5.

[11] Malek Abbaslou R.M., Soltan J and Dalai A.K. The effects of carbon concentration in the precursor gas on the quality and quantity of carbon nanotubes synthesized by CVD method, *Applied Catalysis A: General* 372(2): 147-52.

[12] Liao X.Z., Serquis A., Jia Q.X., Peterson D.E., Zhu Y.T. and Xu H.F., Effect of catalyst composition on carbon nanotube growth, *Appl Phys Lett* 2003; 82(16): 2694-6.

[13] Qi H., Qian C., Liu J., Synthesis of High-Purity Few-Walled Carbon Nanotubes from Ethanol/Methanol Mixture, *Chem Mater* 2006; 18(24): 5691-5.

[14] Liu Q., Ouyang Y., Zhang L., Xu Y., and Fang Y, Effects of argon flow rate and reaction temperature on synthesizing single-walled carbon nanotubes from ethanol, *Physica E: Low-dimensional Systems and Nanostructures* 2009; 41(7): 1204-9.

[15] Biris A.R., Carbon Nanotubes Grown by RF Heating and Their Morphological and Structural Properties, *Particulate Science and Technology,* 2008; 26: 521–8.

[16] Biris A.R., Lupu D., Dervishi E., Li Z., Xu Y., Trigwell S., et al., Multiwall carbon nanotubes synthesized by RF-CCVD on novel CaO supported catalysts, *Physics Letters A* 2008; 372(42): 6416-9.

[17] Biris A.S., Schmitt T.C., Little R.B., Li Z., Xu Y., Biris A.R. et al., Influence of the RF Excitation of the Catalyst System on the Morphology of Multiwalled Carbon Nanotubes, *J Phys Chem C* 2007; 111(48): 17970-5.

PREPARATION OF PURE AND Al DOPED ZnO NANO POWDER USING SIMPLE HYDROTHERMAL METHOD

R. Raj Mohan, K. Rajendran[1] and K. Sambath

Department of ECS, Sri Krishna Arts and Science College, Coimbatore
[1]Department of Electronics, Government Arts College for Women, Ramanathapuram
E-mail: rrajmohan.teacher@gmail.com

ABSTRACT

In this paper, pure and aluminium doped ZnO nanopowders have been synthesized by low temperature hydrothermal method. Zinc nitrate and HMT were used as precursors for preparing pure ZnO and aluminium nitrate is used as doping material for AZO with 1M, 0.5M and 0.25M%. The fine nanostructural characterization of doped and un-doped ZnO powders has been analyzed by X-ray diffraction (XRD), Scanning electron microscopy (SEM) and EDAX studies. The SEM and XRD results show that by comparing of un-doped and doped hydrothermal synthesis it is possible to obtain ZnO nano phases with good composition and morphology.

INTRODUCTION

Zinc oxide (ZnO), a direct wide band gap (3.4eV at Room temperature) *II-VI* compound n-type semiconductor, has a stable wurtzite structure and composed of a number of alternating planes with tetrahedrally-coordinated O^{2-} and Zn^{2+} ions, stacked alternately along the c-axis [1]. It is an important multifunctional material with wide ranging applications in varistors [2], surface acoustic wave (SAW) devices [3], transparent conducting oxide electrodes [4], solar cells [5], blue/UV light emitting devices [6], gas sensors [7, 8], etc. It has been planned to be a more capable UV emitting phosphor than Gallium Nitride (28 meV) because of its larger

V. Rajendran, K. Saminathan, P. Paramasivam and K.E. Geckeler (eds.)
Nano Materials Synthesis and Characterisation, pp. 245-250 (2012).

exciton binding energy (50 meV) [9]. The most significant hurdle in developing and exploiting zinc oxide based materials in electronic and photonic applications is the difficulty in carrier doping (achieving a p-type material); n-type conductivity of ZnO is relatively easy to realize using excess Zn or by doping zinc oxide with Al, Ga, In [10].

So far, various methods such as sol–gel [11], chemical bath deposition (CBD) [12], electrochemical deposition (ECD) [13], hydrothermal (HT) [14], etc., have been utilized to fabricate ZnO nanorods. Among these methods, the hydrothermal method offers advantages of being environmental friendly, low synthesis temperature, and low production cost. Moreover, Bai and coworkers have prepared highly dense structure of Al-doped ZnO nano wires by hydrothermal method followed by the thermal diffusion of Al [15]. For the most part, previous reports have focused on the Al-doped ZnO nano grain thin films, whereas there are few on the nano structured Al-doped ZnO with various morphologies. Till now, the morphology of pure ZnO nanorods and nano tubes synthesized by the hydrothermal deposition was extensively investigated by many research groups [16–17]. In this study, we report ZnO nanorods synthesized via a simple and low temperature hydrothermal method for pure and AZO. Also the influences of dopant ratio and structural property were investigated.

EXPERIMENTAL

The ZnO and AZO (Al**:** ZnO) nanostructures have been synthesized using hydrothermal method. All chemicals used in this experiment are analytical grade and used without any further purication.

Synthesis of ZnO Nanostructures

Preparation of Pure ZnO

Pure ZnO Synthesis was carried out by using zinc nitrate hexahydrate [Zn $(NO_3)_2.6H_2O$] and hexamethylenetetramine (HMT) [$(CH_2)_6N_4$] as precursors. In a classic synthesis process, zinc nitrate was dissolved in 100 ml of deionized water and aqueous solution of 0.1 M hexamethylenetetramine was added to it.

The mixed solution was stirred for 15 min. The pH value of the solution was adjusted by using ammonia solution. The final solution was transferred into a 500 ml RB flask and reuxed at 70°C for 1 hr. The white products were separated from the solution by centrifugation at 6000 rpm for 15 min at 30°C. The final product of ZnO was washed several times with deionized water and dried at 60°C. Finally the powder is annealed in hot air oven at 200°C for 1 hr.

Preparation of Al Doped ZnO

Synthesis of AZO nano particles have been prepared by adding drop wise aluminium nitrate [$Al(NO_3)_3.9H_2O$] solution with different molar concentration 1m%, 0.5% and 0.25m% to the zinc nitrate and HMT mixture The final powder samples S1, S2, and S3 were received for further characterization.

RESULT AND DISCUSSIONS

X-Ray Diffraction (XRD)

Figure 1. shows the XRD pattern of pure ZnO and Al doped ZnO samples at different ratios. As seen from this figure the samples have wurtzite structure with (002) a preferred orientation. To understand the crystalline mechanism of AZO, the average grain size of the nano particles has been estimated from the FWHM of (002) diffraction peak using the Scherrer eq.(1).

$$D_{avg} = 0.9\lambda/\beta \cos \theta_B \qquad \ldots (1)$$

Where, D_{avg} = average grain size, λ = wavelength of incident X-ray beam (1.5406Å), θ_B Bragg diffraction angle in degree, and β = FWHM of θ_B in radians. The grain size of the AZO nano particles increased with increase of Al doping in ZnO, as shown in the Table 1.

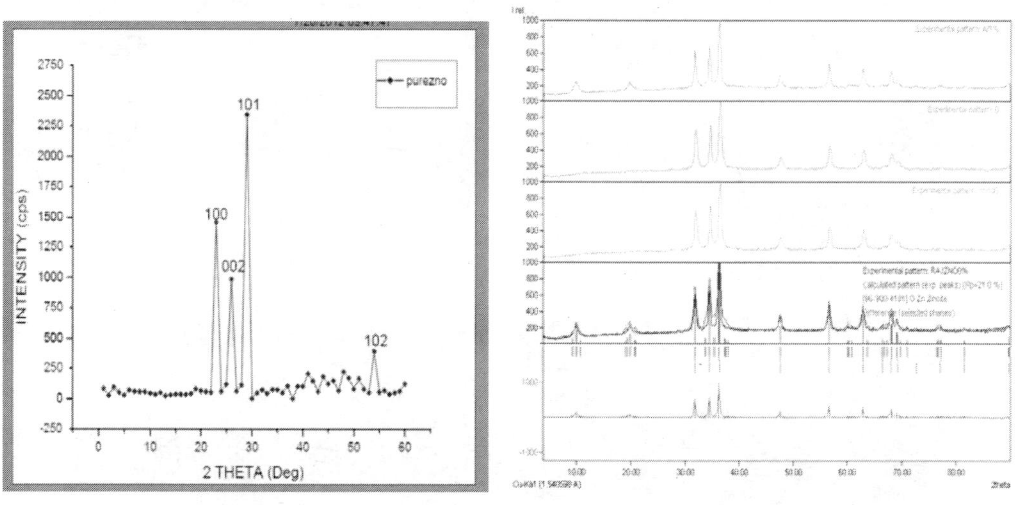

Fig. 1: XRD Pattern of (a) Pure ZnO (b) AZO: Al 0%, 0.25%, 0.5% and 1%

The detected (h k l) peaks were at 2θ value of different samples corresponding to the lattice planes (100), (002) (101), (102), (110), (103), and (112) respectively. The data are matched with the JCPDS card (JCPDS 036–1451/900-4181) for hexagonal wurtzite ZnO. The grain size of AZO increased with increase of Al doping ratio. The peaks of (002) plane of ZnO is less than that of AZO; therefore the crystalline degree of AZO is less than pure ZnO. According to the calculation of FWHM at (002) plane, the highest crystalline of pure ZnO nanorod is achieved because of narrow width. However the doping causes the broadening of the peak and shifting towards lower degree. Hence the arrangement of AZO was randomly irregular, with a variety of morphologies.

Table 1: Grain Size of AZO Samples Prepared by Hydrothermal Method

Samples	*2-Theta (deg)*	*I (cps)*	*D (A)*	*FWHM (deg)*	*Grain Size D (nm)*
Pure ZnO	40.1297	0985	2.60719	0.15180	14.87
S1 (0.25%)	36.4212	2824	2.60116	0.50124	16.2
S2 (0. 5%)	34.6000	2512	2.59471	0.43650	17.4
S3 (1.0%)	34.3708	2104	2.60708	0.36042	28.8

SEM and EDAX

The Fig. 2 shows the SEM images of pure, 0.25%, 0.5% and 1% Al doped ZnO nano structured powder. It is observed that the pure ZnO formed as nanorod clusters whereas the SEM images of AZO show the variation in grain size increasing with Al concentration.

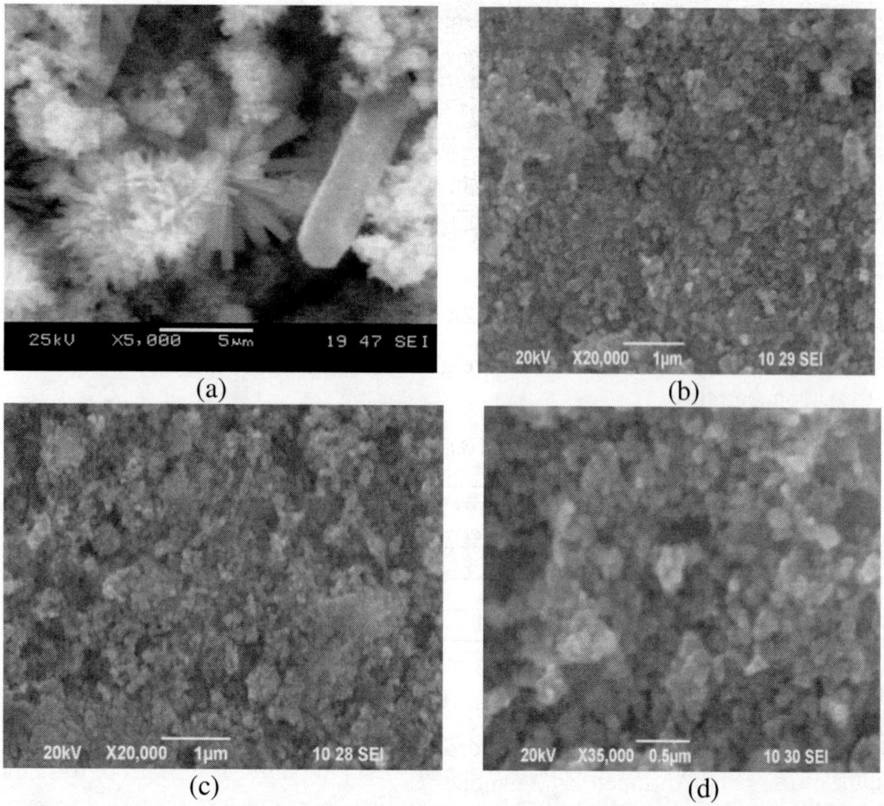

Fig. 2: SEM Images of ZnO: AZO (a) Pure ZnO, (b) 0.25%, (c) 0.5% and (d) 1%

Comparing the crystallization of ZnO with AZO; the presence of Al dopant resulted in lattice disorder which is connected with stress, that allow the grains to develop easily.

The Fig.3.shows the EDAX image of AZO used for the *elemental* composition analysis. The AZO powder contains only Zn, O and Al elements at % of 70, 27 and 3 respectively.

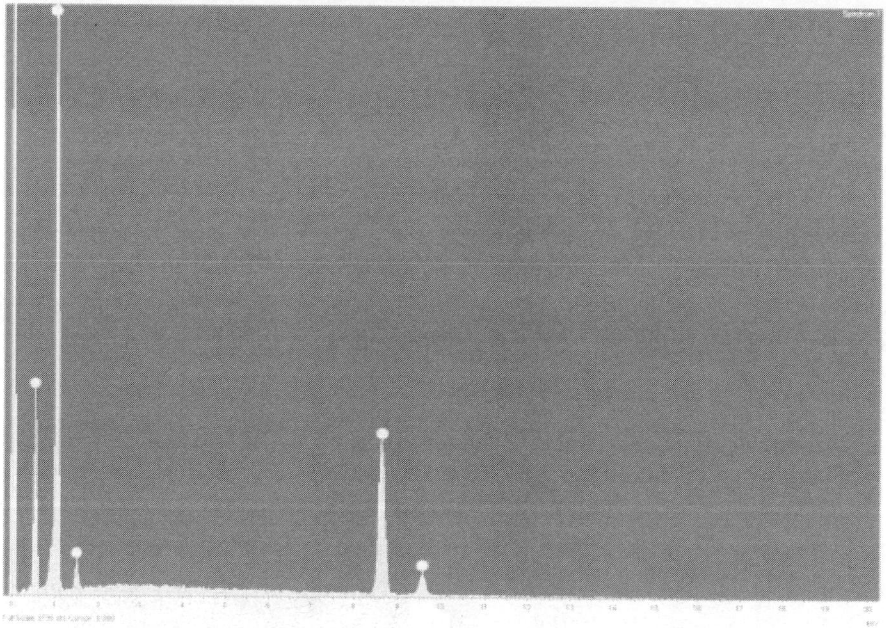

Fig. 3. EDX of AZO Prepared with Zn $(NO_3)_2.6H_2O$, HMT and Al $(NO_3)_3.9H_2O$

The Table 2 gives the details about weight and atomic % of aluminium in ZnO with respect to its concentration.

Table 2: Summary of Wt % and At % of Al in AZO Observed by EDX

Samples	*Wt % Al added*	*Wt % Al as observed by EDAX*	*At % Al as observed by EDAX*
S1	0.25	0.12	0.14
S2	0. 50	0.21	0.25
S3	1.00	0.30	0.35

CONCLUSION

In this present work, low temperature hydrothermal method was used for the successful preparation of pure and Al doped ZnO nanopowders with different molar ratio. The samples were characterized by XRD, SEM and EDX. It is found that the morphology for this material can be changed and the agglomerated nano grains of the AZO powders have different shapes, due to the variation of aluminium concentration. With increasing Al ratio the crystalline quality of ZnO degenerated with stress resulted in lattice disorder.

REFERENCES

[1] C. Klingshirn, *Chem. Phys. Chem.* 8 (2007) 782.

[2] R.N. Viswanath, S. Ramasamy, R. Ramamoorthy, P. Jayavel and T. Nagarajan, Preparation and characterization of nanocrystalline ZnO based materials for varistor applications, *Nanostruct. Mater.* 6 (1995) 993–996.

[3] M.S. Wu, A. Azuma, T. Shiosaki and A. Kawabata, Low-loss ZnO optical waveguides for SAW-AO applications, *IEEE Trans. Ultrasonic Ferroelec. Freq. Control* 36 (1989), 442–445.

[4] M.K. Jayaraj, A. Antony and M. Ramachandran, Transparent conducting zinc oxide thin film prepared by off-axis RF magnetron sputtering, *Bull. Mater. Sci.* 25 (2002), 227–230.

[5] K. Keis, C. Bauer, G. Boschloo, A. Hagfeldt, K. Westermark, H. Rensmo and H. Siegbahn, Nanostructured ZnO electrodes for dye-sensitized solar cell applications, *J. Photochem. Photobiol. A: Chem.* 148 (2002), 57–64.

[6] P. Yang, H. Yan, S. Mao, R. Russo, J. Johnson, R. Saykally, N. Morris, J. Pham, R. He, H. Choi, Controlled growth of ZnO nanowires and their optical properties, *Adv. Funct. Mater.* 12 (2002), 323–331.

[7] S. Roy and S. Basu, Improved zinc oxide film for gas sensor applications, *Bull. Mater. Sci. 25* (2002), 513–515.

[8] A.R. Raju and C.N.R. Rao, Gas sensing characteristics of ZnO and copper impregnated ZnO, *Sens. Actuators B* 3 (1991), 305–310.

[9] Z.K. Tang, G.K.L. Wong, P. Yu, M. kawasaki, A. Ohtomo, H. Koinuma and Y. Segawa, *Appl. Phys. Lett.* 72 (1998), 3270.

[10] Pearton, S. J., Norton, D. P., Ip, K., Heo, Y. W. and Steiner, T., *Superlattices Microstruct,* 2003, 34, 3–32.

[11] K. Prabakar and Heeje Kim, Growth control of ZnO nanorod density by sol–gel method, *Thin Solid Films* 518 (2010), e136–e138.

[12] Qingwei Li, Jiming Bian, Jingchang Sun, Jingwei Wang, Yingmin Luo, Kaitong Sun and Dongqi Yu, Controllable growth of well-aligned ZnO nanorod arrays by low-temperature wet chemical bath deposition method, *Appl. Surf. Sci.* 256 (2010), 1698–1702.

[13] Chien-Te Hsieh, Shu-Ying Yang and Jia-Yi Lin, Electrochemical deposition and Super hydrophobic behavior of ZnO nanorod arrays, *Thin Solid Films* 518 (2010), 4884–4889.

[14] Seong-Jong Kim, Han-Hyoung Kim, Joo-Beom Kwon, Jong-Geun Lee, O. Beom-Hoan, Seung Gol Lee, El-Hang Lee and Se-Geun Park, Novel fabrication of various size ZnO nanorods using hydrothermal method, *Microelectron. Eng.* 87 (2010), 1534–1536.

[15] S.N. Bai, H.H. Tsai and T.Y. Tseng, *Thin Solid Films* 516 (2007), 155.

[16] M. Guo, P. Diao and S. Cai, J. *Solid State Chem.* 178 (2005), 1864.

[17] J.P. Cheng, X.B. Zhang, and Z.Q. Luo, *Surf. Coat. Technol.* 202 (2008), 4681.

STRUCTURAL AND OPTICAL STUDIES OF ZnO NANOPARTICLES SYNTHESIZED BY SOLUTION COMBUSTING METHOD

M. Silambarasan and S. Saravanan

Centre for Photonics and Nanotechnology, Sona College of Technology, Salem, Tamil Nadu
E-mail: silambu.physicist@gmail.com

ABSTRACT

In this paper we report the structural and optical studies of ZnO Nanoparticles synthesized by simple solution combusting method. The XRD pattern of the ZnO Nanoparticles indicted hexagonal wurtzite structure. The average particles size is around 50 nm which is confirmed by HR-SEM and HR-TEM. SAED pattern of ZnO Nanoparticles shows the diffraction rings of the polycrystalline nature of ZnO. EDX analysis indicates Zn and O are present in pure form of the synthesized sample. The FTIR spectrum of the main absorption band is due to Zn-O stretching of ZnO at 446 cm^{-1}. Raman scatting of ZnO Nanoparticles has first and second orders optical modes which are the characteristic bonds of the wurtzite ZnO. The optical properties were studied through UV-Vis and PL spectrum. The UV-Vis absorption was observed at λ_{Max} = 373 nm due to ZnO Nanoparticles. At room temperature PL emission was observed at 419 nm.

INTRODUCTION

Synthesis of semiconductor Nano sized particles has a topic of great attraction in the recent years due to their unique optical, electrical, catalytic, photochemical and bio properties. Among the semiconductor nanocrystals, Zinc Oxide (ZnO) is one of the most important II-VI group elements with wide band gap (3.37 eV) possessing size tunable optical transitions in solar

V. Rajendran, K. Saminathan, P. Paramasivam and K.E. Geckeler (eds.)
Nano Materials Synthesis and Characterisation, pp. 251-256 (2012).

cell, optoelectronics and wider range of application [1-3] such as gas sensors, piezoelctronic transducers, photonic crystals, light emitting devices, photodiodes, photodetectors, optical modulator waveguides, transparent conductive film. ZnO is also receiving a lot of attention because of its antimicrobial property. The bactericidal efficiency of ZnO Nanoparticles (NP) has been reported to increase as the particles size decreases [4, 5]. ZnO is one of five zinc compounds that are currently listed as generally recognized as safe (GRAS) by the U.S Food and Drug Administration (21CFR 182.8991). Different Nanostructures of ZnO have been reported such as Nanorods, Nanocomb, Nanorings, Nanobelts etc, also, many methods have been developed to prepare ZnO NP including wet chemical, sol-gel, template-assisted growth, gas phase reaction, ball milling, etc. [2, 6-8].

The above mentioned methods required complex equipment and complicated operation. So, it is still a challenge to search for simple method for prepared metal oxides NP with a high yield. In 2007, Y. Ni *et* al [9] reported the simple solution combustion route for the rapid and facile preparation of metal oxides in air at ambient pressure. In this paper we report the successful synthesis of ZnO NP by simple solution combusting method.

EXPERIMENTAL

In a typical synthesis, Zinc acetate (0.05M) was dissolved in 100 ml mixed solvents of ethanol and ethyleneglycol with the volume ratio of 70:30. The solution was transferred into a spirit lamp with an absorbent cotton lampwick and then the sprite lamp was fired. After the lampwick was extinguished, the substance became white. The white powder was repeatedly dispersed into distilled water to wash and remove the impurity by ultrasonic and centrifugal process. Finally, the sample was dried at 150°C in hot air oven for 24 h.

Powder x-ray diffraction (XRD) of the ZnO NP was carried out by a SEIFERT-Analytical x-ray diffractometer equipped with Cu-Ká radiation (λ = 0.1540598 nm) at the scanning rate of 0.04°/sec from 20° to 70°. High Resolution-Scanning Electron Microscopy (HR-SEM) and Energy Dispersive X-ray spectrometry (EDX) images were analyzed through Quanta-200F SEM with AMETEK EDX. High Resolution-Transmission Electron Microscopy (HR-TEM) images were obtained by a FEI-Tecnai F20 TEM. Fourier Transform Infrared (FTIR) spectra were recorded in the range of 400cm^{-1} to 4000cm^{-1} by Perkin-Elmer instrument in KBr pelletizes method. Raman spectra was measured with an Ar$^+$ laser line (λ = 513.4 nm). The Ultra Violet-Visible (UV-Vis) absorbance spectra were obtained by a double-beam biospectrophotometer ELICI-BL-198. Room temperature photoluminescence (PL) spectra were recorded to study the optical characterization of the ZnO NP using a Horiba Jobin Spectroflurometer with (*Source*: Xenon lamp 450W) excitation wavelength of 325 nm.

RESULTS AND DISCUSSION

Figure 1(a) shows the powder XRD of synthesized ZnO NP. All the diffraction lines are assigned well to the hexagonal phase of wurtzite structure of ZnO with a reference pattern JCPDS No:

89-1397. The strong and narrow diffraction peaks represents the good crystallinity of pure ZnO. The crystalline size of ZnO is around 6 nm calculated by Scherrer formula. Also, the lattice parameters (a = 3.28Å; c = 5.27Å), Volume of unit cell (147.94(Å)^3), Atomic packing factor (75.51%) were calculated by XRD data. The Nanoscale lattice strain can be expected on the ZnO NP due to very small in crystalline size. The insert of Fig.1(a) shows β *cos* θ is plotted versus *sin c* for ZnO NP. It can be exhibits the negative slope of fitted lines indicate the presence of compressive strain in ZnO NP. Hence, prominent surface and strain resulting from the compression of the prepared ZnO lattice can be expected which would have important effect on the optical properties of ZnO NP.

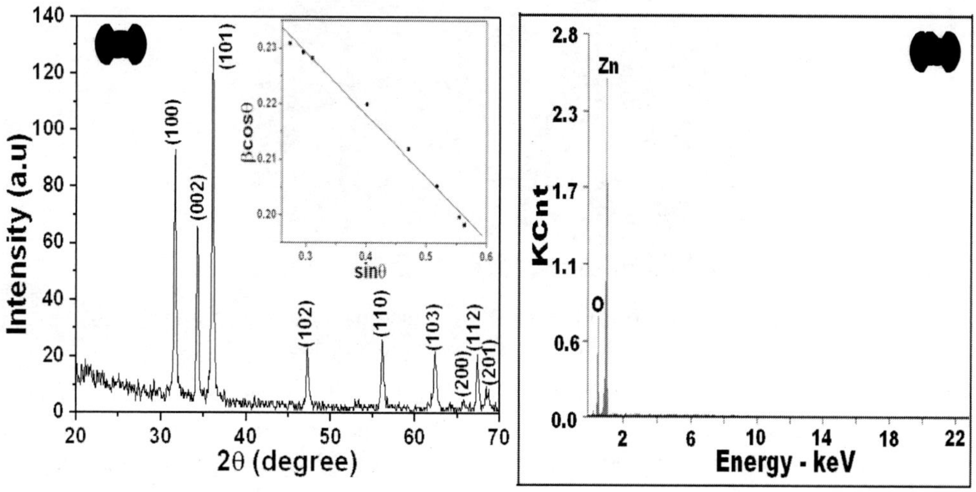

Fig. 1: (a) Powder XRD Pattern (the Insert Shows α cos θ versus sin θ and (b) EDX spectra of ZnO Nanoparticles

Figure 1(b) shows the EDX analysis of ZnO NP. The peaks of Zn and O can be founded easily. Hence, the pure ZnO NP (Zn:O (at %) = 51.23:48.77) have been achieved by using simple solution combustion method. The morphology of the ZnO Nanoparticles was characterized by HR-SEM and HR-TEM. Figure 2 (a) gives a representative SEM image of the ZnO NP. The average particle size is around 50 nm which is confirmed by HR-SEM and HR-TEM (Fig. 2(b)) Presence of some bigger particles might be attributed to the aggregating or overlapping of smaller particles. Figure 2(c) indicates the lattice plane formation of obtained ZnO NP. The corresponding Selected Area Electron Diffraction (SAED) of ZnO NP is shown in Figure 2(d). The diffraction rings show the polycrystalline nature of ZnO. The ZnO NP is structurally uniform and the well-oriented lattice fringe of d-spacing is around 0.23 Å, this indicates wurtzite ZnO crystalline structure. Figure 3(a) shows the typical FTIR spectra of synthesized ZnO NP. In the FTIR spectra at 3425 cm^{-1} indicates O-H bonding from water, the peaks at 1634 and 1029 cm^{-1} are due to the vibration of ZnO, the peaks at 1380 cm^{-1} have indicate stretching of carbolated group (COO^-). The FTIR spectrum of the main absorption band is due to Zn-O stretching of ZnO at 446 cm^{-1}. Prominent IR peaks are analyzed and assigned well with the reported literature [10, 11]. Absorption bands show the presence of resonance interaction between vibrational modes of oxide ions in the crystal.

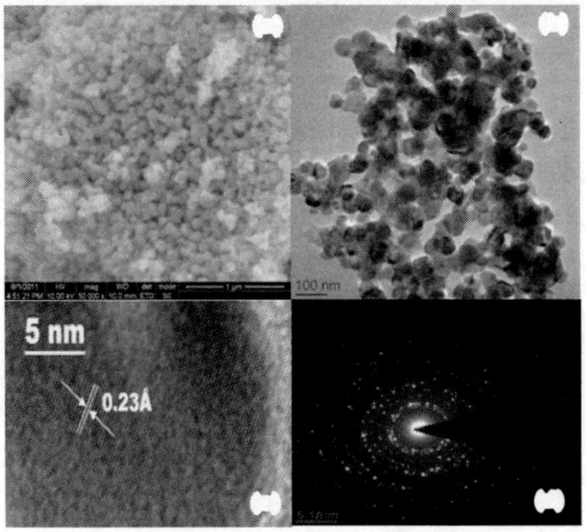

Fig. 2: (a) HR-SEM (b) HR-TEM (c) Lattice plan Formation and (d) SAED Pattern of ZnO Nanoparticles

Since the Raman spectra are sensitive to crystallization, structural disorder and defects in micro and nanostructures, vibration properties of the hexagonal shaped ZnO NP were investigated by Raman scattering technique. The characteristic optical phonon modes at the centre of Brillouin Zone (Γ point, $\Gamma = 1A_1 + 2B_2 + 1E_1 + 2E_2$). Where A_1 and E_1 are two polar branches, which spilt into longitudinal optical (LO) and transversal optical (TO) components with different frequencies due to macroscopic electric fields associated with the LO phonons. The E_2 modes are nonploar with low frequency mode (E_{2L}) associated to the heavy Zn sublattice and high frequency mode (E_{2H}) involving only oxygen atoms. While the A_1, E_1 and E_2 modes are active in Raman and infrared spectra, the B_1 modes are generally inactive in Raman spectra and are so-called silent modes [7, 12].

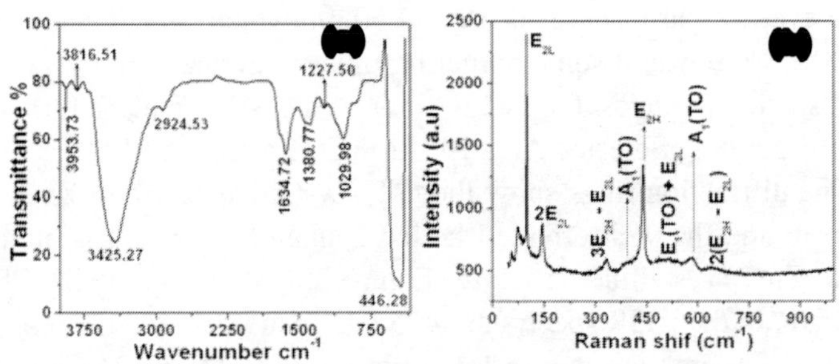

Fig. 3: (a) FT-IR and (b) Raman Spectra of ZnO Nanoparticles

Figure 3(b) presents the room temperature Raman spectrum of the ZnO NP in the 50-1000 cm^{-1} spectral range. The spectrum consists of four peaks located at about 100, 386, 440 and 580 cm^{-1}, which correspond to the E_{2L}, A_1 (TO), E_{2H} and A_1 (LO) fundamental phonon modes of hexagonal ZnO, respectively. The Raman peak located at about 144 cm^{-1} was assigned to $2E_{2L}$ second order phonon mode. Other three Raman peaks located at about 331, 508 and 664 cm^{-1} could be assigned to the $3E_{2H} - E_{2L}$, $E_1(TO) + E_{2L}$ and $2(E_{2H} - E_{2L})$ multi phonon scattering modes, respectively. The optical studies of ZnO NP were carried out by UV-Vis absorption and PL emission spectrum. Figure 4(a) is the UV-Vis absorption spectrum. An absorption peaks centered at $\lambda_{Max} = 373$ nm (3.32 eV) was found, which is in good agreement with the literature [9]. The PL spectrum of the ZnO NP is shown in Fig 4(b). The spectrum of the ZnO NP mainly consist of five emission bands: a strong UV emission band at 419 nm (2.96 eV), a weak blue band at 467 nm (2.65 eV), a weak blue-green band at 494 nm (2.51 eV), a weak green band at 558 nm (2.22 eV) and a narrow yellow-orange band at 608 nm (2.04 eV). The strong UV emission corresponds to the exciton recombination related near-band edge emission of ZnO [13]. The blue and weak blue-green emissions are possibly due to surface defects in the ZnO NP. The weak green band emission corresponds to the singly ionized oxygen vacancy and this emission results from the recombination of a photo-generated hole with the single ionized charge state of specific defects [14]. The narrow yellow-orange band is frequently assigned with excess oxygen as well as interstitial zinc atoms. The low intensity of green emission may be due to the low density of oxygen vacancies during the preparation of ZnO NP, whereas the room temperature PL emission intensity should be attributed to high purity with perfect crystallinity of synthesized ZnO NP.

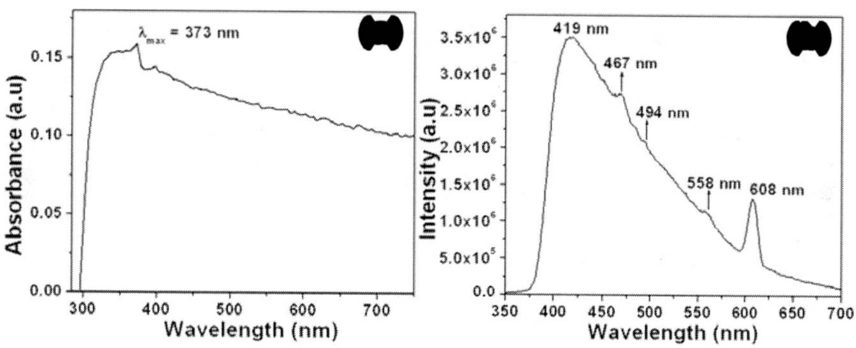

Fig. 3: (a) UV-Vis Absorption and (b) PL spectra of ZnO Nanoparticles

CONCLUSION

In summary, the structural and optical characteristics of ZnO NP synthesized by simple solution combusting method were studied. The XRD, EDX, HR-SEM, HR-TEM, SAED, FT-IR and Raman studies are performed to investigate the structural and morphologies of synthesized

ZnO NP. These results indicate high purity hexagonal wurtzite structure with the average particles size of around 50 nm can be achieved by simple solution combusting method. Raman scatting of ZnO NP has first and second orders optical modes which are the characteristic bonds of the wurtzite ZnO. The optical properties were studied through UV-Vis and PL. Due to Nanometer particle size, detailed information can be expected at the surface of ZnO NP, which influences this oxygen-vacancies related luminescence process greatly.

REFERENCES

[1] A.B. Djurisic *et al.*, *Prog. Quantum Electron.* 34, 191-259, 2010.
[2] Z.L.S. Seow *et al*, *Nanotechnology* 20, 045604, 2009.
[3] J.J. Cole et al., *Nano Lett.* 8, 1477-1481, 2008.
[4] T. Jin et al., *J. Food Sci.* 74, 46-52, 2009.
[5] N. Padmavathy et al., *Sci. Technol. Adv. Mater.* 9, 035004, 2008.
[6] E. Pal et al., *Colloids Surf.*, A. 340, 1-9, 2009.
[7] J. Zhao et al., *Mater Lett.* 64, 569-572, 2010.
[8] Q. Zhang et al., *Adv. Mater.* 21, 4087-4108, 2009.
[9] Y.Ni et al., *Nanotechnology* 18, 155603, 2007.
[10] B.N. Dole et al., *Curr. Appl Phys.* 11, 762-766, 2011.
[11] S. Maensiri et al., *J. Cryst. Growth.* 289, 102-106, 2006.
[12] C.X. Xu et al., *Nanotechnology* 15, 856-861, 2004.
[13] L. Bergman et al., *J. Appl. Phys.* 96, 675-683, 2004.
[14] R.S. Zeferino et al., *J. Appl. Phys.* 109, 014308, 2011.

SYNTHESIS AND CONDUCTIVITY STUDY ON TiO_2 INCORPORATED PVCAC BASED POLYMER ELECTROLYTE SYSTEM

T. Subbu Lakshmi and S. Rajendran

School of Physics, Alagappa University, Karaikudi, Tamil Nadu
E-mail: sraj54@yahoo.com

ABSTRACT

Polymer electrolytes based on PVCAC were prepared using solvent casing technique. XRD and FTIR analyses confirm the comlexation of polymer and salt. From the conductivity measurement, it is observed that the composition PVCAC/ $LiBF_4/EC/TiO_2$ (15 wt%) shows maximum ionic conductivity 7.82×10^{-5} S/cm at room temperature (303 K). The non-linear behaviour of the temperature dependent plot indicates that the ion transport in polymer electrolyte depends on polymer segmental motion, which obeys the Vogel-Tamman-Fulcher (VTF) equation.

INTRODUCTION

Rapid growth in technologies requires the development of a new generation of high performance energy sources. Polymer-based ion conducting materials have generated remarkable interest in the field of lithium batteries due to their application as electrolyte. In this application, the polymer electrolyte provides both the mechanical function of a separator between two electrodes and the ionic conduction medium of an organic solvent for the lithium salt. The incorporation of solid inorganic fillers seems to be an alternative for increasing the ionic conductivity [1-3]. Some fillers like TiO_2 show strong capability to provide higher conductivity, cationic transference number and mechanical strength [4,5]. The high mechanical stability and thermal stability of Poly (vinyl chloride-co-vinyl acetate) PVCAC

V. Rajendran, K. Saminathan, P. Paramasivam and K.E. Geckeler (eds.)
Nano Materials Synthesis and Characterisation, pp. 257-262 (2012).

copolymer is superior to related homopolymers. Polymer electrolyte films are prepared from PVCAC and $LiBF_4$ with different ratio of TiO_2 by using solution casting technique.

EXPERIMENTAL

Poly(vinyl chloride-co-vinyl acetate) (PVCAC; Aldrich, USA) and Lithium tetrafluoroborate ($LiBF_4$; Aldrich, USA) were used as received. All the solid polymer electrolyte films were prepared by solution casting technique. Appropriate weight of polymer and salt were dissolved in Tetrahydrofuran (THF; Merck, India) separately at room temperature. After the complete dissolution of polymer and salt they were mixed together and then the plasticizer and TiO_2 (Aldrich, USA) added to it. This mixture was stirred for 24 hours and then it was allowed to evaporate slowly until a homogeneous solution was obtained. The obtained highly viscous solutions were cast onto clean glass plates to obtain the desired thin film membranes. The films were further dried for 5 hours in vacuum to remove the residual solvent THF.

The Fourier Transform Infrared (FT-IR) spectra of the samples were taken with Perkin Elmer 577-IR spectrophotometer in the range of 400-4000 cm^{-1} in transmittance mode at ambient conditions. The X-ray diffraction patterns were recorded with X'pert PRO PANalytical diffractometer in the scanning range $2\theta = 10$ to $80°$ using Cu-Kα (wavelength, $\lambda = 1.541$Å). The ionic conductivity of polymer electrolytes were carried out with the help of stainless steel blocking electrodes by using a computer controlled μ-Autolab type-III Potentiostat/ Galvanostat in the frequency range 40Hz–300 KHz. The conductivity (σ) was calculated using the formula $\sigma = l/R_bA$, where l is the thickness of the polymer electrolyte film, R_b is the bulk resistance, and A is the surface area of the polymer electrolyte film.

RESULTS AND DISCUSSION

XRD Studies

The XRD pattern of the pure and complex are shown in Fig. 1. From Fig. 1, pure polymer exhibits amorphous nature and it shows a broad diffraction peak around ($2\theta = 23°$) and the XRD pattern of $LiBF_4$ shows high crystallinity which is revealed from the sharp peaks. These sharp peaks are almost found to be disappeared in the polymer complex. The absence of crystalline peaks may be due to the addition of plasticizer, hence reduces crystallinity and enhances amorphous nature. This in turn increases the free volume of the system and thus permitting the free flow of ions from one site to the other [6].

In case of systems S2, S3 and S5, fillers are added in the composition of (x = 5 wt%, 10 wt%, 20 wt %,) which show small crystalline peaks. Complex S4 reveals more amorphous nature compared to other films. This is in accordance with the conductivity data. Further the addition of inorganic filler into the complex would increase the dissolution of the charge carriers in the matrix and hence ionic conductivity has been improved [7].

On the other hand TiO_2 filler is crystalline, and is showing its characteristic crystalline peaks at $2\theta = 25, 27, 35$ and $54°$. So TiO_2 containing micro composites partially suppress the amorphous nature of the sample and exhibit some small crystalline peaks.

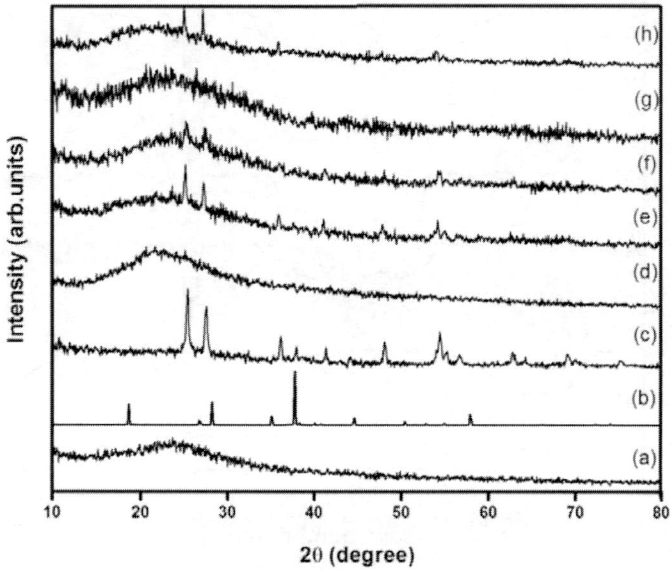

Fig. 1: XRD patterns of (a) Pure PVCAC (b) Pure $LiBF_4$ (c) Pure TiO_2 PVCAC/$LiBF_4$/EC/TiO_2 [x= 0wt% (d), 5wt% (e), 10wt% (f), 15wt% (g), 20wt% (h)]

FT-IR Studies

The FTIR spectrum of pure polymer, salt and complexes PVCAC/$LiBF_4$/EC/TiO_2 are shown in fig.2. The absorption peaks of the pure copolymer at 2970 cm^{-1} is assigned to (-C-H) stretching mode and 1736 cm^{-1} to C=O stretch. And the vibrational bands at 1439 cm^{-1} is due to CH_3 rocking and the peaks at 1372, 1237, and 1024 cm^{-1} correspond to (C-O) stretch coupled with (C-C) vibration. The vibrational peak at 996 cm^{-1} exhibits skeletal vibration and at 740, 700, 635 and 612 cm^{-1} infer the C-Cl stretch of PVCAC [8]. The characteristic vibrational peak of $LiBF_4$ at 532 cm^{-1} is assigned to BF_4^- ion. This peak is found to disappear in the complex revealing the interaction between the polymer and salt [9]. The characteristic peak of TiO_2 at 1650 cm^{-1} is found to shift to 1688 cm^{-1} in the complex [10]. By comparing the band spectra of pure polymers and polymer salt complex, it is clear that the band assignments of polymer salt complex are found to shift from their pure spectra. In addition to these, some new peaks are found in the complexes. The formation of new peaks along with changes in the existing peaks in the FTIR spectra confirms the complexation.

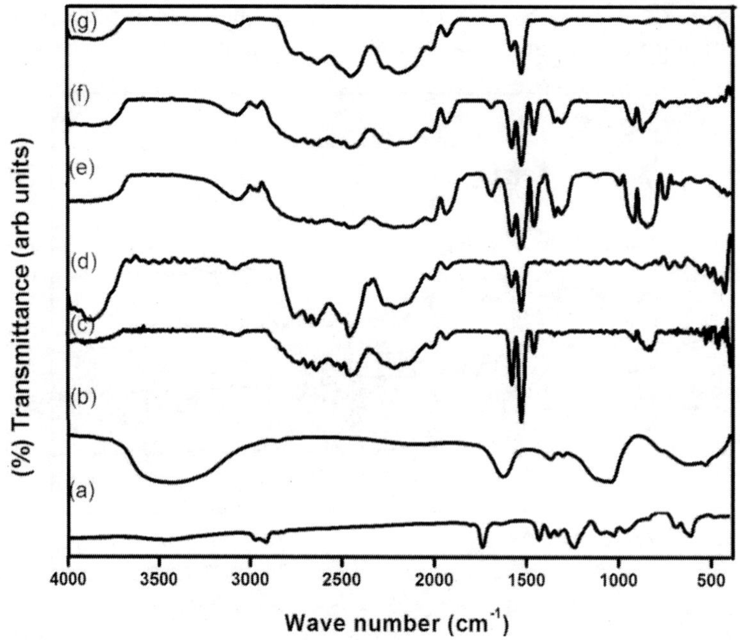

Fig. 2: FTIR Spectra of (a) Pure PVCAC (b) Pure $LiBF_4$, PVCAC/$LiBF_4$/EC/TiO_2 [x = 0wt% (c), 5wt% (d), 10wt% (e), 15wt% (f), 20wt% (g)]

Conductivity Measurements

Impedance plots of PVCAC/$LiBF_4$/EC/TiO_2 (TiO_2 = 0, 5, 10, 15, 20 wt%) for various temperatures were taken. From the complex impedance plot, the intercept at the high frequency range on the Z' axis gives the bulk resistance value of the electrolyte. The conductivity of the electrolyte is calculated from the measured bulk resistance, area and thickness of the polymer electrolyte system.

It is observed that the system with TiO_2 (15 wt%) shows maximum ionic conductivity 7.816×10^{-5} S/cm at room temperature (303 K). The ionic conductivity of the electrolyte systems with varying filler concentration is depicted in Fig.3. From the figure it has been observed that as temperature increases the conductivity also gets increased for all the compositions. This is in agreement with the free volume theory. According to free volume theory as temperature increases the polymer can expand easily and produce free volume. Thus ions and solvated molecules can move into the free volume and the resulting conductivity is exhibited by the overall mobility of ions, supporting the disappearance of high frequency semi circular region. It is observed that the conductivity of the electrolyte increases linearly for the increasing filler concentration from (x = 0 to 15 wt %) and at higher filler concentration (x = 20 wt %) again the conductivity gets decreased. The decrease in conductivity at higher TiO_2 contents is consistent with other reports [11].

This may be due to agglomeration of primary particles to bigger particles which then leads to lower surface area. This mechanism assumes that the surface area of the filler is the key factor in determining high conductivity [12].

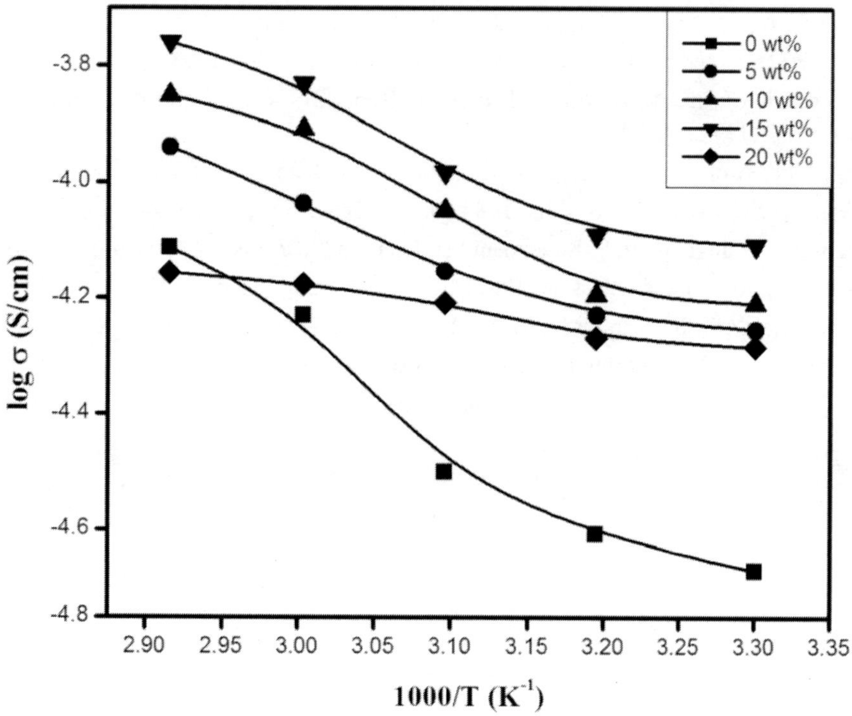

Fig. 3: Impedance Plot for the Compositions PVCAC/$LiBF_4$/EC/TiO_2 [x = 0 wt% (S1), 5wt% (S2), 10wt% (S3), 15wt% (S4), 20wt% (S5)]

CONCLUSIONS

PVCAC based composite polymer electrolyte system was prepared by solvent casting technique. The amorphicity nature of the sample was revealed from XRD studies and it also gave evidence for the complex formation of polymer and salt. FTIR spectrum confirms the complexation (i.e.) the formation of homogeneous polymer salt electrolyte system. Electrical conductivity of the samples was studied at different temperatures. Maximum ionic conductivity of 7.8163×10^{-5} S/cm at 303 K was obtained for the compositions PVCAC/$LiBF_4$/EC/TiO_2 (15wt%). As the temperature is increased, the conductivity gets increased which supports the free volume theory.

REFERENCES

[1] B. Scrosati, F. Croce and S. Panero, *J. Power Sources* 100 (2001) 93.

[2] A.S. Best, J. Adebahr, P. Jacobsson, D.R. MacFarlane and M. Forsyth, *Macromolecules* 34 (2001) 4549.

[3] C. Karlsson, A.S. Best, J. Swenson, W.S. Howells and L. Borjesson, *J. Chem. Phys.* 118 (2003), 4206.

[4] F. Croce, G.B. Appetecchi, L. Persi and B. Scrosati, *Nature* 394 (1998), 456.

[5] F. Croce, R. Curini, A. Martinelli, l. Persi, F. Ronci, B. Scrosati and R. Caminiti, *J. Phys. Chem.B* 103 (1999), 10632.

[6] R.E. Hollingsworth and J.R. Sites (1982), *J. Appl. Phys.*, 53: 5357–5358.

[7] H. Van Ryswyk and Arthur B. Ellis (1986), *J. Am. Chem. Soc.*, 108: 2454–2455.

[8] K. Uosaki, Y. Shigematsu, S. Kaneko and H. Kita (1989), *J. Phys. Chem.*, 93: 6521–6526.

[9] R.R. Chang, R. Iyer and D.L. Lile (1987), *J. Appl. Phys.*, 61: 1995–2004.

[10] H.J. Frenck, W. Kulisch and R. Kassing (1989), *Proc. SPIE* , 1144: 250–258.

[11] S. Rajendran, O. Mahendran, R. Kannan, (2002) *J. Phys. and Chem. Solids*, 63, 303-307.

[12] S. Rajendran, O. Mahendran and R.Kannan (2002), *J. Solid State Electrochem* 6: 560-564.

CeO_2 NANOCRYSTALINE THIN FILMS: PREPARATION AND CHARACTERIZATION

D.L. Monika, C.J. Shilpa, D.V. Sunitha, J. Malleshappa, H. Nagabhushana, K.L. Narasimha Rao[1] and B.M. Nagabhushana[2]

Prof. C.N.R.Rao Advanced Material Research, Tumkur University, Tumkur
[1]Department of Instrumentation and Applied Physics, Indian Institute of Science, Bangalore
[2]Department of Chemistry, M.S Ramaiah Institute of Technology, Banglore
E-mail: bhushanvlc@gmail.com; 0810monika@gmail.com

ABSTRACT

Cerium oxide (CeO_2) thin films were prepared by sol-gel spin coating technique on quartz and glass substrates. The stable sol was prepared by cerium nitrate as a precursor, 2 methoxy ethanol as a solvent, acetyl acetone, PEG as an acid catalyst and stabilizer respectively. The coated films are characterized by X-ray diffraction (XRD), scanning electron microscopy (SEM) and UV-vis spectroscopy. The particle size obtained by Debye Scherer's formula was found to be 35-155 nm. Further, dislocation density, micro strain and texture co-efficient were calculated from XRD peaks. The optical band gap energy estimated using Tauc relations and was found to be 3.60eV.

INTRODUCTION

Cerium oxide (CeO_2) thin films have received considerable interest because of their high transparency in the visible, near IR region, chemical stability [1], high refractive index, good transmission and wide band gap[2-3] and electro-optical properties. Therefore they find potential application in insulating buffer layers, ion conducting layers, ultra-precise polishing, gas sensors, catalysis, SOFC, optical filters, electrical shielding and protective coatings, etc., [4-6]. Therefore, pure and doped CeO_2 thin films have been widely used in the field of semiconductors, photo

V. Rajendran, K. Saminathan, P. Paramasivam and K.E. Geckeler (eds.)
Nano Materials Synthesis and Characterisation, pp. 263-268 (2012).

catalysis, fuel cells and high-temperature superconductors [3-7]. Among these applications, the optical properties of the prepared films such as the band gap energy play a significant role regarding the device quality and range of applications [8-9].

In recent years some attempts have been performed for the preparation of rare earth ion doped CeO_2 by various techniques such as pyrolysis, solvethermal, sol gel, solid state ceramic reactions, etc.[10]. The sol-gel method of fabricating thin films offers potential advantages over traditional technique as because of two temperature processing, easy coating of large surface, small thickness, high optical quality and purity. In the present work we reporting sol-gel derived spin coated nanostructured CeO_2 thin films deposited on glass and quartz substrate and Characterized by different techniques.

EXPERIMENTAL

Preparation

The CeO_2 films are prepared by sol gel spin coating method. The precursor cerium nitrate was mixed with 2-methoxy ethanol solvent and solution was stirred vigorously for 10 min at room temperature in a closed beaker. Then 30 ml of PEG and 10 ml of acetyl acetone is added as stabilizing agent and acid catalyst respectively, solution was stirred for 24 hours. After stirring the stock solution was filtered and kept in air tight beaker for aging. Substrates were cleaned with HCl, acetone, distilled water and ultrasonic cleaner. Once the solution is prepared film was deposited on glass and quartz substrate using spin coating method with spin sped 3000 rpm in 50 secs. Obtained films are dried at 80°C for 2 hrs. The formation of CeO_2 stock solution can be represented by the following flowchart.

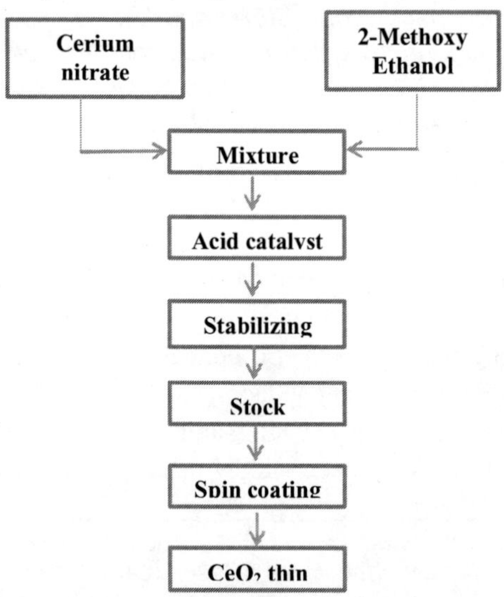

Flow chart for the preparation of CeO_2 thin films

Measurements

The structural properties were studied by X-ray diffraction measurements with Cu ká radiation with wavelength 1.5406 Å. The average dimension of crystallite size was determined by Debye Scherer method. The morphological features of CeO_2 thin films were studied by Scanning electron microscope (SEM) and optical measurements of the crystalline films were carried out at room temperature using UV-Vis Spectrophotometer in the wavelength range 400 to 800 nm. Swanepole's envelope method was employed to evaluate the optical constants.

RESULT AND DISCUSSION

XRD

Fig. 1 shows the XRD patterns of CeO_2 thin films of different thickness annealed at 500°C on glass substrate. Peaks assigned to (111), (200), (220) and (311) reflexes at an angle 28.4°, 33.4°, 47.6°, 56.3° which are well matched with JCPDS card No. 81-0792 and corresponds to cubic fluorite structure with lattice parameter ~ 5.4 Å. The crystallite size were estimated using Scherer formula

$$D = \frac{0.9\lambda}{\beta cos\theta} \quad ... (1)$$

where λ is wavelength of X-rays, β is full width at half maximum and θ is angle of diffraction. The value found for the average grain size corresponding to the (111) peak for film t_1 nm is 34 nm and for the film t_2 is 155 nm. Dislocation density was calculated using grain size D, given by the Williamoson and Smallmans relation [11].

$$\delta = \frac{1}{D^2} \quad ... (2)$$

where D is crystallite size. The possible dislocation density and the variation of texture coefficient for the peaks are given in Table 1. The *TC(hkl)* represents the texture of particular plane deviation[12]. A sample with randomly oriented crystallite presents *TC(hkl)* = 1. while the larger this value, the larger abundance of crystallites oriented at *(hkl)* direction.

Table 1: Structural Parameters of CeO_2 Thin Films.

Sample	*2θ degree*	*(hkl)*	*$\delta x10^{14}$ m^{-2}*	*I/I_0*	*TC*
t_1 = 49 nm	28.69	(111)	8.50	0.3120	0.4996
	32.98	(200)	46.0	0.5150	1.1004
	47.72	(220)	4.30	0.7628	3.2540
t_2 = 62 nm	28.67	(111)	22.1	0.3093	0.5128
	32.69	(200)	260	0.7243	1.0982
	47.40	(220)	6.30	0.7243	3.2025

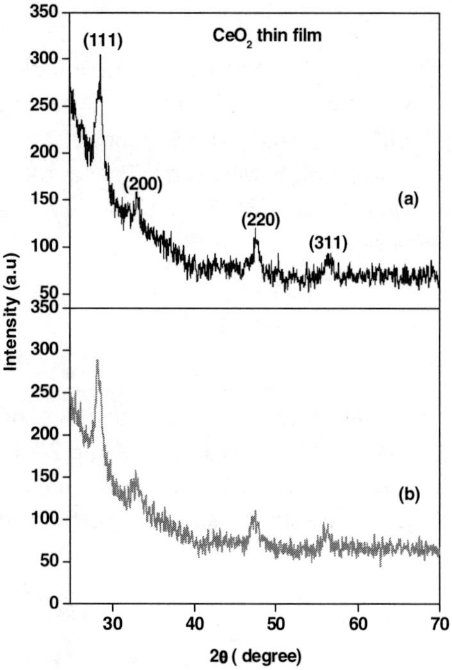

Fig. 1: X-ray Diffraction Patterns of (a) t_1, (b) t_2 CeO_2 thin Films.

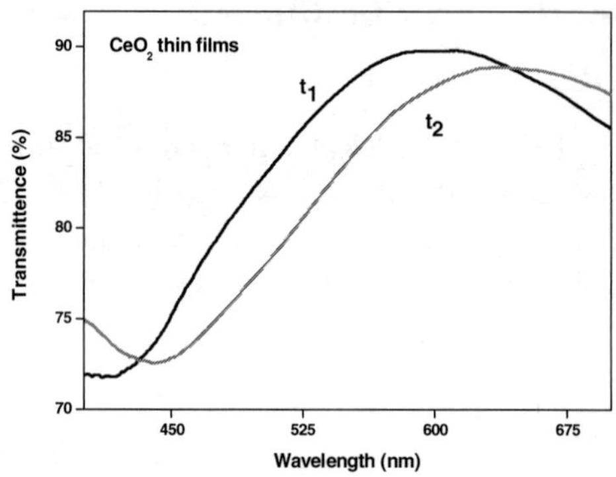

Fig. 2: Transmittance Spectra of CeO_2 thin Films

Optical Properties

Figure 2. shows the transmittance spectra for CeO_2 thin films. The measurements were conducted in the room temperature using glass substrate as background. The obtained films are largely transparent in the visible region. The refractive index and thickness of the CeO_2 thin films were estimated by transmittance spectra [13]. Thickness of the film was found to be 49 nm (t_1) and 62 nm (t_2). Further the maximum transmittance reaches to 90% for the film with thickness of 49 nm. It is clear from the Fig. 2 the transmittance decreases with increase of film thickness [14].

The absorption coefficient (á) of the thin films were calculated from the respective transmittance spectra using relation

$$\alpha = \frac{ln\left(\frac{1}{T}\right)}{t} \qquad \text{... (3)}$$

where *T* is the transmittance of the film and *t* is the thickness of the film coated on glass substrates at given wavelength. Optical band gap (E_g) of nanostructured CeO_2 film was estimated using wood-Tauc relation given by

$$(\alpha h\nu)^n = B\ (h\nu - E_g) \qquad \text{... (4)}$$

where B is constant, $n = 2$ for direct band gap and $n = 1/2$ for indirect band gap. $(\alpha h\nu)^n$ vs $h\nu$ curve for two films of different thickness have been plotted. The band gap value changed from 3.5 to 3.6 eV for the film t_2 to t_1 it is observed that decrease in the band gap value may be due to the result of improvement in the crystalline nature and stress relaxation of the film with increase in thickness t_2 to t_1[15-16]. The refractive index (n) has been estimated by envelope technique. The observed refractive index values for CeO_2 films annealed at 500° C are found to be 2.5 for t_1 and 2.4 for t_2 at 550 nm. The refractive index values found to be decreases with increasing wavelength as shown in Fig. 4 [17].

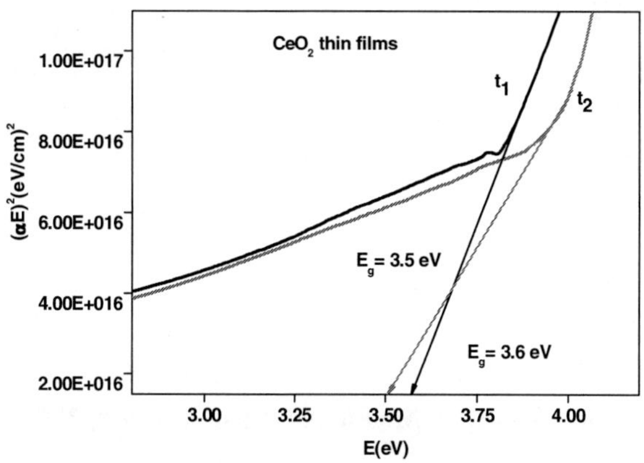

Fig.3: Optical Band Gap of CeO_2 Thin Film

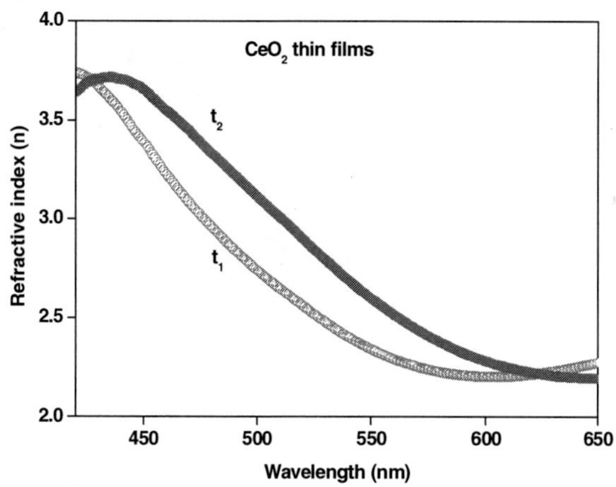

Fig. 4: Plot of Refractive Index as a Function Wavelength of CeO_2 Film

CONCLUSION

CeO_2 nanocrystaline thin films have been deposited on glass substrates by sol gel spin coating method. The X-ray diffraction patterns of film reveal the existence of cubic fluorite structure.

The average grain size was found to be 35 nm and 155 nm for t_1 and t_2 films respectively. The optical constants such as refractive index (n) and extension coefficient k were determined from transmittance spectrum in UV-vis regions using envelope method. The thickness of the film t was calculated from interference of transmittance spectra, also an energy band gap value (E_g) was calculated which is found to be 3.5 and 3.6eV for t_1 and t_2 respectively. The deposited films CeO_2 thin films was suitable for many optical devices like solar cells, etc., because of wide crystallized high transmittance in the visible region and wide band gap value.

ACKNOWLEDGEMENTS

The authors gratefully acknowledge NRB-DRDO, New Delhi for providing the financial assistance and Dr. M. Vishwas, Government science college, Bangalore for fruitful discussion regarding optical studies.

REFERENCES

[1] R.G. Schwab, R.A. Steiner, G. Mages and H.J. Beie,. Vol. *Thin Solid Films* 207 (1992) 283.

[2] D.E.Zhang, X.J.Zhang, J.M Song H.G Zheng.. 7-2468, 2006, Vol. Chem.Phys.Chem.

[3] S.R.Foltyon, P. Tiwari,, R.C Dye, M.Q le X.D. Wu. 63-1848, 1993, Vol. Appl.Phys.Lett. .

[4] H. Guo, Y. Qiao (2008), Vols. App. Surf. Sci., 254- 1961–1965.

[5] T. Masui, K. Machida, T. Sakata, H. Mori G. Adachi,. (1997), *Vols. J. of Alloys and Compds* 256 -97-101.

[6] M.A.A. van Wijick, M.A.J. Verhoeven, E.M.C.M. Reuvekamp, C.J. Gerritsma, D.H.A. Blank, H. Rogalla,. *Vol. Appl. Phys. Lett.* 67 (1996).

[7] W.D.Xiao, Q.L Guo. E.G wang. *Vol. Chem.Phys. Lett.* 368(2003)527.

[8] A.Yoshikawa, H. Matsunami, Y. Nanshi, *Widebandgap semiconductors,* s.l. : Springer Berlin Heidelberg, 2007.

[9] L.G Devi B.N. Murthy. *Vol. Catal. Lett.* 125 (2008) 320.

[10] Anees A. Ansri, S.P singh, B.D. Malhotra. *Vols. J. alloys and comp* 509(2011) 262-265.

[11] D. P. Padiyan, A. Marikini, K.R. Murli. *Vol. Mat. Chem. And Phys.* 78 51 (2002).

[12] *Structure of metals.* C.S Barret. T.B massalski. 1980, S.l. : Pergamon press, Oxford.

[13] Swanepoel, R. *Vol. J.Phys. E* 16(1983) 1214.

[14] Yinzhu Jiang, Naoufal Bahlawane, 2009, *Vols. J. of Alloys and Comp* 485, L52-L55.

[15] D.Yuvaraj K Narimha Rao. Vols. Vaccum 82(2008)1274-1279.

[16] Tan ST, Chen BJ, Sun XW, Fan WJ, Kwok HS Zhang XH et al. Vol. J. Appl. Phys. 2005; 98:013505.

[17] M. Vishwas, Sudhir Kumar Sharma, K. Narasimha Rao, S. Mohan, K.V. Arjuna Gowda and R.P.S. Chakradhar. *Vols. Spectrochimica Acta Part* A 74 (2009) 839–842.

PREPARATION AND CHARACTERIZATION OF NANOPOROUS ACTIVATED CARBON AND ITS APPLICATION AS A HETEROGENEOUS FLUIDIZED BED CATALYST FOR THE TREATMENT OF MUNICIPAL WASTEWATER

S. Karthikeyan, R. Boopathy, P. Maharaja, J. Sharmila, P. Kumara Sashidara and G. Sekaran

Environmental Technology Division, Council of Scientific & Industrial Research (CSIR) – Central Leather Research Institute (CLRI), Adyar, Chennai
E-mail: ganesansekaran@gmail.com

ABSTRACT

In the present investigation an attempt was made to treat the municipal waste water in Fluidized Immobilized Catalytic Carbon Oxidation (FICCO) reactor. In this study rice husk carbon (nanoporuos Activated Carbon, NAC_{800}) is used as the catalyst, in which the contaminants adsorb the catalyst. The main objective of the present study is to assess the efficiency of reducing the organic pollutant in the waste water through fluidized immobilized carbon catalytic oxidation cell reactor with minimum sludge production. The effective removal of organic pollutants through various concentrations of starch, protein, surfactant, catalyst loading, and time was found to be most successful in reducing the COD and BOD removal efficiencies of the FICCO reactor.

INTRODUCTION

Increasing discharge and improper management of municipal waste water have created a great concern among the environment. Municipal waste water consists of toxic compounds or

V. Rajendran, K. Saminathan, P. Paramasivam and K.E. Geckeler (eds.)
Nano Materials Synthesis and Characterisation, pp. 269-274 (2012).

compounds which may be potentially mutagenic of carcinogenic. There is a need for the protection of the public health in a manner commensurate with environmental, economic, political and social concerns (Metcalf and eddy). The removal efficiencies of suspended solids (SS) and chemical oxygen demand (COD) are around 50% and 30% respectively in a typical primary treatment plant.

Hence application of biological methods such as fluidized bed reactors has generated a lot of interest in the recent past in the field of waste water treatment processes due to its high performance efficiency compared to the conventional suspended growth and fixed-film waste water treatment processes. The major advantage of fluidized bed reactor over other biodegradation systems is a higher biomass concentration, and a higher mass transfer, resulting in a higher rate of biodegradation.

Therefore the main aim of the present study is to evaluate the effectiveness of reducing the organic pollutant in the waste water through fluidized immobilized carbon catalytic oxidation cell reactor with minimum sludge production.

EXPERIMENTAL

The activated carbon samples from rice husk were prepared in two sequential steps: pre-carbonization at 400°C and chemical activation (Kennedy et al. 2004). The carbon samples have been characterized for surface area, pore volume and pore size, X-ray diffraction technique (XRD) and a SEM.

RESULTS AND DISCUSSION

Fig. 1 presents nitrogen adsorption/desorption isotherms at 77°K for NAC_{700}, NAC_{800} and NAC_{900} respectively. The presence of nanoporous is justified by the hysteresis effect and the slope of the plateau increased to yield type IV isotherms with a significant increase in the nitrogen uptake through the entire pressure range. XRD pattern of MAC_{800} (Fig. 2) contains a mixture of amorphous and crystalline phases of silica. The Low angle XRD spectrum of MAC_{800}, which was a characteristic of nanoporous material, is shown in the Fig. 3. It shows a very intense diffraction peak (100) and three obvious peaks (110), (200) and (210), which suggest the preservation of the hexagonal structure (regularity) with a thicker wall. The thermo gravimetric analysis of NAC_{800} samples were carried out in the presence of nitrogen as shown in Fig. 4. NAC_{800} exhibits three-step weight loss in the presence of nitrogen atmosphere. The SEM image of NAC_{800} sample, shown in Fig. 5, depicts that the activated carbon matrix is highly porous in nature with major distribution of nanoporous, which is the characteristic feature of activation using phosphoric acid as the activant.

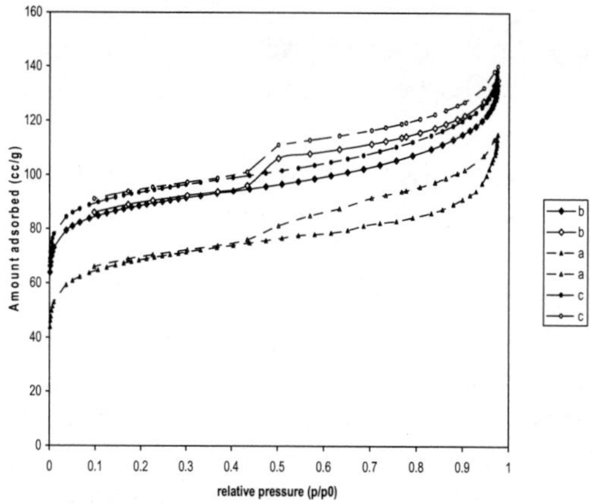

Fig. 1: N_2 Adsorption/desorption Isotherms at 77 K Onto (a) NAC_{700} (b) NAC_{800} and (c) NAC_{900}

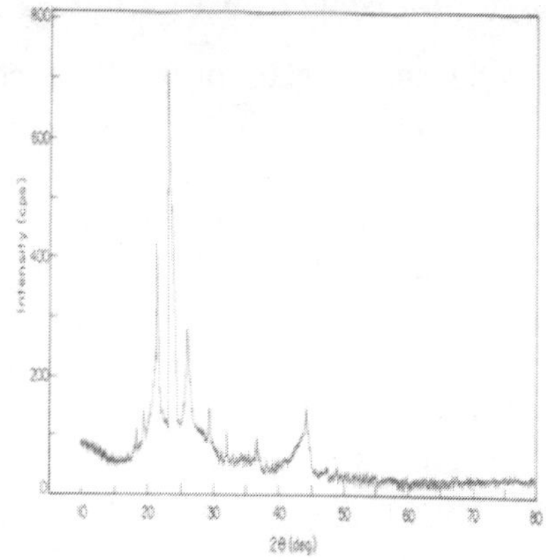

Fig. 2: XRD pattern of NAC_{800}

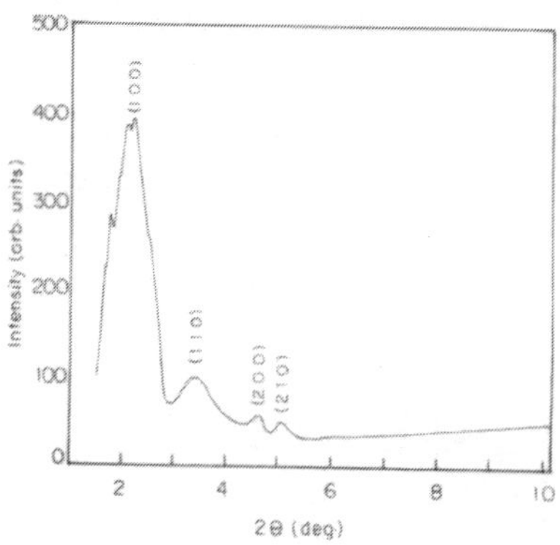

Fig. 3: Low angle XRD spectrum of NAC_{800},

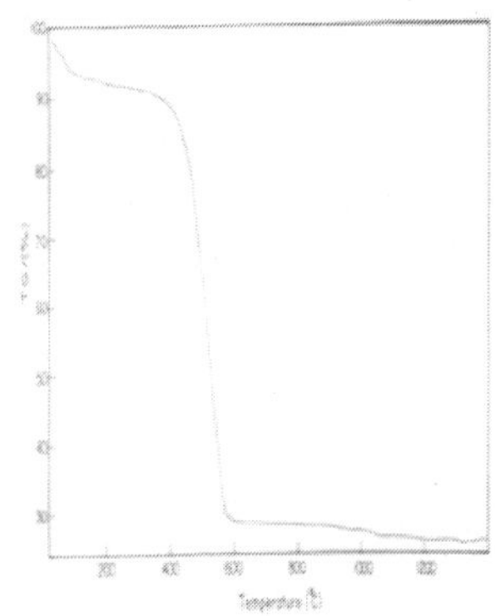

Fig. 4: TGA spectrum of NAC

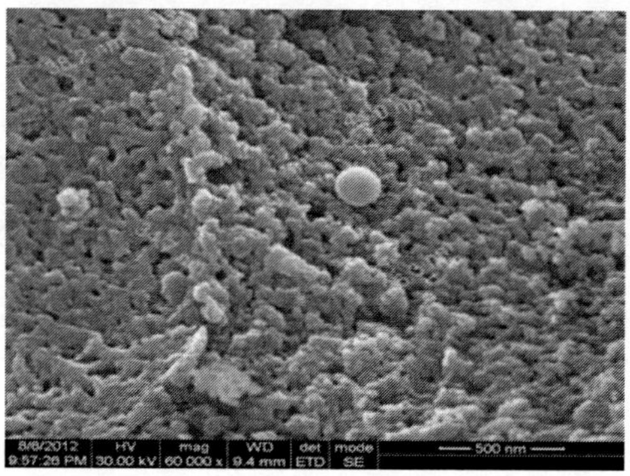

Fig. 5: SEM

Design and Selection of FICCO Reactor

An acrylic sheet reactor of height 12 cm and width 9 cm was fabricated. such reactors were fabricated [Fig 6] with different dimensions of upper hopper and differing in the provision for aeration. In addition to these the raw water and FICCO treated water were analyzed for characteristics which include pH, sulphates, chloride, COD, BOD, total solids, dissolved solids and suspended solids level [Table 1]. Figure 7 shows the percentage removal of COD and BOD with and without adding catalyst varying with time. Oxidation removed BOD by 46% and 84% with and without carbon respectively. Figure 8 shows the effect of catalyst mass loading on COD during the catalytic oxidation of domestic waste water. The percent reductions of COD increased with the increase in catalyst loading from 1 to 12 g/620ml. With further increase in catalyst mass loading from 12 to 15 g/620ml, there has been no increase in percent COD removal. The performance of the FICCO reactor was studied with the concentration of anionic surfactants. The average inuent concentration of surfactants was 23.8 mg/L, of which the FICCO reactor removed 80%. The COD and BOD removal efficiencies reached 82.7% and 81.5% respectively which is shown in the graph (Fig. 9), leading to average efuent concentrations of 4.8mg/L. When the reactor was operated at sucient aeration condition with the addition of starch, the removal eciencies of COD and BOD reached 88.1% and 87.7% (Fig. 10) respectively. The performance of the FICCO reactor was studied with the concentration of adding oil and protein to the waste water in addition to surfactant and starch as shown in the Fig. 11 and Fig. 12.

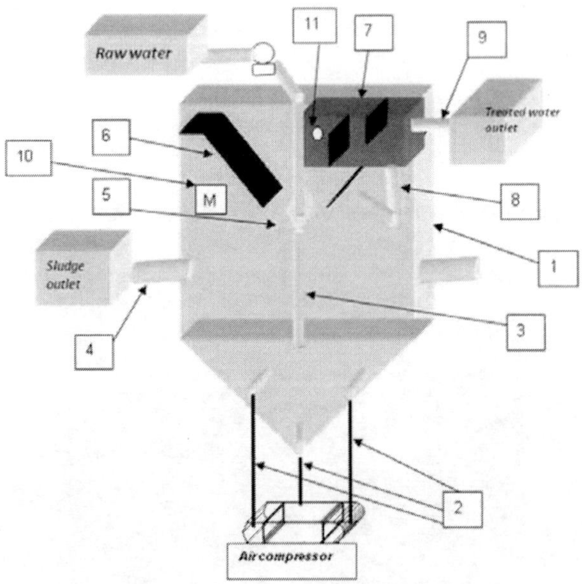

Fig. 6: Schematic Flow Diagram for Fluidized Bed Oxidation for the Treatment of Municipal Wastewater

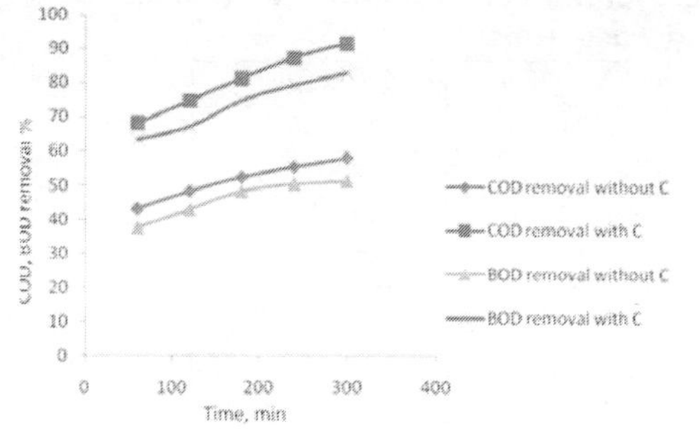

Fig. 7: Shows the Percentage Removal of COD and BOD with Function of Time

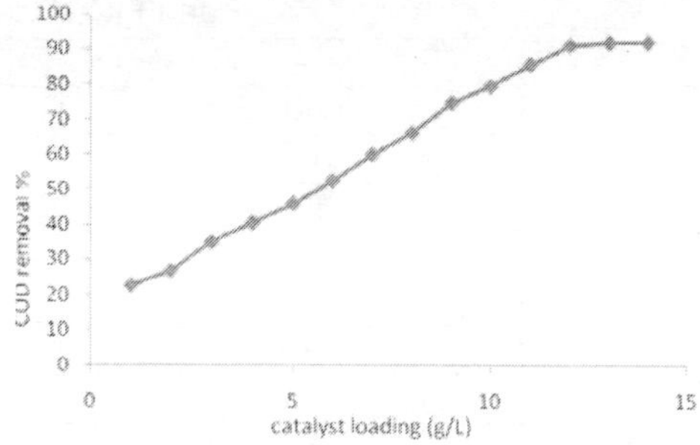

Fig. 8: Percentage Removal of COD as a Function of Catalyst Loading

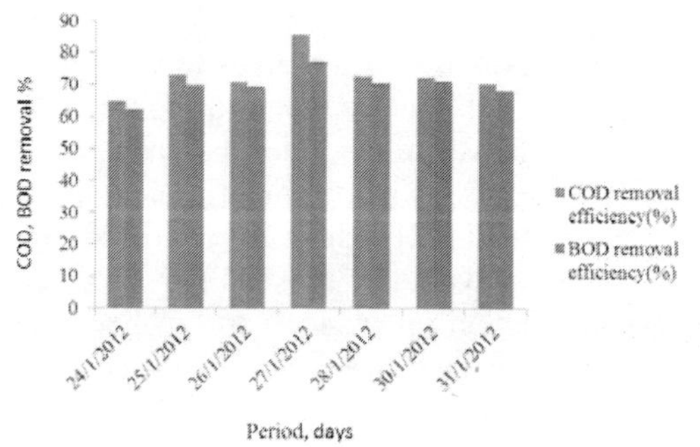

Fig. 9" Effect of Surfactant Loading on the COD Removal

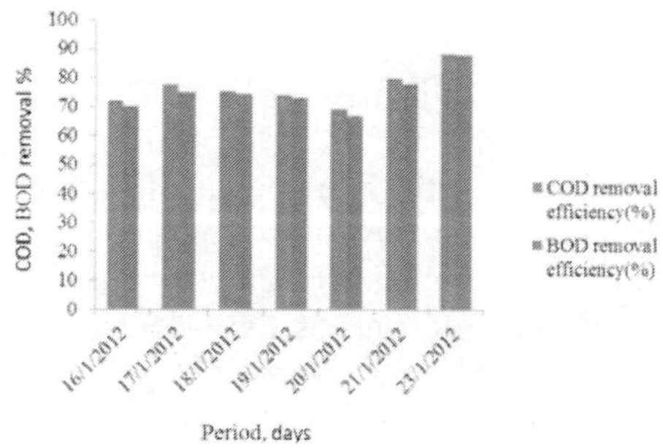

Fig. 10: Effect of Starch Loading on the COD Removal

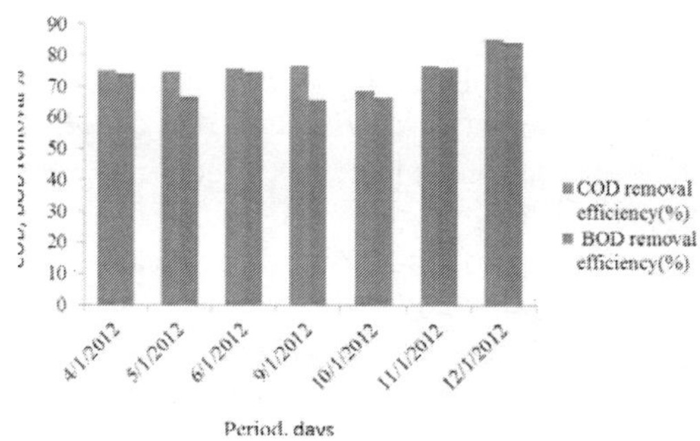

Fig. 11: Effect of oil Loading on the COD Removal,

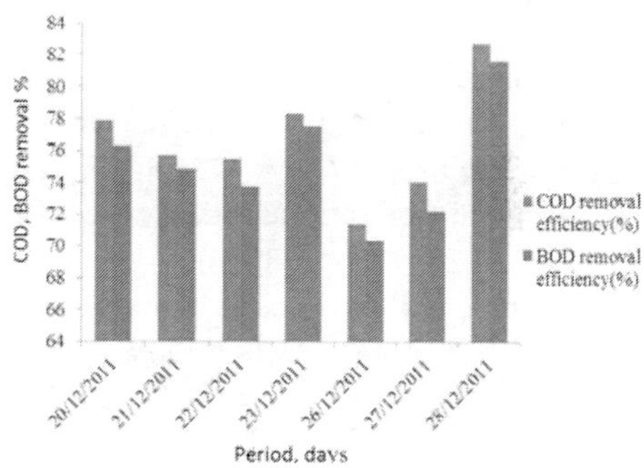

Fig. 12: Effect of protein loading on the COD removal

Table 1: Physicochemical parameters for raw domestic waste water and FICCO treated waste water

S. No	*Parameters*	*Raw Waste Water*	*FICCO Treated*
1.	pH	7.6 ± 0.6	8.2 ± 0.2
2.	BOD	171.1 ± 51.7	79.3 ± 5.2
3.	COD	514.7 ± 154.6	238.1 ± 16.5
4.	Total suspended solids	187.1 ± 60	82.6 ± 9.6
5.	Nitrogen	50.2 ± 21.6	31.1 ± 5.8
6.	Sulphate	36.1 ± 10.3	16.3 ± 5.9
7.	Chlorides	40 ± 6	13.2 ± 3.7
8.	Surfactant	11.1 ± 3.1	4.47 ± 0.4
9.	Oil and Grease	60 ± 8.9	16 ± 1.8

CONCLUSION

The rice husk based carbon was used as catalyst for the oxidation of organics in wastewater discharged from domestic origin. The rice husk based was efficient enough to reduce COD, BOD by 91.6% and 84% respectively. The consistent removal of pollution parameters was accompanied with the removal of pathogenic and antibiotic resistant bacteria. The sludge production is only 60% of the conventional sewage treatment plant. The investment cost towards Sewage treatment plant by employing rice husk based carbon would be drastically reduced and thereby the operational cost towards electrical energy consumption will also be very much reduced. The reactor C with upper hopper length of 3.5cm is considered to be the desired optimum condition for running the reactor with high COD and BOD removal efficiencies. The outlet of the fluidized reactor was then transferred to the algal pond for further treatment through algal process to make the water suitable for release into the outer environment.

REFERENCES

[1] Ahmadna A.L., Chong M.F., Bhatia S., Ismail S. (2006). Drinking water reclamation from palm oil mill effluent (POME) using membrane technology, *Desalination*, 191: 35-44.

[2] B.S. Girgis and A.A.M. Daifullah. Removal of some substituted phenols by activated carbon obtained from agricultural waste, *Water Res*., 32 (1998), pp. 1169–1177

[3] Eckenfelder W., Argaman A., 1991. Principles of biological and physical/chemical nitrogen removal[M]. In: Phosphorus and nitrogen removal from municipal wastewater (Sedlak R., ed.). New York: Lewis Publishers.

[4] Faust, S.D., and Hunter, J.V. (1967). Principles and applications of water chemistry. New York: John Wiley and Sons.

[5] Hasler, B., Lundhede, T., Martinsen, L., Neye, S., Schoi, J.S. (2005): "Valuation of groundwater protection versus water treatment in Denmark by Choice Experiments and Contingent Valuation" NERI Technical Report No. 543.

[6] Izrail S. Turovskiy and P.K. Mathai, Wastewater Sludge Processing, 2006, pp. 42.

[7] John. L. Kennedy, Mohan Das.K, Sekaran. G. (2004). Integrated biological and catalytic oxidation of organics/inorganics in tannery wastewater by rice husk based mesoporous activated carbon-*Bacillus sp, Carbon*, 42: 2399-2407.

SYNTHESIS OF CARBON NANOTUBES FROM ETHANOL USING RF-CCVD AND Fe-Mo CATALYST

Ramakrishnan S., Jelmy E.J., Dhakshnamoorthy M.[1], Murali Rangarajan, Vasanthakumari R.[1] and Nikhil Kothurkar

Department of Chemical Engineering and Materials Science, Amrita Vishwa Vidyapeetham, Coimbatore

[1]Polymer Nanotechnology Centre, B S Abdur Rahman University, Chennai, Tamil Nadu

E-mail: nikhil.kothurkar@gmail.com

ABSTRACT

Multiwalled carbon nanotubes (MWCNTs) were synthesized by RF heating at 850°C using an MgO-supported bimetallic Fe-Mo catalyst. Ethanol was used as the carbon source. The MWCNTs were tested using Raman spectroscopy, High resolution transmission electron microscopy (HRTEM) and thermo gravimetric analysis (TGA). HRTEM studies show nanotubes with an average diameter 12.2 ± 3.23nm (95 % confidence interval). In Raman spectroscopy, the ratios of intensities of the G and D band and the 2D and G bands were 0.90 and 0.47 respectively. These ratios are generally indicative of high crystallinity and purity of the MWCNTs. The MWCNT purity as calculated from TGA analysis.

INTRODUCTION

Since their discovery in 1991, multiwalled carbon nanotubes [1] (MWCNTs) have been at the forefront of the nanomaterials research. These nanometer size materials can think to be arising from rolling of graphene sheets into a cylindrical form. Soon after discovery, extensive research efforts on these tubular graphite structures revealed that CNT are the materials of great interest from fundamental point of view as well as for future applications in the areas of hydrogen

V. Rajendran, K. Saminathan, P. Paramasivam and K.E. Geckeler (eds.)
Nano Materials Synthesis and Characterisation, pp. 275-280 (2012).

storage, electronics, catalysis and polymer composites [5, 6]. These carbon nanotubes were synthesized by arc discharge method, laser ablation method and catalytic chemical vapor deposition (CCVD) method. Of these, CCVD is the most widely used because of the advantages that it presents in controlling the growth morphology of these nanostructures. The catalyst system and the reaction parameters like temperature and the flow rates of the carrier and hydrocarbon gases can be directly controlled [11, 12]. Moreover, by using RF-CCVD [15-17], high qualities CNTs (high crystallinity and defect free) are possible. Inductive heating in the CVD process can significantly reduce the energy consumption and enable better control on the heating rate compared to external furnace methods. Alcohols are particularly attractive as the hydrocarbon source in CVD synthesis of CNTs because they are cheap, easily available and safe to handle. They also produce high purity CNTs due to the etching effect by oxygen-containing radicals generated during alcohol decomposition, which oxidize the amorphous carbon [13, 14]. We propose to characterize in detail the morphologic and structural properties of the nanotubes synthesized at 850°C by RF-CCVD from ethanol on MgO supported Fe – Mo catalysts.

EXPERIMENTAL

The catalyst used was Fe $(NO_3)_3$ $9H_2O$, $(NH_4)_6Mo_7O_{24}$. $4H_2O$ supported on MgO (Sigma Aldrich, 99.99%, purity). Ethanol (Laboratory Grade) was used as the carbon source.

The Fe-Mo/MgO catalyst system was prepared in the weight ratio of 1:0.1:12. The melt salts Fe $(NO_3)_3.9H_2O$ and $(NH4)_6Mo_7O_{24}$. $4H_2O$ were dissolved separately in distilled water. These metal solutions were mixed with prepared MgO suspension and sonicated for 1 hr. The solvent was removed using a rotary evaporator at 90°C. The powder was further dried overnight at 150°C for 15 hours and calcinated in air at 700°C for 7 hours [2]. About 100 mg of the Fe-Mo/MgO catalyst was uniformly spread into a thin layer in a graphite boat and placed in the 2" quartz tube of the RF-CCVD setup. Heat was applied by adjusting the RF power of the generator (350 kHz) to maintain a constant temperature of 850°C for the duration of the reaction. The tube was purged with 800 mL/min flow of Ar for 10 min to remove the ambient air. Ethanol was introduced using a syringe pump at 4 mL/h for about 30 min. The tube was rinsed with Ar and allowed to cool to room temperature. The reaction product (powder) was collected from the graphite boat. It was sonicated in aqueous hydrochloric acid (v/v ratio: 1:1) for 30 min to remove the catalyst and support, followed by washing in water to neutral pH. The obtained CNTs were filtered and dried overnight at 120°C.

CHARACTERIZATION

High resolution transmission electron microscopy (HRTEM) was carried out on a JEOL JEM 2100 instrument. Thermogravimetric analysis (TGA) was conducted on a EXSTAR6200 (TG/DTA) thermal analyzer at 10 °C/min from room temperature to 900°C in air. Raman analysis carried out using a Renishaw (RM 1000) instrument with an Argon Ion Laser source of wavelength 514.5 nm. The exposure time for the each measurement is 100 second and 2 accumulations for the each measurement.

RESULT AND DISCUSSION

The Raman spectrum presented in the Figure 1 shows the characteristic bands corresponding to multiwall carbon nanotubes. The D band (between 1305 and 1345 cm^{-1}) is associated with the vacancies, defects, and carbonic impurities (amorphous carbon, glassy carbon, etc) that destroy the symmetry of the graphene sheet. The G band (between 1500 and 1600 cm^{-1}) corresponds to the splitting of the E_2g stretching mode for graphite, and the 2D band (between 2450–2650 cm-1) corresponds to the second harmonic of the D band and is associated to the degree of crystallinity of the graphitic layers. The calculated ratios of the intensities of the G to D and 2D to G bands in the Raman spectra of carbon nanotubes were 0.90 and 0.47 respectively [7,8,9,10].

Morphology of the formed MWCNTs was analyzed by using HRTEM. HRTEM analyses of 15 nanotubes show an average diameter of 12.2 ± 3.23nm (95 % confidence interval) for the MWCNTs. Figure 2A and 2B show the presence of the agglomerated MWCNTs while Figure 2C and 2D show the parallel arrangement of MWCNTs. The arrows in the HRTEM images indicate crystalline domains and also small amount of amorphous carbon have been observed.

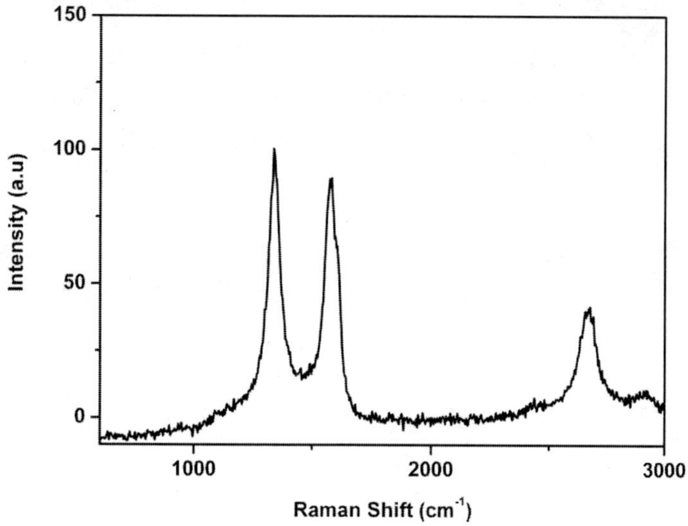

Fig. 1: Raman Spectra of MWCNTs

TGA analysis was performed on the formed MWCNT in order to find out the quality and purity of the formed MWCNTs. Figure 3 shows that three distinct weight losses were observed in the TGA analysis. The first weight loss (3%) below 150°C was due to the removal of moisture present in the MWCNTs. The second weight loss (4.46 %) in the interval of 200 – 400°C was due to the removal of the amorphous carbon present in the material [3]. A large weight loss (66%) in the interval of 400°C – 670°C was due to the decomposition of crystalline carbon present in MWCNTs[4].The TGA analysis thus shows the formation of over 66% of MWCNTs.

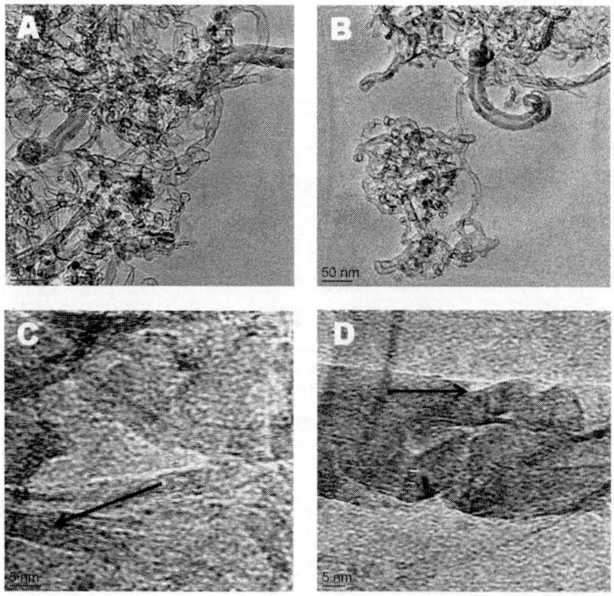

Fig. 2: (A, B, C and D) HRTEM Image of the MWCNTs

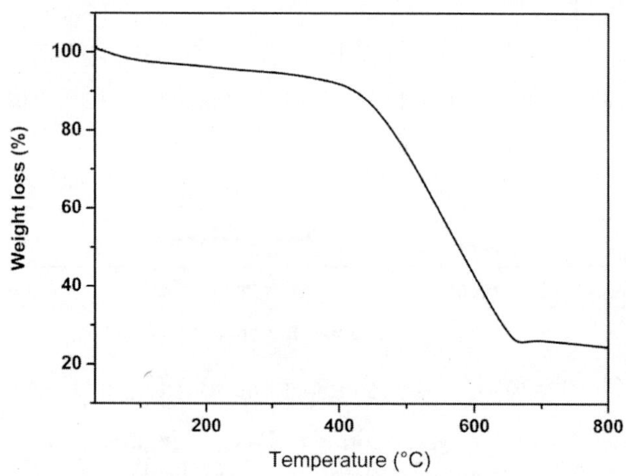

Fig. 3: TGA curve of the MWCNTs

CONCLUSIONS

Multiwalled carbon nanotubes were synthesized successfully using RF induction heating method and using ethanol source and a bimetallic catalyst of Fe-Mo on MgO support material. The main advantage of using ethanol as the precursor for the preparation of highly crystalline MWCNTs, is its ready availability and low cost. Raman spectroscopy revealed a G to D band intensity ratio of 0.90. The product contained over 66% MWCNTs with average outer diameters of 12.2 ± 3.23nm (95 % confidence interval) out of a population of 15 nanotubes.

ACKNOWLEDGEMENTS

We would like to express our gratitude for the valuable inputs of Dr. Alexandru Biris in doing this research. We are also grateful to DRDO (Grant No. ERIP/ER /0703668/M/01/2796/D (R&D)/1071) for their generous funding.

REFERENCES

[1] Iijima S, Ichihashi T. Single-shell carbon nanotubes of 1-nm diameter. Nature, 363(6430), 603-5. 1993.

[2] Biris AR, Li Z, Dervishi E, Lupu D, Xu Y, Saini V, et al. Effect of hydrogen on the growth and morphology of single wall carbon nanotubes synthesized on a Fe-Mo/MgO catalytic system. Phys. Lett. A, 372(17), 3051-7, 2008.

[3] Biris AR, Biris AS, Lupu D, Trigwell S, Dervishi E, Rahman Z, et al. Catalyst excitation by radio frequency for improved carbon nanotubes synthesis. Chem Phys Lett, 429(1-3), 204-8, 2006.

[4] Alexandru R. Biris. High-Quality Double-Walled Carbon Nanotubes Grown by a Cold-Walled Radio Frequency Chemical Vapor Deposition Process. Chem Mater., 20, 3466–72, 2008.

[5] Derycke V, Auvray S, Borghetti J, Chung C-L, Lefèvre R, Lopez-Bezanilla A, et al. Carbon nanotube chemistry and assembly for electronic devices. Comptes Rendus Physique,10(4), 330-47, 2009

[6] Faris Y, inodot, lmaz, Zuhal K. Conducting polymer composites of multiwalled carbon nanotube filled doped polyaniline. J Appl Polym Sci, 111(2), 680-4 2009.

[7] Santangelo S, Messina G, Faggio G, Lanza M, Milone C. Evaluation of crystalline perfection degree of multi-walled carbon nanotubes: correlations between thermal kinetic analysis and micro-Raman spectroscopy. J Raman Spectrosc, 42(4), 593–602. 2011

[8] Tzong-Ming W, Shih-Hsiang L. Synthesis, characterization, and electrical properties of polypyrrole/multiwalled carbon nanotube composites. J. Polym. Sci., Part A: Polym. Chem., 44(21), 6449-57, 2006

[9] Dresselhaus MS, Jorio A, Hofmann M, Dresselhaus G, Saito R. Perspectives on Carbon Nanotubes and Graphene Raman Spectroscopy. Nano Lett., 10(3):751-8, 2010.

[10] Yoshida A, Kaburagi Y, Hishiyama Y. Full width at half maximum intensity of the G band in the first order Raman spectrum of carbon material as a parameter for graphitization. Carbon, 44(11), 2333-5. 2006.

[11] Malek Abbaslou RM, Soltan J, Dalai AK. The effects of carbon concentration in the precursor gas on the quality and quantity of carbon nanotubes synthesized by CVD method. Appl Catal A-Gen, 372(2), 147-52, 2010.

[12] Liao XZ, Serquis A, Jia QX, Peterson DE, Zhu YT, Xu HF. Effect of catalyst composition on carbon nanotube growth. Appl Phys Lett., 82(16), 2694-6. 2003

[13] Qi H, Qian C, Liu J. Synthesis of High-Purity Few-Walled Carbon Nanotubes from Ethanol/ Methanol Mixture. Chem Mater., 18(24), 5691-5, 2006.

[14] Liu Q, Ouyang Y, Zhang L, Xu Y, Fang Y. Effects of argon flow rate and reaction temperature on synthesizing single-walled carbon nanotubes from ethanol. Physica E Low Dimens Syst Nanostruct , 41(7), 1204-9. 2009.

[15] Biris AR. Carbon Nanotubes Grown by RF Heating and Their Morphological and Structural Properties. Part. Sci. Technol., 26, 521–8, 2008.

[16] Biris AR, Lupu D, Dervishi E, Li Z, Xu Y, Trigwell S, et al. Multiwall carbon nanotubes synthesized by RF-CCVD on novel CaO supported catalysts. Phys. Lett. A, 372(42), 6416-9, 2008.

[17] Biris AS, Schmitt TC, Little RB, Li Z, Xu Y, Biris AR, et al. Influence of the RF Excitation of the Catalyst System on the Morphology of Multiwalled Carbon Nanotubes J Phys Chem C, 111(48), 17970-5, 2007.

STRUCTURAL ANALYSIS OF BULK AND NANO $La_{0.7}Pb_{0.3}MnO_3$ PEROVSKITE MANGANITE MATERIALS

K. SAKTHIPANDI AND V. RAJENDRAN

Centre for Nano Science and Technology, K S Rangasamy College of Technology, Tiruchengode, Tamil Nadu
E-mail: veerajendran@gmail.com

ABSTRACT

Bulk and nanostructured $La_{0.7}Pb_{0.3}MnO_3$ perovskite manganite samples were synthesised using solid-state reaction and sonochemical methods, respectively. The X-ray diffraction studies confirmed that the prepared bulk and nanoperovskite manganite samples were crystalline in nature. The particle size of bulk perovskite samples is in the range of 300–400 nm, whereas it is 47 nm in nanoperovskite samples. In addition, relative percentage change in ultrasonic velocity shows an interesting anomaly in both bulk and nanoperovskite samples. The ferromagnetic-to-paramagnetic phase transition temperature (T_C) was marked from the observed softening and hardening in sound velocity. In nanoperovskite samples, the existence of diffused phase transition took place at a lower temperature compared with the corresponding bulk samples, which was attributed to the coexistence of orbitally ordered and disordered phases.

INTRODUCTION

In recent years, the $La_{1-x}A_xMnO_3$ (where A is a divalent element such as Ca, Sr, Pb, Na, and Rb) perovskite structure has been widely studied due to its unique electrical and magnetic properties [1–4]. The structural/electric behaviour of $LaMnO_3$ insulator was observed while doping the divalent ions instead of La in an $LaMnO_3$ insulator [4]. $La_{1-x}Pb_xMnO_3$ (LPMO)

V. Rajendran, K. Saminathan, P. Paramasivam and K.E. Geckeler (eds.)
Nano Materials Synthesis and Characterisation, pp. 281-286 (2012).

manganite shows a ferromagnetic (FM) metallic phase at room temperature along with large magnetoresistance (MR). LPMO manganites in their bulk phase are used in a wide range of applications, such as in magnetic field sensors, hard-disk read heads, and microwave active components, due to their unique physicochemical and many giant magnetoresistance properties [5–8].

The double-exchange interaction between pairs of Mn^{3+} and Mn^{4+} ions through an oxygen atom is responsible for the FM and metallic properties of these manganese oxides [9]. Therefore, knowledge on the structural properties of perovskites such as surface spin, structure disorders, half-metallicity, and their possible influence on magnetism, is an essential requirement [10,11] while studying the use of manganite oxides for a wide range of applications. Nanostructured manganites with large grain boundaries are known to show pronounced low-field magnetoresistance properties [12]. Nanostructured perovskites show unique size-dependent optical, electronic, magnetic, thermal, mechanical, and chemical properties, which are distinct from those of their corresponding bulk perovskites [12,13]. The nanostructured LPMO shows promising applications such as piezoelectric transducers and actuators, ultrasonic devices, infrared devices, magnetic sensors, magnetic reading heads for memory devices, and non-volatility memory devices [14,15].

Among the different methods available to characterise the manganite materials, on-line ultrasonic characterisation is a novel and promising technique to examine the structural/phase transitions on-line and to correlate the differences between FM and paramagnetic (PM) phase physical properties [16]. The lattice degrees of freedom are directly related to the measured velocities and attenuation, which in turn are used to explore the structural and phase transition changes in materials [17].

Therefore, in this study, an attempt has been made to synthesise the bulk and nanocrystalline $La_{0.7}Pb_{0.3}MnO_3$ perovskite manganite using solid-state reaction and sonochemical methods. A systematic study on both bulk and nano LPMO perovskites samples has been conducted using on-line ultrasonic studies as a function of temperature to gain insight into the behaviour of transition and the microscopic origin of the colossal magnetoresistance effect.

MATERIALS AND METHODS

Bulk and nanostructured $La_{0.7}Pb_{0.3}MnO_3$ were prepared using, respectively, the solid-state and sonochemical reaction methods, as prepared elsewhere [18]. X-ray diffraction (XRD) patterns of the prepared bulk and nano LPMO samples were determined using a powder X-ray diffractometer (X'Pert Pro; PANalytical, the Netherlands) at a scan range of 10–80° at a scanning rate of 1° min^{-1} using a monochromatised CuKá radiation source operated at 40 kV and 30 mA. Transmission electron microscope (TEM; CM 200; Philips, USA) was used to obtain substructural information and determine particle size of the prepared samples. Ultrasonic longitudinal velocity measurements were carried out from 300 to 400 K using the experimental facility available in the author's laboratory. The relative percentage variation of measured velocity and its first derivative were calculated using standard formula [18].

RESULTS AND DISCUSSION

The XRD pattern of bulk and nano LPMO perovskites is shown in Fig. 1. The observed diffraction peaks in the XRD pattern of bulk and nano LPMO samples are used to study the crystalline nature of the prepared samples. The observed XRD peaks of bulk LPMO perovskite manganite samples are indexed and well matched with the triclinic (JCPDS files no. 85-1788) as reported elsewhere [19]. The observed triplet around 32.5° for triclinic structure is also given in Fig. 1 as an insert. However, the diffracted peaks of all nanostructured LPMO perovskite manganite materials matched well with the rhombohedral structure (JCPDS file no. 4-095-1256).

Figure 2 shows the TEM images of bulk and nano LPMO perovskite samples. It is evident from the images that particles show spherical morphology with fine and clear grain boundaries and narrow range. The range of particle size distribution is obtained considering the minimum and maximum particle size. The estimated particle size distribution in bulk LPMO as observed from micrographs is 300–450 nm whereas the average particle size for nanoperovskite sample is 47 nm. The selected area electron diffraction (SAED) of bulk and nanoperovskite samples is shown as an insert in Fig. 2. The circular lines in SAED pattern indicate that the BLPMO and NLPMO perovskites are crystalline in nature. In addition, the diffraction dots and rings observed in SAED pattern of bulk sample reveal the single and polycrystalline nature of the bulk sample.

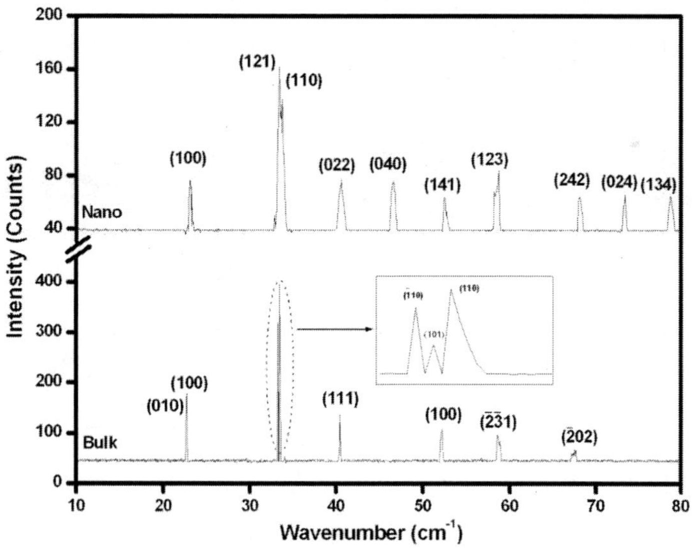

Fig. 1: XRD Pattern of Bulk and Nano LPMO Perovskite Manganite Materials

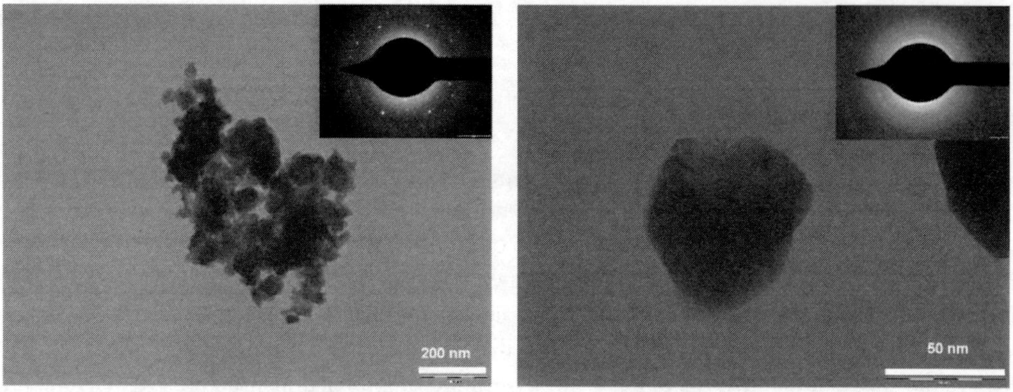

(a) Bulk smple (b) Nano sample

Fig. 2: TEM Image of Bulk and Nano LPMO Perovskite Manganite Materials

The phase transition temperature of bulk and nanoperovskite samples is examined by on-line ultrasonic studies. The relative percentage variation in measured longitudinal velocity is calculated from the measured ultrasonic parameters. The temperature-dependent relative percentage variation in measured longitudinal velocity and its first derivative are shown, respectively, in Figs. 3 and 4. The observed anomalous behaviour in relative percentage variation in velocity is used to explore the FM–PM magnetic transition of prepared perovskites sample. The anomalous behaviour occurs during aging of sample due to the existence of phase/structural transition in materials [16].

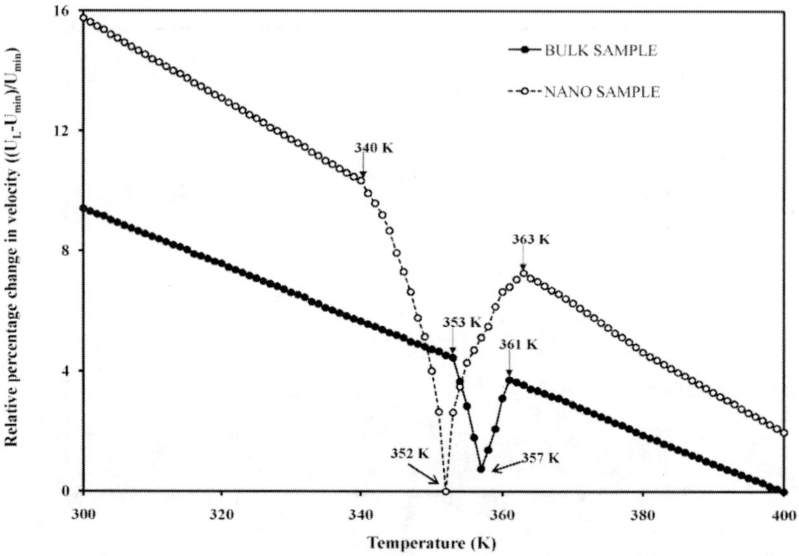

Fig. 3 Relative Percentage Variation in Ultrasonic Longitudinal Velocity

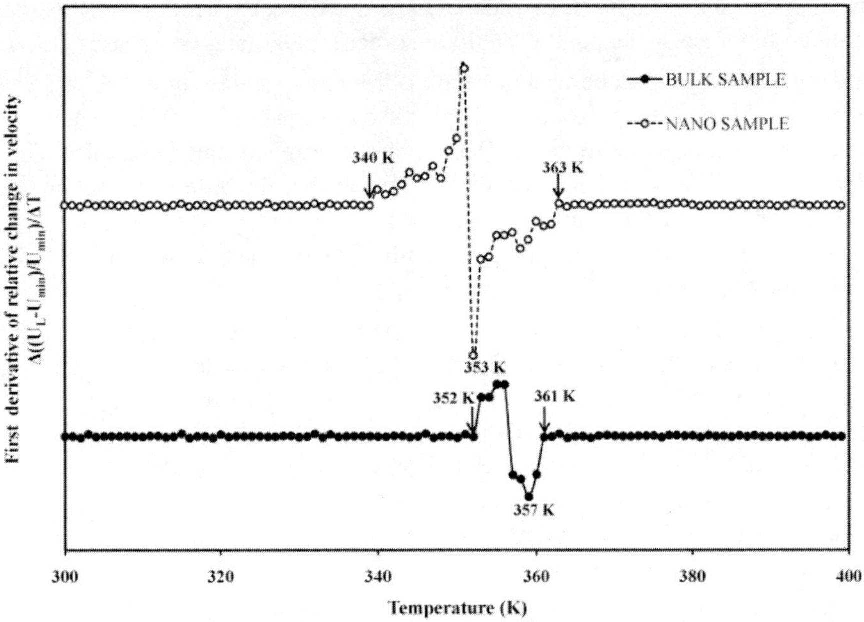

Fig. 4: First Derivative of Relative Variation in Longitudinal Velocity

The temperature-dependent different phase transitions are identified by correlating the temperature at which an anomalous behaviour is observed in ultrasonic velocities/attenuation with the phase diagram of LPMO perovskites reported earlier [16]. In the bulk sample, a sudden fall in relative change in velocity starts from 353 K and reaches a minimum at 357 K, followed by an increase in velocity up to 361 K. However, in nanoperovskite sample, it starts at 340 K and then reaches a minimum at 352 K, followed by an increase in velocity up to 363 K. The temperature at which the observed minimum change in relative velocity is correlated with the phase diagram of LPMO samples [15,16]. The observed anomalous temperature in ultrasonic measurements is in agreement with that at which FM–PM phase transition takes place in LPMO perovskite samples [15]. Hence, the observed anomalous temperature is marked as FM–PM phase transition temperature.

Further, the observed anomalous behaviour is used to reveal more insight about the FM–PM transition. The first derivative of relative change in longitudinal velocity (Fig. 4) is used to reveal precise information about the temperature-dependent ultrasonic velocity. It shows the predominate peaks used to reveal the precise starting, ending, and anomalous temperature (T_C) of phase transition. The observed T_C for nanoperovskite sample was lower than that for the corresponding bulk perovskite samples. Further, the observed sharp transition in bulk sample with a single peak is due to the coarse grains of the sample, which occurs due to the presence of orbitally ordered (OO) phases [20]. In contrast, the observed broad transition with large peaks in nanoperovskite samples is considered to be due to the existence of small/softer crystal size attributed to the coexistence of orbitally disordered (OD) and OO phases [13,20].

From Figs. 3 and 4, it can be seen that bulk LPMO perovskite samples show a sharp FM–PM transition, whereas nanoperovskite samples show a broad FM–PM transition. This is in line with the observation made from the XRD pattern for nanocrystalline samples. The induced stress during the synthesis of nanocrystalline is responsible for observed peak broadening in XRD pattern. An abrupt change in coupling between spin, charge, orbital, and lattice degrees of freedom and interactions at transition temperature are responsible for the observed anomaly in ultrasonic measurements [13].

The observed larger magnitude in nanoperovskite samples is not only attributed to a conventional FM spin fluctuation near FM–PM transition but is also related to the existence of strong electron–lattice interactions via Jahn–Teller (JT) distortion of $Mn^{3+}O_6$ and the spontaneous spin ordering below FM–PM transition [21]. The observed large softening below T_C is attributed to the significant change in lattice parameters or the Mn–O bond length [13]. Thus, it leads to a significant reduction in the JT distortion of the $Mn^{3+}O_6$ octahedral due to the spin–lattice coupling below the transition temperature [21]. This indicates a direct evidence for weak magnetic coupling between the grains below T_C in nanoperovskite manganite materials [21]. Thus, it leads to low-field magnetoresistance in the nanostructured perovskite manganites than that in the bulk perovskite samples. Thus, on-line ultrasonic measurements are more sensitive in exploring the structural/phase transition behaviour in bulk and nanostructured perovskite manganite materials.

CONCLUSION

In this investigation, bulk and nanostructured $La_{0.7}Pb_{0.3}MnO_3$ perovskite manganite samples were synthesised, respectively, using solid-state reaction and sonochemical methods. The crystallinity and phase purity of both bulk and nanoperovskite samples were confirmed by XRD. TEM studies have confirmed that the particle size of bulk LPMB perovskite samples is in the range of 300–450 nm, whereas it is 47 nm in case of nanoperovskite samples. On-line ultrasonic measurements show FM–PM transition (T_C) at 357 K for bulk sample, whereas for nanoperovskite sample, it is at 352 K. Further, diffused phase transition in nanoperovskite sample occurs at lower T_C than the corresponding bulk sample. The broad FM–PM transition in nano LPMO perovskites confirms the coexistence of OO and OD phases.

ACKNOWLEDGMENTS

KS acknowledges the for Senior Research Fellowship grant (8/570(0001)/2011 dt. 29.03.2011) from the Council of Scientific and Industrial Research (CSIR), New Delhi, for this work.

SYNTHESIS AND CHARACTERIZATION OF MULTI-WALLED CARBON NANOTUBES BASED COMPOSITE MATERIALS DERIVED VIA ELECTROSTATIC INTERACTION OF DENDRIMER/ METAL NANOPARTICLES

EAGAMBARAM MURUGAN AND SIVARANJANI ARUMUGAM

Department of Physical Chemistry, School of Chemical Sciences, University of Madras, Maraimalai Campus, Guindy, Cheinnai, Tamil Nadu
E-mail: dr.e_murugan68@gmail.com

ABSTRACT

Five types of Multiwalled carbon nanotubes (MWCNT) functionalized with poly (propylene imine) (G2) dendrimer (PPI-G2) encapsulated/stabilized silver nanoparticle (AgNPs) composites viz., MWCNT-PPI (G2)-AgNPs has been prepared by varying the [Ag^+] load and keeping the amount of MWCNTs and PPI(G2) as a constant. Initially, MWCNT was oxidized with HNO_3/H2SO_4 and thus produced MWCNTs-COOH. In a separate procedure, 5 different load of [Ag^+] was complexed with PPI (G2) dendrimer and thus formed PPI (G2)-Ag complex. Each PPI (G2)-Ag complex was functionalized individually onto MWCNT-COOH via electrostatic interaction. The resulting mixture was reduced by $NaBH_4$ and thus produced the corresponding MWCNTs based composite materials (MWCNT-PPI (G2)-AgNPs). Then, the resulting 5 types of MWCNT-PPI (G2)-AgNPs were characterized with FT-IR, UV-Vis, FESEM and EDS analyses.

INTRODUCTION

In recent years, carbon nanotubes have been attracted considerable attention due to their unique structural, mechanical and electrical properties. Particularly, carbon nanotubes can exhibit

V. Rajendran, K. Saminathan, P. Paramasivam and K.E. Geckeler (eds.)
Nano Materials Synthesis and Characterisation, pp. 287-294 (2012).

metallic conductivity, chemical and thermal stability, extremely high tensile strength and elasticity, the ability to absorb gas molecules as nanocapillaries, solubility when treated with surfactants and potential for further chemical functionalization [1]. However, a major constraint experienced in CNTs is the poor solubility and processibility of carbon nanotubes [2]. Hence, functionalization of CNTs have received greater interest so as to enhance the dispersion, processing and chemical modifications [3-5]. The hybrid nanostructures composed of carbon nanotubes and nanoparticles have great concern and used in electronic, magnetic or catalytic applications [6]. Generally, nanoparticles are attached to the carbon nanotubes by covalent, noncovalent and electrostatic interaction [7]. For example, the attachment of organic molecules onto CNTs have improved greatly the solubility of CNTs and brought more extensive applicability [8,9]. But electrostatic self-assembly [10] of organic molecules with CNTs is the most versatile and facile method. The periphery amine groups of poly(ethylene imine) provide a reactive handle for immobilizing the dendrimer-encapsulated nanoparticles (DENs) onto surfaces has been reported [11]. Nanometer-scale metal and semiconductor particles sometimes exhibit unique physical or chemical properties that are interesting in their own right but that also hold out the promise of technological advances in the fields such as catalysis, sensors, electronics, and biotechnology [12-14].

However, to the best of our knowledge, there is no report available for preparation of MWCNTs based composite material via functionalization of MWCNTs with poly (propyleneimine) (G2) (PPI-G2) dendrimer encapsulated silver nanoparticles by electrostatic assembly. In this paper, we report new MWCNTs based composite materials by integration of MWCNT, PPI (G2) dendrimer and metal nanoparticles. The said composite material has been developed in 3 different stages. Initially, MWCNT was oxidized with HNO_3/H_2SO_4 and thus produced MWCNTs-COOH. In step 2, by using commercial PPI (G2) with NH_2-terminal group dendrimer encapsulated/stabilized AgNPs was prepared. In step 3 the PPI (G2) encapsulated/ stabilized AgNPs was attached to MWCNTs-COOH surface via electrostatic interaction and thus produced MWCNT-PPI dendrimer encapsulated/stabilized silver nanoparticle (MWCNT-PPI (G2) AgNPs) composites. The introduction of metal nanoparticles to the CNTs surface can generate new nanostructures with excellent functions in the field of optics [15], electronics [16] and electrocatalysis [17].

EXPERIMENTAL SECTION

MWCNT Functionalized with PPI (G2) Dendrimer Encapsulated/Stabilized Silver Nanoparticles Composite (MWCNT-PPI (G2)-Ag)

Reagents and Materials

MWCNTs with purity greater than 95% were purchased from SRL, PPI (G2) dendrimer (Sigma), Silver nitrate (SRL), sodium borohydride ($NaBH_4$, SRL), H_2SO_4 (SRL) and HNO_3 (SRL) were

AR grade of 99% purity and used as received. The functionalization of –COOH onto MWCNTs was performed by treating the 50mg of pristine MWCNTs with H_2SO_4 and HNO_3 (3:1 ratio) and refluxed the mixture at 80 ^{0}C for 4 h, washed with water until attain a neutral state and they were dried under vacuum at 60 ^{0}C. Then, the -COOH functionalized MWCNTs were dispersed in water, obtaining a stable dispersion of MWCNTs-COOH with negative charge on the surface of the MWCNTs.[18] Subsequently, by adopting the separate procedure, the $AgNO_3$ (2 mM) solution was slowly added into 2 mg/mL of commercial PPI (G2) solution maintain the pH at 9.0-9.1 under stirring condition, this mixture was stirred for 1 h, after driving completion of the coordination reaction, the resulting PPI (G2)-Ag complex was transferred to the above MWCNT-COOH dispersed aqueous solution (0.03 mg/mL) and subsequently subjected to ultrasonication treatment for 15 min. The resulting product was mixed with 0.3 mL of 0.1 M $NaBH_4$ solution. Several minutes later, the obtained product (MWCNT-PPI (G2)-AgNPs) was centrifuged and washed for two cycles. Similar types of another 4 composites were prepared by varying the [Ag^+] load in constant amount of PPI (G2) and keeping the MWCNTs-COOH as a constant and thus produced 4 types of MWCNT-PPI (G2)-AgNPs composites (Scheme 1 next page).

RESULTS AND DISCUSSION

The functionalization of –COOH onto MWCNTs was established through FT-IR result. The observed FT-IR (Fig. 1a) shows a weak absorption band at 1628 cm^{-1} and this is related to the stretching vibration of C=C in pristine MWCNTs. In the case of acid functionalized MWCNTs (MWCNTs-COOH) a new peak at 1716.70 cm^{-1} was noticed and this is due to the $C=O_{(str)}$ and similar broad band in the range of 2500 to 3300 cm^{-1} was also noticed and this is related to the $OH_{(str)}$ (Fig. 1b), the appearance of these two characteristic peaks confirms the functionalization of –COOH on MWCNTs and this observation also agreed well with the literature values.

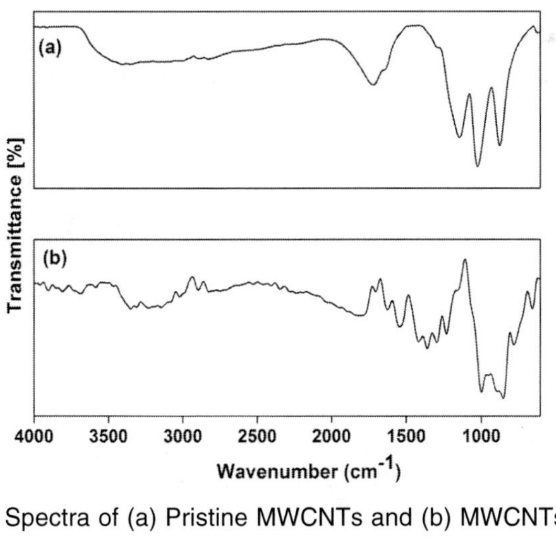

Fig. 1: FT-IR Spectra of (a) Pristine MWCNTs and (b) MWCNTs-COOH

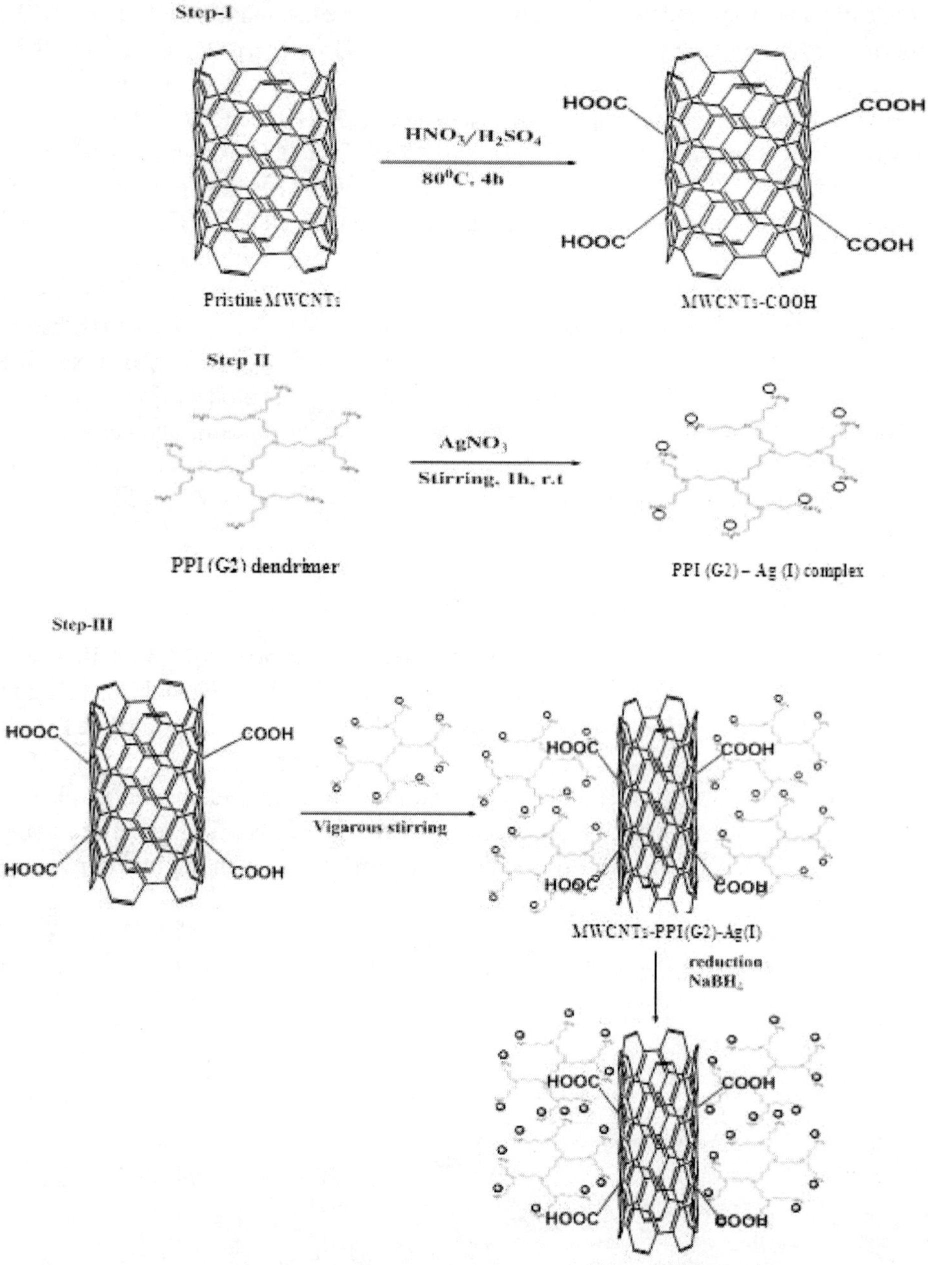

Scheme 1. MWCNT-PPI (G2) Dendrimer Encapsulated/Stabilized Silver Nanoparticles (MWCNT-PPI-G2-AgNPs)

The functionalization of MWCNTs with dendrimer encapsulated/stabilized metal nanoparticles has been followed periodically by UV-visible spectroscopy. That is, although we prepared 5 types MWCNTs-PPI(G2)-AgNPs composites by varying the [Ag^+] load and characterized the same with UV-Vis, FESEM and EDS analyses, but as a representative sample, we are shown here spectra derived from low (2mM) and higher (6mM) load of AgNPs immobilized composite materials. The complexation of Ag^+ with commercial PPI (G2) was known from UV-Vis spectra. That is, a characteristic absorption peak was noticed around 260 nm irrespective of [Ag^+] load, and observed further that the intensity of the same peak was gradually increased from 2mM (Fig. 2a) to 6mM (Fig. 2d). Then these 5 types of complexed solution were mixed individually with the solution dispersed in identical load of MWCNT-COOH in aqueous phase, the resulting MWCNT-PPI (G2) Ag^+ complex obtained in each load of [Ag^+] was analyzed with UV-Vis and observed that the intensity of the characteristic peak noticed at 260 nm was again increased gradually when compared with their corresponding PPI (G2)-Ag^+ complexed sample. Therefore, gradual increment of characteristic peak intensity at 260 nm noticed from complex of MWCNTs-PPI (G2)Ag^+ derived 2mM [Ag^+] load (Fig. 2b) to 6mM [Ag^+] load (Fig. 2e) indicates the characteristic absorption band of assembled CNTs [19]. Then, each MWCNT-PPI (G2) Ag^+ complex derived irrespective of [Ag^+] load was allowed for reduction through addition of $NaBH_4$ and thus produced corresponding composite, viz., MWCNT-PPI (G2)-AgNPs. To ensure the reduction of Ag^+ or immobilization/formation of AgNPs on MWCNT-PPI (G2), the obtained resulting mixtures were analyzed with UV-Vis. The spectra derived from the composites of 2mM of [Ag^+] and 6mM of [Ag^+] were shown in Fig. 2c and Fig. 2f respectively. The observed UV-Vis spectra reveals that irrespective of the composite, the surface plasmon resonance (SPR) peak was noticed above 390 nm and thus confirms the reduction of Ag^+ or immobilization/formation of AgNPs onto the MWCNTs-PPI (G2) matrix.

Element	*Weight%*	*Atomic%*
C K	73.53	93.23
Na K	0.75	0.50
S K	7.91	3.76
Ag L	17.81	2.51
Totals	100.00	

The PPI (G2)-Ag complexes are positively charged complex ions, which can be adsorbed on the surface of negatively charged CNTs. After reduction by $NaBH_4$, the heterogeneous nanostructures were received. In order to know the surface morphology of the MWCNTs-PPI (G2)-AgNPs composite, all the composites were studied with FESEM and the observed images are compared. Here also, as a representative, the images derived from 4mM [Ag^+] load of MWCNT-PPI (G2) composite are shown in Fig. 3a. The obtained images reveal that the smooth surface of pristine-MWCNT has been drastically changed into heterogeneouse surface with

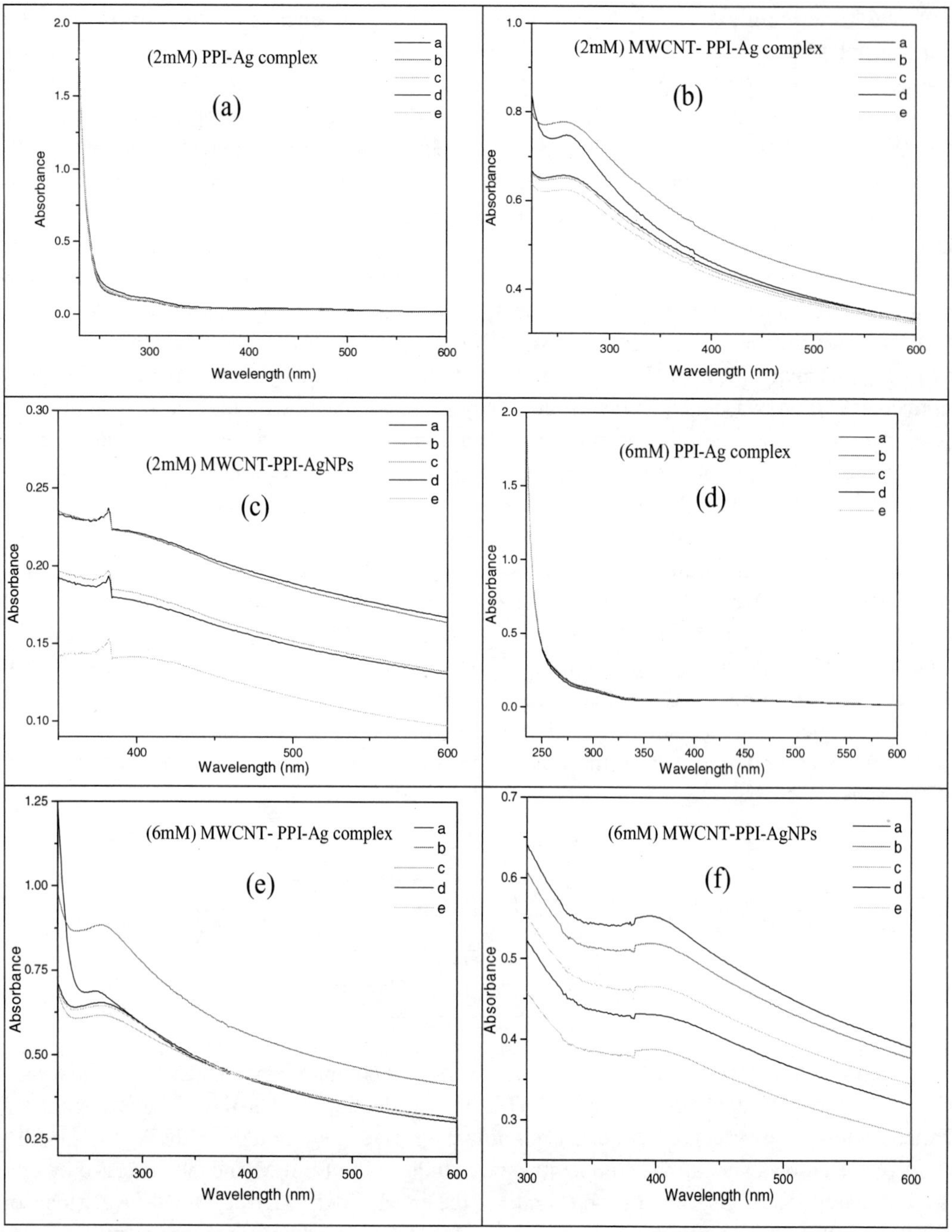

Fig. 2: UV-Visible Spectra of MWCNTs-PPI (G2)-AgNPs (2mM and 6mM)

homogeneous distribution of tiny nodules. The appearance of heterogeneous surface is due to the functionalization of PPI (G2) on MWCNTs and similarly availability of tiny nodules is the contribution of AgNPs. Since, we recorded FESEM images for MWCNTs-PPI (G2)-AgNPs composite derived from 4mM [Ag^+] load of MWCNTs and AgNPs are possible to determined. That is from the image, it is understand that the size of single MWCNTs was 20-50 nm and the size of the AgNPs was approximately 22 nm. The presence of AgNPs in the MWCNTs-PPI (G2) composite are also evaluated through EDS analysis (Fig. 3b), and the weight percentage of AgNPs was found to be 17.81%.

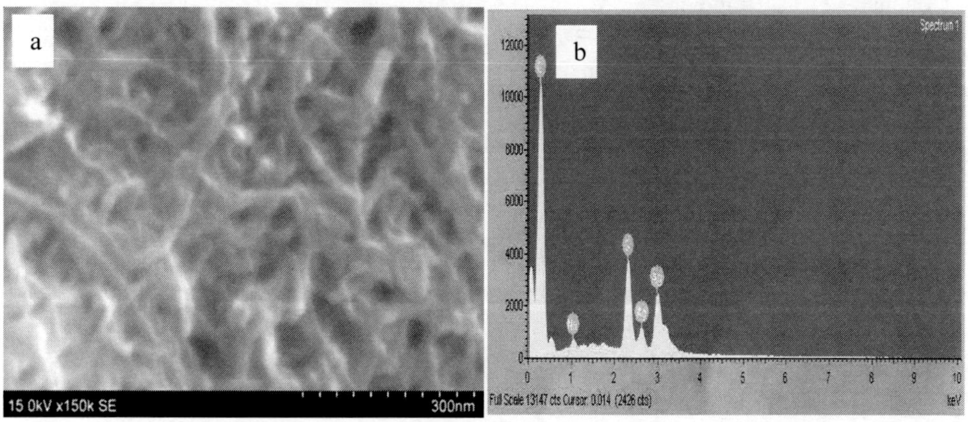

Fig. 3: (a) FESEM Spectrum of MWCNTs-PPI (G2)-AgNPs (4 mM metal load) and (b) EDS Spectrum of MWCNTs-PPI (G2)-AgNPs (4 mM metal load)

CONCLUSION

In conclusion, 5 types of MWCNT functionalized with PPI (G2) dendrimer encapsulated/stabilized silver nanoparticles composite viz., MWCNT-PPI (G2)-AgNPs were developed by varying the [Ag^+] load. The resulting MWCNT-PPI (G2) stabilized silver nanoparticle composites were characterized with FT-IR, UV-VIS, FESEM and EDS analyses. The functionalization of –COOH onto MWCNTs was confirmed from the appearance of characteristic peaks such as $C{=}O_{(str)}$ at 1716.70 cm^{-1} and, $OH_{(str)}$ from 2500 to 3300 cm^{-1}. The appearance of SPR peak at 390 nm from UV-Visible spectra confirms the formation of AgNPs onto MWCNT-PPI (G2) matrices. From the FESEM results, it is understand that the size of the single MWCNTs produced in composite derived from 4mM [Ag^+] load was observed as 20-50 nm and the size of the AgNPs deposited on MWCNTs-PPI (G2) was approximately 22 nm. The EDS results explain that weight percentage of AgNPs was found to be 17.81%. The incorporation of metal NPs onto the CNTs surface forms a new functional nanostructure and thus possesses some novel properties, viz., optics, catalysis, etc., hence, this technique can be extended further for various applications in future.

ACKNOWLEDGMENT

The authors acknowledge the NCNSNT (National Centre for Nanoscience and Nanotechnology), University of Madras for providing financial support.

REFERENCES

[1] Hirsch, A., Functionalization of Single-Walled Carbon Nanotubes, *Angew. Chem. Int. Ed.*, 41, 853, 2002.

[2] Luong, J.H.T., Hrapovic, S., Wang, D.S., Bensebaa, F., Simardc, B., Solubilization of Multiwall Carbon Nanotubes by 3-Aminopropyltriethoxysilane Towards the Fabrication of Electrochemical Biosensors with Promoted Electron Transfer, *Electroanalysis.,* 16, 132-139, 2004.

[3] Fu, K.F., Sun, Y.P., Dispersion and Solubilization of Carbon Nanotubes, *J. Nanosci Nanotechnol.*, 3, 351, 2003.

[4] Sun, Y.P., Fu, K.F., Lin, Y., and Huang, W.J., *Acc. Chem. Res.*, 35, 1096, 2002.

[5] Liu, J., Rinzler, A.G., Dai, H.J., Hafner, J.H., Bradley, R.K., and et al., *Science.*, 280, 1253, 1998.

[6] Xing, Y., *J. Phys. Chem. B,* 108, 19255-19259, 2004.

[7] Kim, B., and Sigmund, W. *Langmuir.,* 20, 8239-8242, 2004.

[8] Vostrowsky, O. and Hirsch, A., *Angew. Chem. Int. Ed.,* 43, 2326, 2004.

[9] White, C.T., and Mintmire, J.W., *J. Phys. Chem. B.*, *190*, 52, 2005.

[10] Zhang, M.N., Yan, Y.M., Gong, K.P., Mao, L.Q., and Guo, Z.X., Chen, Y. *Langmuir*, 20, 8781, 2004.

[11] Zhou, Y., Ma, R., Ebina, Y., Takada, K., and Sasaki, T., *Chem. Mater.*, 18, 1235, 2006.

[12] Murray, C.B., Kagan, C.R., and Bawendi, M.G., *Annu. Rev. Mater. Sci.*, 545-610, 2000.

[13] Templeton, A.C., Wuelfing, W.P., and Murray, R.W. *Acc. Chem.Res.*, 33, 27-36, 2000.

[14] Shenhar, R., and Rotello, V.M., *Acc. Chem. Res.,* 36, 549-561, 2003.

[15] Kamat, P.V., *J. Phys. Chem. B,* 106, 7729, 2002.

[16] Neuendorf, R., Quenten, and M., Kreibig, U., *J. Chem. Phys.*, 104, 6348, 1996.

[17] Sasaki, T., Ebina, Y.; Fukuda, K., Tanaka, T., Harada, M., et al., *Chem. Mater.*, 14, 3524, 2002.

[18] Xiaoge Hu, Tie Wang, Xiaohu Qu, and Shaojun Dong, *J. Phys. Chem. B,* 110, 853-857, 2006.

[19] Rouse, J.H., and Lillehei, P.T., *Nano Lett., 3*, 59, 2003.

SURFACE CHARACTERIZATION OF COBALT PROMOTED MoO_3/SiO_2 SOLID ACID CATALYSTS BY XRD AND OTHER RECHNIQUES

Vinod B. Mane, Mahind Lalaso and Sharda P. Dagade

Department of Chemistry, Yashwantrao Mohite College,
Bharati Vidyapeeth Deemed University, Pune
E-mail: dagade@rediffmail.com

ABSTRACT

Molybdenum-silica binary oxide supported cobalt promoted (loading 0.5, 5, 10%) solid acid catalysts was prepared by sol-gel method using ammonium heptamolybdate and ETS-40 (silica source). Synthesized gel is dried at room temperature and calcinations at 773K, 873K and 973K. The resulting materials were characterized by means of X-ray powder diffraction, thermal analysis and FT-IR spectroscopic methods. The XRD measurements further reveal the presence of MoO_3, amorphous SiO_2 and Cobalt oxide phases were observed. The IR stretching mode spectra of pure SiO_2 treated at 773K and compared with the cobalt promoted solid acid catalysts. For pure SiO_2, the band observed at 975 cm^{-1} is assigned to the Si-O stretching mode of free silanol group. In the mixed oxide this frequency is shifted to lower frequency due to Si-O-Co linkage formation.

INTRODUCTION

The molybdenum oxide as such or in supported form is widely used as catalysts in a great number of reactions. Particularly, molybdenum oxide supported on silica is a well-known solid acid catalysts which possesses both strong Lewis and Bronsted acidity [1, 2]. On various supports it has been successfully studied by the newly developed low temperature oxygen chemisorptions technique.

V. Rajendran, K. Saminathan, P. Paramasivam and K.E. Geckeler (eds.)
Nano Materials Synthesis and Characterisation, pp. 295-302 (2012).

Supported molybdenum oxide catalysts are widely used in various catalytic processes [3-7]. These catalysts are normally obtained by impregnating the catalytically active molybdenum oxide species on an inorganic oxide support for the purpose of increasing the catalytic activity, selectivity and life of the catalysts. It is also well known that the type of support plays an important role on the catalytic properties and for a given reaction the activity and selectivity of the catalyst can be improved by the use of an support oxide.

Molybdenum oxide without support or supported on SiO_2 or alumina is a well-known solid acid catalyst useful for acid catalyzed reactions such as the oxidation of ammonia to elemental N_2 [8]. In recent years, promoted molybdenum on silica have gained much attention for transesterification reactions, nitration and many acid catalyzed reactions due to their super acidity, non-toxicity and a high activity at low temperatures. Dongare et al observed that molybdenum promoted silica exhibits strong acidity and excellent catalytic properties for various organic synthesis and transformation reactions in the liquid phase.

Recently supported molybdenum oxide has attracted much attention for its use as solid acid catalysts, especially for oxidation, hydroxylation and transesterification [9-13]. The catalytic activity of supported molybdenum oxide depends on the degree of dispersion and structure of the supported phase. Therefore, the activity has been related to the nature of the active phase, which itself depends on the nature of the active phase, which itself depends on the nature of the support, the molybdenum loading and also the preparation method [14-16].

Most of the studies reported the impregnation method but recently some groups report the use of the sol-gel method as an alternative and superior route to the impregnation method. Smith et al has reported the effect of the nature of surface species on the activity and selectivity of catalysts.

The molecular structure of molybdenum oxide on different supports has also been extensively studied by various techniques over the past decade. Considering the importance of the catalyst, we have prepared Co(II) supported on MoO_3/SiO_2 composites by sol-gel method [17]. B.M. Reddy et al . reported that the various mixed oxide supports TiO_2-ZrO_2,TiO_2-SiO_2. TiO_2-Al_2O_3, SiO_2-ZrO_2 and SiO_2-TiO_2-ZrO_2 used in investigation and study were prepared by homogeneous coprecipitation method and were deposited with MoO_3 by adopting a wet impregnation method [18].

The need for new and improved catalysts with different distribution of catalytic functionalities has increasingly inspired the research in the use of mixed oxides as supports. We have prepared various compositions of cobalt oxide (0.5, 5 & 10%) on Mo/SiO_2 catalyst by sol gel method and then it is characterized by using various techniques to understand the surface properties of the catalysts.

EXPERIMENTAL

Catalyst Preparation

Co (II) supported on Mo/SiO_2 catalyst was prepared by sol-gel process to obtain the catalyst with high surface area as well as for uniform distribution of Co/Mo on silica support. Ethylsilicate-40 (CAS Registry No. 18954-71-7) was used as a silica source for the first time because of its higher silica content as well as low cost as compared to tetraethyl orthosilicate, which is generally preferred silica source for sol-gel synthesis of silica support in catalysis [15]. 250 g of ethylsilicate-40 was mixed with 250 ml of dry isopropyl alcohol under constant stirring. 5.225 g of cobalt acetate was dissolved in 100 ml of isopropyl alcohol and was added to ethyl silicate solution under vigorous stirring. 20.31 g of ammonium Heptamolybdate (AHM) was dissolved in 100 ml of water and was added to the above solution under constant stirring. A viscous gel was obtained which was further stirred for another two hours and was kept overnight. A transparent solid product was obtained which was air dried and heated in an oven at 110°C for 12 hours. The product was further calcined at 450°C for 12 hours. The product was ground to fine powder, pressed in the form of pellet and granulated by crushing the pellet to –10 + 20 mesh size for use as a catalyst for reaction.

Catalysts Characterization

XRD was carried out on a Rigaku Miniflex X-ray diffractometer with CuKα radiation soure (λ = 1.54178 Å) scanning rate of 0.02 ^{0}/s was applied to record the pattern in the 2θ range of 20–80^0. The FT-IR spectra obtained with a Nicolet iD5 spectrometer (4000-400 cm^{-1}) using KBr pellet technique. Thermogravimetric analysis (TGA) was conducted on a Shimadzu (Simultaneous DTA-TGA System, DTG-60) under nitrogen flow and then it was heated up at a rate of 10 K/min to 1100°C. Scanning Electron Microscope image and EDX (JEOL-JSM 6360A) were obtained at accelerating voltage 20 KeV. Sample were deposited on a sample holder with an adhesive carbon foil and sputtered with gold.

RESULTS AND DISCUSSION

FT-IR Analysis

The FTIR spectra of the $Co/Mo/SiO_2$ calcined at 500^{0}C for 5 h are shown in Fig. 1. A strong and wide absorption band, caused by the stretching vibration of the O-H bond, appears at 3488 cm^{-1} (Fig. 1). However a weak absorption band appears at 1611.54 cm^{-1} due to flexural vibration of the O-H bond. Absorbed water and hydroxyl on the sample surface contribute to these absorption bands. A strong absorption band, caused by the vibration of the chemical bond of Si-O-Si, is present at approximately 1085.50 cm^{-1}. In the above sample, the vibration of the chemical bond of Si-O-Si moves towards the low intensity band causes a strong band at approximately 1084 cm^{-1}. This condition also indicates a certain interaction between molybdenum and supported matrix. From the spectrum (Fig. 1), the vibration of the chemical bond of Si-O-Si causes a strong absorption band at approximately 1085.50 cm^{-1} [17]. However, in the 10% Co/Mo/SiO_2 sample, the vibration of the chemical bond of Si-O-Si moves toward the low intensity and causes a strong absorption band observed at around 1084 cm^{-1}. Same effect is observed in 0.5 and 5% samples. This condition also indicates the interaction between cobalt and molybdenum, this phenomenon indicates the effect of Co addition. Upon loading with increasing amounts of metal, some new features appear that show the effect of cobalt content as presented in Fig. 1. The band due to MoO_3 (950 cm^{-1}) did not appear exactly at the usual frequency in the case of low loading molybdenum. This may be the due to interference of silica band. This interference vanished with the increase in Mo loading. The simultaneous and consecutive increase in the bands assigned at 990, 860 and 650 cm^{-1} may be due to the decomposition of Mo=O=Si bonds and the formation of MoO_3.

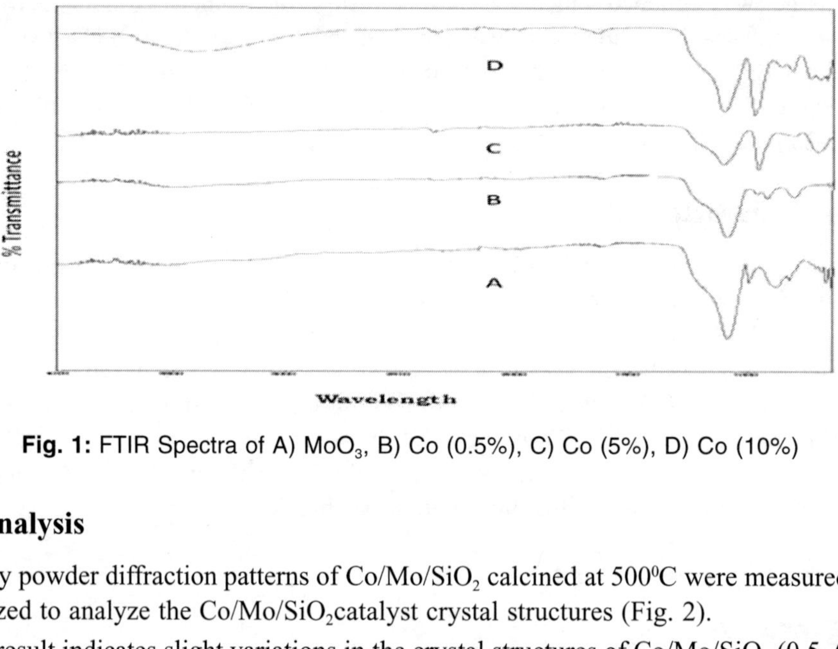

Fig. 1: FTIR Spectra of A) MoO_3, B) Co (0.5%), C) Co (5%), D) Co (10%)

XRD Analysis

The X-ray powder diffraction patterns of Co/Mo/SiO_2 calcined at 500°C were measured. XRD was utilized to analyze the Co/Mo/SiO_2catalyst crystal structures (Fig. 2).

The result indicates slight variations in the crystal structures of Co/Mo/SiO_2 (0.5, 5, 10%) compared with Mo/SiO_2 [17,18]. The Mo/SiO_2 catalyst dispersion characteristic peak appears at approximately 2θ =27.3°, which is the characteristic peak of MoO_3. The characteristic peak of the Co/Mo/SiO_2 dispersion appears at approximately $2\theta = 23°$. This peak is almost unchanged compared with undoped SiO_2. However, the peak of the Co/Mo/SiO_2 catalyst is slightly sharper than that without Co dopant. No obvious diffraction peak appears in the spectrum, implying an incomplete crystal structure of catalysts and a degree of crystal defects. This condition indicates a high level of catalytic activity. A Powerful tool to look at the phase composition is X-ray diffraction. The resulting diffraction pattern can be matched with a crystal structure database and the phases present can be identified. $CoMoO_4$ phase is observed at 33.5° and 39.1° (JCPDS file no. 21-0868). As we increase the Co content then intensity of the band also increases.

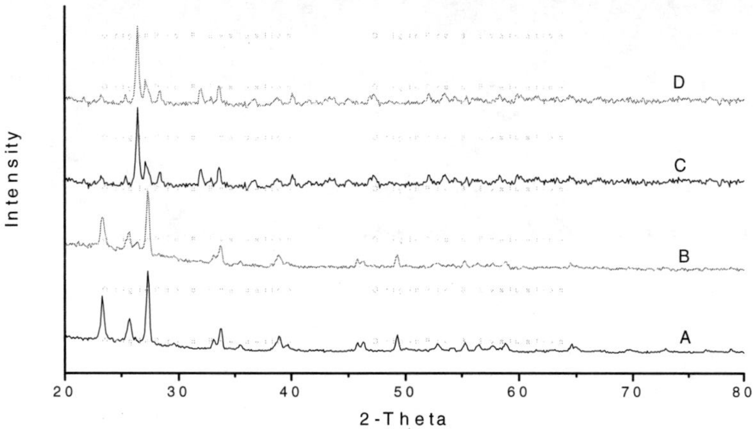

Fig. 2: XRD Graph of (A) MoO_3, (B)Co (0.5%), (C) Co (5%), (D) Co (10%)

From a more detailed analysis of the XRD data it is also possible to estimate the average crystallite size using the Debye-Scherrer equation: the FWHM (full width at half maximum) is inversely proportional to the crystallite size. Below table showed the crystallite size is around 30 nm in all compositions of the catalysts.

Table 1: XRD Data of $Co/Mo/SiO_2$ (0.5%; 5%; 10%) Catalyst

Sr.No.	*Catalysts*	*Crystallite Size, nm*	*FWHM*
1.	Mo/SiO_2	30.83	0.26
2.	$Co/Mo/SiO_2$ (0.5%)	29.34	0.27
3.	$Co/Mo/SiO_2$ (5%)	29.80	0.27
4.	$Co/Mo/SiO_2$ (10%)	30.48	0.26

SEM and EDX Analysis

The SEM images and EDX analysis of the prepared mixed oxide catalysts were carried out and the results were shown in Fig. 3. EDX analysis showed the presence of Co, Mo, and Si ($Co/Mo/SiO_2$) was detected in the synthesized catalysts. The scanning electron micrograph analysis depicted the cubic nature of the particles with mean particle size of about 500 nm.

This image shows global and uniform particles which are coherent together. The EDX data of $Co/Mo/SiO_2$ in Fig. 3 shows two intense and prominent peaks of Molybdenum and silica and other comparatively smaller peak of cobalt at 6.9 keV. These results confirmed the existence of Co atoms in the solid catalysts. Therefore, it may be concluded that Co ions are uniformly dispersed among the crystallites of molybdenum and silica.

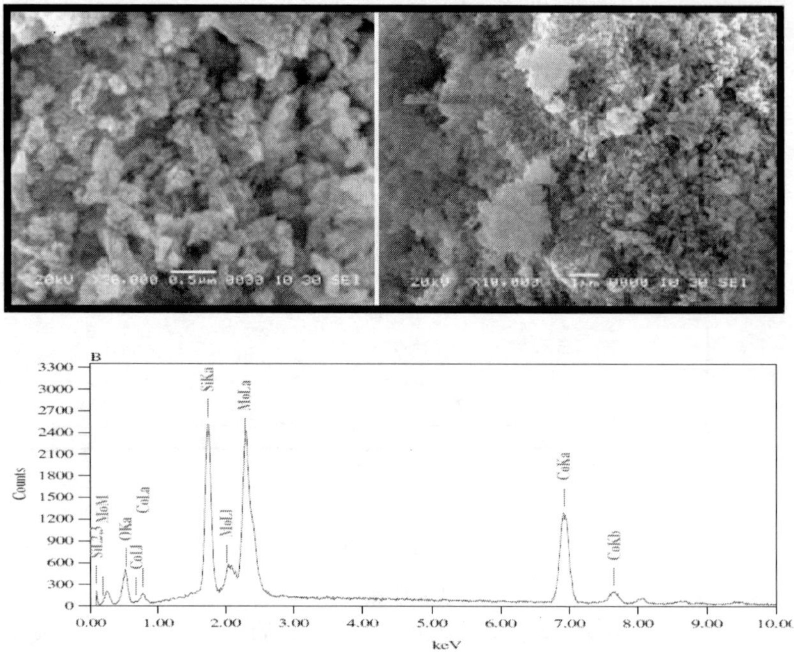

Fig. 3: SEM Images and EDX Graph

Thermal Analysis

In the 0.5% $Co/Mo/SiO_2$ the step I start from 29.57-170.14⁰C which showed 12.288% weight loss due to vaporization of physically adsorbed water. Above 170.14⁰C and upto 547.21⁰C gradually weight loss was observed. The weight loss in this region can be considered as a result of dehydroxylation of hydroxyl group present in the catalyst. In step III corresponds to decomposition of AHM via intermediates into MoO_3. In case of 5% $Co/Mo/SiO_2$, temperature start with 34.55-207.02⁰C where weight loss observed due to loss of adsorbed water and silica precursor, which extends upto 508.97⁰C. After 508.97⁰C there is formation of MoO_3 from AHM to various intermediates. In case of 10% $Co/Mo/SiO_2$ the heating upto 421.28⁰C, the major weight loss was occurred, i.e. upto 24.165%. The weight loss in this region can be considered as a result of loss of water and decomposition of Molybdate species to MoO_3.

By comparing the weight loss of three samples (0.5, 5, 10%) $Co/Mo/SiO_2$, the 5% $Co/Mo/SiO_2$ sample indicates the major weight loss *i.e.* around 47.467% as compare to other samples. This might be due to the increase in Cobalt oxide percentage in the catalyst. The result showed that after calcination temperature i.e. 500⁰C the remarkable weight loss occurs. Therefore utilizing 500⁰C as the catalyst calcination temperature in the present study is suitable.

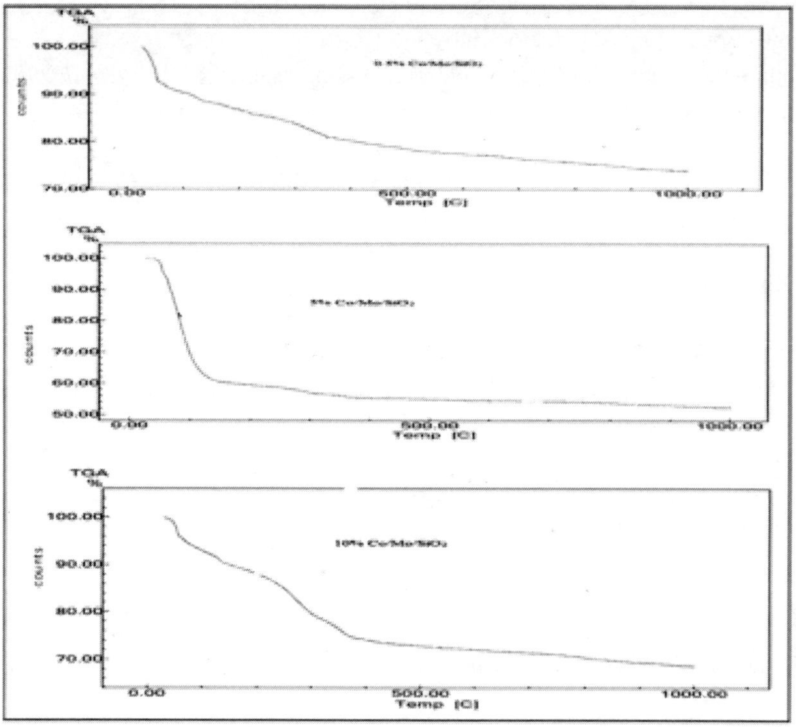

Fig. 4: TGA Curves of (0.5%, 5%, and 10%) $Co/Mo/SiO_2$

CONCLUSION

The catalysts were prepared by sol gel method and surface properties were evaluated by various techniques such as FT-IR, XRD, SEM and EDX, and TGA. XRD results indicate that the formation of two major phases of MoO_3 and $CoMoO_4$ which is favorable for the catalytic reactions. Other techniques support to the formation of catalysts with dispersion of cobalt oxide on the Mo/SiO_2.

ACKNOWLEDGEMENT

We thankful to Prin. K.D. Jadhav and Head, Department of Chemistry for their valuable support and help. Authors also thanks to UGC, New Delhi for financial support under MRP Scheme.

REFERENCES

[1] Auroux, A. and Gervasini, A., Microcalorimetric study of the acidity and basicity of metal oxide surface, *J. Phys. Chem.*, 94, 6371–6379, 1990.

[2] Kawai, M., Tsukuda M, Tamaru K., Surface electronic structure of binary metal oxide catalyst ZrO_2/SiO_2, *Surf. Sci.*, 111, L176, 1981.

[3] Massoth, F.E., Handbook of Spent Hydroprocessing Catalysts. *Adv. Catal.* 27, 265, 1978.

[4] Haber, J., The Role of Molybdenum in Catalysts, Climax Molybdenum Co., Ann Arbor, MI, 1981.

[5] Gates, B.C., Katzer, J.R., and Schuit, G. C. A., McGraw-Hill, Chemistry of Catalytic Processes. (Chapter 5),New York, 1979.

[6] Hu, H., Wachs, I.E. 3rd World Congress on Oxidation Catalysis Proceeding, *J. Phys. Chem.*, 99, 10911, 1995.

[7] Knozinger , H., Taglauer, E., Preparation of Solid *Catalysts. Catal. Roy. Soc. Chem.* Combridge, 10, 1-40, 1993.

[8] Lietti, L., Ramis, G.; Busca, G., Bregani, F., Forzatti, P. Characterization and reactivity of MoO_3/SiO_2 catalysts in the selective catalytic oxidation of ammonia to N_2, *Catal. Today*, 61, 187–195, 2000.

[9] Williams,C.C., Ekerdt, J.G., Jehng, J.M., Hardcastle, F.D., Turek, A.M., Wachs, I.E., A Raman and ultraviolet diffuse reflectance spectroscopic investigation of silica-supported molybdenum oxide, *J. Phys. Chem*, 95, 8781, 1991.

[10] Carbucicchio, M., Trifiro, F., Redox processes at the surfaces of $Fe_2O_3.MoO_3/SiO_2$ catalysts. *J. Catal.* 62, 13, 1980.

[11] Bruckman, K., Grzybowska, B., Che, M., Tatibouet, J.M., Methanol oxidation on MoO_3/TiO_2 catalysts, *Appl. Catal.* A, 96, 279, 1993.

[12] (a) Ono, T., Miyata, H., Kubokawa, Y., J. Chem. Soc. Faraday Trans., 183,1761,1987.

(b) Ma,X., Gong, J., Wang, S., Gao, N., Wang, D., Yang, X., He, F., Reactivity and surface properties of silica supported molybdenum oxide catalysts for the transesterification of dimethyl oxalate with phenol, *Catal. Commun.*, 5, 101, 2004.

[13] Lietti, L., Ramis, G., Busca, G., Bregani, F., Forzatt, P., Characterization and reactivity of MoO_3/SiO_2 catalysts in the selective catalytic oxidation of ammonia to N_2, *Catal. Today.*, 61, 187, 2000.

[14] Dongare, M.K., Bhagwat, V.V., Ramana, C.V., Gurjar, M.K., Silica supported MoO_3: a mild heterogeneous catalyst for the Beckmann rearrangement and its application to some sugar derived ketoximes, *Tetrahedron Lett.*, 45, 4759, 2004.

[15] Umbarkar, S.B., Biradar, A.V., Mathew, S.M., Dagade, S.P., Dongare, M.K., Vapor phase nitration of benzene using mesoporous MoO_3/SiO_2, solid acid catalyst, *Green Chem.* 8, 488, 2006.

[16] Suzuki, K., Hayakawa, T., Shimizu, M., Takehira, K., Partial oxidation of methane over silica supported molybdenum oxide ctalysts, *Catal. Lett.*, 30, 159, 1995.

[17] Reddy, B.M., Chowdhury, B., Smieniotis, P.G., An XPS study of the dispersion of MoO_3 on TiO_2- ZrO_2, TiO_2-SiO_2, TiO_2-Al_2O_3, SiO_2-ZrO_2, and SiO_2- TiO_2-ZrO_2 Mixed oxides, *Appl. Catal. A*, 211, 19-30, 2001.

SYNTHESIS OF ZnO NANORODS AND PARTICLES: CHARACTERIZATION, GAS SENSING PROPERTIES

B.S. RAVIKUMAR, H. NAGABHUSHANA, KURIEN MATHEW[1], B.M. NAGABHUSHANA[2] AND R.P.S. CHAKRADAR[3]

Centre for Nano Research (CNR), Tumkur University, Tumkur
[1]Space Application Centre, ISRO, Ahmedabad
[2]Dept. of Chemistry, M.S. Ramaiah Institute of Technology, Bangalore
[3]National Aerospace Laboratory (CSIR), Bangalore
E-mail: bhushanvlc@gmail.com; bsrkumar7@gmail.com

ABSTRACT

Uniform ZnO nanorods and particles were synthesized by hydrothermal (180°C for 18hrs) and combustion (350°C) techniques respectively. The synthesized nano products were characterized by X-ray diffraction (XRD), Scanning electronic microscope (SEM), Fourier transform Infrared Spectroscopy (FTIR) and Photoluminescence (PL) properties. The obtained ZnO show polycrystalline nature with preferential peak (101) having an average crystallite size 50-85 nm and nanorods have diameters of ~70-90 nm and length upto 1.5 μm. The IR spectrum confirms the presence of characteristic bands of ZnO. Photoluminescence spectrum of the synthesized nanopowders exhibit a dominant, sharp and strong ultraviolet (UV) emission with a suppressed deep-level emission indicating good crystal quality and optical properties. Gas sensitivity experiments on ZnO nanorods and nanoparticles were carried out under different temperatures. The sensitivity of ZnO nanorods and particles were studied towards different gases and found nanorods have good response at ~ 300°C for CO_2 gas.

INTRODUCTION

Nanostructured materials are considered as good candidates for toxic and flammable gas sensing applications due to their large surface area-to-volume ratio and the size factors. Nanostructures

V. Rajendran, K. Saminathan, P. Paramasivam and K.E. Geckeler (eds.)
Nano Materials Synthesis and Characterisation, pp. 303-310 (2012).

of well-established gas sensing materials like tin oxide [1-3], tungsten oxides [4-6], zinc oxide[7,8], titanium oxide [9,10], and indium oxide [11,12] have shown higher sensitivity, faster response, lower operating temperature, and/or enhanced capability to detect low concentration gases compared with the thin lm counterparts. Zinc oxide is a class of versatile material that offer manifold technological applications including gas sensors, opto-electrochronic and optical modulation devices, photocatalysis, hydrophilic surface design, etc. Gas sensors based on zinc oxide are sensitive to a variety of gases such as NO_2, C_2H_5OH, H_2S, LPG, CO and NH_3 [13]. On the other hand, nanostructure zinc oxide such as nanorods [14], nanoparticles [15] etc. have been synthesized using, hydrothermal reaction, combustion synthesis. These nanostructures provide good blocks for developing high performance gas sensors. Here in, we report the synthesis of zinc oxide nanorods and particles by hydrothermal and combustion methods and characterized by powder x-ray diffraction, scanning electron microscopy, Fourier transform Infrared Spectroscopy and Photoluminescence properties. The gas sensing properties were studied at different operating temperatures.

EXPERIMENTAL

Preparation of Zinc oxide Nanorods/Particles

In hydrothermal process the stoichiometric amount of $Zn(NO_3)_2.6H_20$ was dissolved in 30ml doubly distilled water. At room temperature, the stoichiometric quantities of analytically grade cetyl trimethyl ammonium bromide (CTAB) and NaOH were dissolved in 30ml doubly distilled water to form transparent solution under stirring, the stirred solution was transferred into the above zinc nitrate solution under continues constant stirring. The resulted solution was transferred to a teflon-lined stainless steel autoclave of 100ml capacity and heated at 180^0 C for 18hrs. The obtained white solid product was separated from the solution by centrifugation, washed with distilled water and absolute ethanol for four times, respectively, and then dried in air at 80^0 C. A high yield ZnO nanorod was obtained. The obtained ZnO nanorods were used as the sensitive material in the experiment for CO_2 gas response.

During combustion synthesis process the stoichiometric quantities of analytical grade $Zn(NO_3)_2$ and fuel sugar are dissolved in minimum quantity of doubly distilled water. These quantities determines the equivalence ratio O/F = 1 (Oxidizing (O) and reduction (R) valencies of the components), due to exothermic reaction maximum energy was released. The aqueous redox mixture in the petri dish is placed in a pre-heated muffle furnace at 350^0 C. Initially the solution boiled with frothing and foaming and underwent dehydration. During complete dehydration, the surface ignited, burning with temperature (~1000^0 C) and yielding a voluminous solid product within 4-5 mins.

CHARACTERIZATION

All the samples were characterized using X-ray diffractometer (Model PW 3710) with Cu Kα incident radiation of wavelength 1.5405Å. XRD patterns were recorded from 20^0-80^0 (2°) with a scanning step of 0.02^0. The morphology of the sample was examined using scanning electron microscopy (SEM). The photoluminescence studies were carried out using luminescence spectrophotometer with Xe lamp as a source of excitation wavelength 254 nm. Typical infrared (IR) spectrums of the materials were recorded using FTIR model.

The paste prepared from a mixture of ZnO nanorods/particles with an organic adhesive in an agate mortar, then the paste was coated on the alumina substrate on which pair of platinum electrodes had been previously printed by screen printing technique, followed by drying at 100^0 C for 60 mins and subsequently calcined at 600^0 C for 90 mins. The gas sensing properties were measured in a stainless steel chamber system with a coil resistance heater. The carrier gas (dry synthetic air) mixed with a desired concentration of CO_2 gas was flowed through the polythene tube. And a multimeter (Agilent 34970A) was used to monitor the change of electrical resistance upon turning the target gas on and off. The sensor response is defined as $(R_a - R_o)/R_o$, where R_o is the resistance in air and R_a is the change in resistance after the CO_2 gas was introduced.

RESULTS AND DISCUSSION

Plots of X-ray diffraction data of zinc oxide samples prepared are shown in Fig. 1. All the diffraction peaks corresponding to (100), (002), (101), (102), (110), (103), (200), (112), (201) planes were matched to the hexagonal wurtzite ZnO in agreement with the standard JCPDS No. 36-1451, having a lattice parameters a = b = 0.3249 nm, and c = 0.5205 nm. No additional characteristic peaks are observed which confirms the phase purity of the sample. The average crystalline size (D) was estimated from the Debye-Scherrer's formula

$$D = \frac{0.9\lambda}{\beta \cos\theta} \quad \ldots (1)$$

Figure 2(a) shows the SEM images of nano crystalline powder synthesized by solution combustion technique. Particles are connected to each other to make large network systems with irregular pore sizes and shapes. Here the porous network with lot of voids is typically due to escaping of gases during combustion synthesis. Figure 2(b) showed typical SEM image of the flower-like 1D ZnO nanorods synthesized by sol-hydrothermal process. These Nano flowers were composed of shorter Nanorods assembles because the orientation growth along the c-axis was inadequate in lacking the driving force of the growing units. It reveals that the well-defined ZnO Nanocrystals synthesized by the CTAB-assisted hydrothermal processes have high aspect ratio and high surface to volume ratio in the presence of surfactant, which serves as a growth controller and an agglomeration inhibitor leading to the high crystallization products.

And the surfactant played an important role in controlling the morphology of the ZnO nanocrystals.

The purity and molecular structure of the product were analysed by the FTIR spectroscopy. Figure 3(a) shows the FTIR spectrum of the ZnO nanopowder which was acquired in the range of 400-4000 cm^{-1}. No impurity peaks are related to other group which shows the purity of the sample. The two peaks at ~3420 cm^{-1} and 1625 cm^{-1} are attributed to O-H stretching vibration and H-O-H bending vibration, which are assigned to small amount of H_2O existing in the ZnO sample. The peak at 425 cm^{-1} is ascribed to Zn-O stretching vibration. A small peak at 1356 cm^{-1} is due to weak nitrate group.

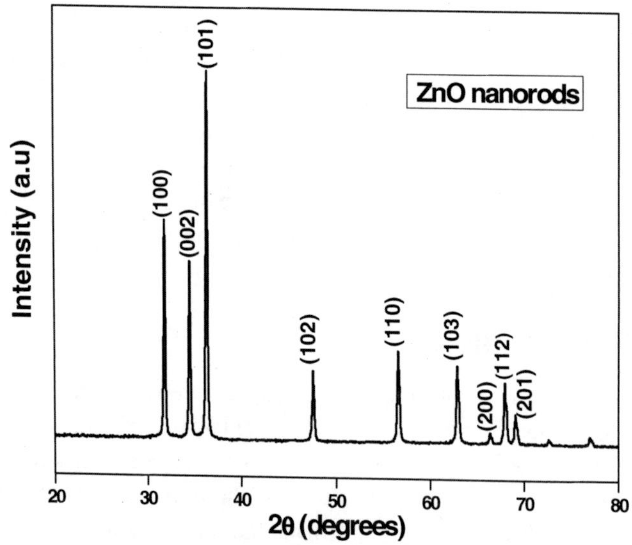

Fig. 1

Fig. 2

Figure 3(b) shows typical room temperature photoluminescence spectrum of ZnO nanopowders. Upon 254 nm excitation, the emission spectrum of ZnO has the asymmetric curve in the visible region, which implies the superposition of multiple emission bands. High intensity sharp peak at 421 nm in violet-region was observed while low intensity peaks at 483 nm and 528 nm were observed in the blue-green and green region. The blue green emission centered at 483 nm and 528 nm may be ascribed to the intrinsic defects such as oxygen vacancies, zinc vacancies, oxygen interstitials and zinc interstitials [16, 17]. For sensing measurement, the concentration of the CO_2 gas was regulated by mixing with dry synthetic air. In present study response (%) is defined as R_g/R_a. where R_a and R_g are the electric resistance in air and in CO_2 gas respectively. Temperature dependence on the gas sensing properties could result from physisorption and chemisorption rates of the oxygen ions on the metal oxide surface [18]. The oxygen molecules adsorbed on the surface of ZnO to form ions (O_2^-, O^-, O^{2-}) from the conduction band of ZnO depending on the temperature in air atmosphere. The stable ions are O_2^- below 100^0 C,O^- between 100 and 300^0 C,O^{2-} above 300°C [19]. By the fallowing reactions:

$$CO_2(_{gas}) + e^- \rightarrow CO_2(_{ads})^- \qquad \text{... (2)}$$

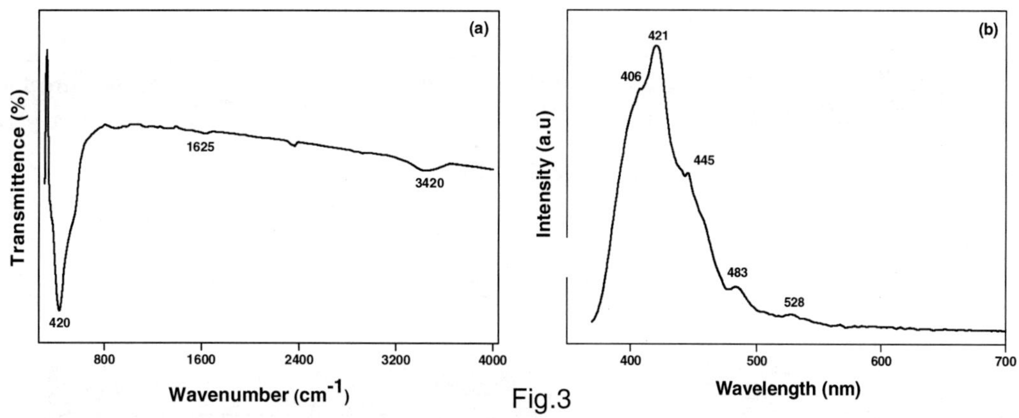

Fig. 3

The above reactions result in the decrease of concentration of electron on the surface of ZnO which results in an increase in resistance of the material.

$$CO_2(_g) + e^- \rightarrow CO_2(_{ads})^- \qquad \text{... (3)}$$

$$CO_2(_{ads})^- + 2O_{ads}^- \rightarrow 2O_{ads}^{2-} + CO_2 \qquad \text{... (4)}$$

These series of reactions result in the further decreased concentration of electrons on the surface of ZnO, which led to the increase in resistivity of the material.

For the nanoscale sensors, high surface to volume ratio are the main reason for the reaction of gases and samples [20]. Sensors fabricated with synthesized materials were exposed to

2000 ppm of CO_2 gas in synthetic air at different temperature to find out optimum operable temperature which was found to be good response at 300^0 C for ZnO nanorod sample. In addition in order to testify the effect of the sample's morphology on its gas response in a gas sensor, the sample of ZnO nanoparticles obtained from combustion technique was also studied under the same condition.

The gas sensitivity –test time in min plots shown in Fig. 4(a) reveals that ZnO nanorods sensors has better response than a nanoparticles based sensor. Even though the big surface area of ZnO particles, there will be more and more particles ZnO particles are connected to each other to make large network systems with irregular pore sizes that will increase the energy barrier of electron transport, but less gas channeling after applying in a gas sensors. Compare to nanoparticles, nanorods has more gas channels lead to increase in effective surface area which can contact with the gas. Higher gas response of the sensor made from ZnO nanorods might be attributed to higher surface areas. We also measured the dynamic response properties of the nanorods to 2000 ppm CO_2 in synthetic air at the operating temperature of 300^0 C. The response time and recovery time of the sensors are quite good. Response time, time required to reach 90% of its steady value and recovery time to 2000 ppm CO_2 in dry air for ZnO nanorods are shown in fig 4(b).

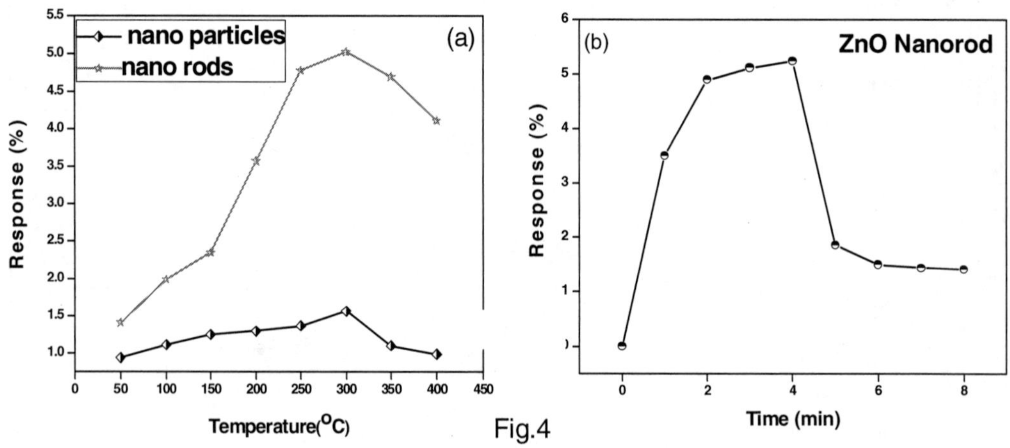

Fig. 4

CONCLUSIONS

In summary, we prepared well-crystallized ZnO nanorods/particles by wet chemical methods and characterized their size and shape by different techniques. The gas sensing properties are investigated using own build chamber system with coil resistive heater. The result demonstrates that the sensors based on ZnO nanorods are good response to 2000 ppm CO_2 at 300^0 C. And response time and recovery time were found to be good for the same concentration of CO_2 gas.

ACKNOWLEDGEMENTS

The authors gratefully acknowledge Indian Space Research Organization (ISRO)-SAC, Ahmedabad, for the financial support to carry out this research work.

REFERENCES

[1] A.Kolmakov, Y.X. Zhang, G.S. Cheng, M. Moskovits. *Adv. Mater.* 15 (12) 2003, 997-1000.

[2] E.Comini, G. Faglia, G. Sberveglieri, D. Calestani, L. Zanotti and M. Zha, *Sens.Actuators* B 111, 2005, 2-6.

[3] Y. B. Shen, T. Yamazaki, Z. F. Liu, D. meng, T. Kikuta, N. Nakatani, *Sens. Actuators B* 135, 2009, 524-529.

[4] A. Ponzoni, E. Comoni, G. Sberveglieri, J. Zhou, S.Z. Deng, N.S. Xu, Y. Ding and Z.L. Wang, *Appl. Phys. Lett.*, 88 (20), 2006, 203101.

[5] K.M. Sawicka, A.K. Prasad and P.I. Gouma. *Sens. Lett.* 3 (1) 2005 31-35.

[6] C. S. Rout, A. Govindaraj and C.N.R. Rao. *J. Mater. Chem.* 16 (40) (2006), 3936-3941.

[7] Yu-zhen Lv, Cheng-rong Li, Lin Guo, Fo-chi Wang, Yue Xu and Xiang-feng Chu. *Sens. Actuators* B 141, 2009, 85-88.

[8] S.G. Ansari, Rizwan Wahab, Z.A. Ansari, Young-Soon Kim and Gilson Khang. *Sens. Actuators B* 137, 2009, 566-573.

[9] O.K. Varghese and C.A. Grimes, *J. Nonosci. Nanotechnol.* 3(4), 2003, 277-293.

[10] A.S. Zuruzi, A. Kolmokov, N.C. Macdonald and M. Moskovits, *Appl.Phys. Lett.* 88 (10), 2006, 102904.

[11] C. Li, D.H. Zhang, X.L. Liu, S. Han, T.Tang, J. Han, C.W. Zhou. *Appl, Phys. Lett.* 82 (10), 2003, 1613-1615.

[12] C. Li, D.H. Zhang, X.L. Liu, S. Han, T. Tang and C.W. Zhou. *Appl. Phys. Lett.* 83 (9), 2003, 1845-1847.

[13] Xu Jiaqiang, Chen Yuping, Chen Daoyong and Shen Jianian. *Sens. Acuators B* 113, 2006, 526-531.

[14] Jiaqiang Xu, Yuan Zhang, Yuping Chen, Qun Xiang, Qingyi Pan and Liyi Shi, *Mater. Sci. Engg.* B 150 (2008) 55-60.

[15] A. Jagannatha Reddy, M.K. Kokila, H. Nagabhushana, R.P.S. Chakradhar, C. Shivakumara, J.L. Rao and B.M. nagabhushana. *J. Alloys and Comp.* 509, 2011, 5349-5355.

[16] V.S. Gorelik, S.N. Mikov, M.I. Sokolovskii T and Tsuzuki, *Inorg. Mater.* 42 (2006) 282-285.

[17] S.A. Shojaee, M.M. Shahraki, M.A. Faghihi Sani, A. Nemati and A. Yousefi, *J. Mater. Sci. Mater*. Electron. 21 (2010) 571-577.

[18] D. Manno, G. Micocci, R. Rella, A. Serra, A. Taurino and A. Tepore, *J. appl. Phys*. 82, 1997, 54-59.

[19] A.Z. Sadek, S. Choopun, W. Wlodarski, S.J. Ippolito and K. Kalantar-zadeh. *IEE Sens.* J.7 2007, 919-924.

[20] X.J. Huang Y.K. Choi, *Sens. Actuators B* 122, 2007, 659-671.

ACID AND BASE MEDIATED CITRIC ACID UREA ASSISTED COMBUSTION PROCESS FOR THE PREPARATION OF NANOCRYSTALLINE $NiFe_2O_4$

R. INDHRAJOTHI, M. VENKATESWARLU[1] AND N. SATYANARAYANA

Department of physics, Pondicherry University, Puducherry
[1]R&D, REC, Amara Raja Batteries Ltd, Tirupati, Andhra Pradesh
E-mail:indhrajothisharini@yahoo.com

ABSTRACT

Nanocrystalline $NiFe_2O_4$ nanoparticles are prepared by citric acid urea assisted combustion process using metal ions, citric acid and urea ratio as 1:1:0.5 and also without and with the addition of excess acid and base. The prepared samples are characterized using XRD, FTIR, DSC and SEM-EDS techniques. The impurity free, nanocrystalline $NiFe_2O_4$ is obtained for the sample prepared with the addition of excess acid in the citric acid urea assisted combustion process.

INTRODUCTION

Nanocrystalline spinal ferrites are very important materials because of their interesting magnetic and electrical properties and find applications in magnetic recording media, catalysis, drug delivery, sensors and electrodes for lithium ion battery, etc. [1-5]. The novel electrical and magnetic properties of ferrites can be tuned by controlling the size of their nanoparticles. Various wet chemical methods are used to prepare nanocrystalline ferrites. Among these, combustion process can be effectively used to prepare high pure, stoichiometric materials at a lower temperature. Various physical and chemical conditions can control the combustion process and in turn its particle size [6-7]. Hence, in the present work, $NiFe_2O_4$ samples are prepared using citric acid urea assisted combustion process for the metal ions: citric acid: urea ratio

V. Rajendran, K. Saminathan, P. Paramasivam and K.E. Geckeler (eds.)
Nano Materials Synthesis and Characterisation, pp. 311-318 (2012).

(M:CA:U) = 1:1:0.5 at three different chemical conditions by adding excess acid and base. The complete process is explained through reaction mechanism with the help of characterization results of XRD, FTIR, DSC and SEM-EDS.

EXPERIMENTAL SYNTHESIS

Three sets of nanocrystalline $NiFe_2O_4$ samples are prepared using citric acid urea assisted combustion process for the metal ions: citric acid: urea ratio (M:CA:U) = 1:1:0.5 at three different chemical conditions by adding excess acid and base. The chemicals used in the process are AR grade nickel nitrate, ferric nitrate nanohydrate, citric acid and urea. The required nickel nitrate and ferric nitrate solutions for the 5 g end product are sonicated for 10 minutes each. Later, the mixture of nickel nitrate and ferric nitrate solutions are also sonicated for 10 minutes and stirred at 353 K. The required citric acid and urea solutions are added to the above solution on continuous stirring after 30 minutes. The samples are labeled as raw. In the other two set of samples excess 2ml of concentrated nitric acid, 2 ml of ammonia solution are added on continuous stirring and labeled as acid and base respectively. The water gets evaporated from the solution mixture on continuous stirring and results in polymeric resin. The polymeric resin is dried in oven at 333 K for 24 h and formed the volume expanded porous polymeric intermediate. The polymeric intermediate of raw and acid showed porous nature with volume expansion. The polymeric intermediate of base showed hard nature. The polymeric intermediate is calcined at different temperatures and formed the nanocrystalline $NiFe_2O_4$. The complete process is monitored by powder X-ray diffraction, fourier transform infrared spectroscopy, differential scanning calorimetry, scanning electron microscopy and energy dispersive X-ray spectroscopy. The XRD spectra are taken using PANalytical XpertPRO diffractometer with Cu $K_{\acute{a}}$ radiation of wavelength 0.15418 nm from 10-80° at a scan rate of 1.5° per minute. Fourier transform infrared spectra were recorded on Shimadzu FTIR/8300/8700 spectrophotometer in the frequency range of 4000–400 cm^{-1} with a 2 cm^{-1} resolution for 20 scans. FTIR spectra were recorded on thin transparent pellets and each pellet was made by pressing a mixture of 0.5 mg sample and 200 mg of spectra-pure KBr. The thermal behavior of the prepared samples was analyzed using TA instrument Q20DSC differential scanning calorimeter. The particle size and morphology and elemental mapping of $NiFe_2O_4$ sample is analyzed using Hitachi SN 3400N SEM instrument.

RESULT AND DISCUSSION

Reaction Mechanism

In the citric acid urea assisted combustion process, citric acid is used as fuel as well as chelating agent and urea is acting as additional fuel.

Scheme 1: When urea is added to the mixture, it reacts with the H^+ ion and release CO_2 gas and form NH_4. NH_4 reacts with nitrates and form ammonium nitrate according to the chemical reaction.

$$NH_2\text{-}CO\text{-}NH_2 + 2H^+ + H_2O \rightarrow 2NH_4^+ + CO_2 \text{ - Hydrolysis of urea}$$

$$NH_4^+ + NO_3^- \rightarrow NH_4NO_3$$

The formed ammonium nitrate is a combustion aid and ignites the combustion reaction at a lower temperature. The heat energy released during the combustion reaction is sufficient to form the required crystalline phase.

Addition of Acid

When concentrated nitric acid added to the metal nitrates, citric acid urea solutions, the nitric acid neutralize the ammonia formed in the step 1 and inhibit the formation condition of ammonium nitrate, instead the ammonia and nitrates liberated as ammonia gas, NO_2 and form the porous polymeric intermediate during evaporation.

Addition of Base

When concentrated ammonia solution is added to the metal nitrates, citric acid urea solutions, the added ammonia mixed with the ammonia generated from step 1 and favor the formation condition of ammonium nitrate and increase the quantity of ammonium nitrate.

X Ray Diffraction

Figures 1-3 respectively show the X ray diffraction patterns for the $NiFe_2O_4$ polymeric intermediate prepared using citric acid urea assisted combustion process using M:CA:U ratio 1:1:0.5 without and with addition of excess acid and base obtained at different temperatures. From Fig. 1, the X ray diffraction patterns for the $NiFe_2O_4$ polymeric intermediate prepared using citric acid urea assisted combustion process without addition of acid and base show peaks corresponds to the formation of ammonium nitrate. This is confirmed by comparing the observed diffraction peaks with the JCPDS data. On further calcinations at 448 and 623 K the ammonium nitrate is removed from the sample during combustion reaction and formed the $NiFe_2O_4$ phase, which is confirmed by comparing the diffraction pattern with the JCPDS data. The crystallite size is calculated using Scherer's formula and it is found to be 21 nm. This confirmed the formation of nanocrystalline $NiFe_2O_4$ phase. From Fig. 2, the X ray diffraction patterns for the $NiFe_2O_4$ polymeric intermediate prepared using citric acid urea assisted combustion process with excess addition of acid shows peak free diffraction pattern, which corresponds to the amorphous phase. On further calcinations at 448 and 623 K, the crystalline $NiFe_2O_4$ phase is formed, which is confirmed by comparing the observed diffraction pattern with the JCPDS Jcpds data. The crystallite size is calculated using Scherer's formula and it is found to be 14 nm. This confirmed the formation of nanocrystalline $NiFe_2O_4$ phase. From Fig.

3, the X ray diffraction patterns for the $NiFe_2O_4$ polymeric intermediate prepared using citric acid urea assisted combustion process with excess addition of base shows peaks corresponds to the formation of ammonium nitrate. This is confirmed by comparing the observed diffraction peak with the JCPDS data. On further calcinations at 448 and 623 K the ammonium nitrate is removed from the sample during combustion reaction and formed the $NiFe_2O_4$ phase, which is confirmed by comparing the diffraction pattern with the JCPDS data. The crystallite size is calculated using Scherer's formula and it is found to be 36 nm. This confirmed the formation of nanocrystalline $NiFe_2O_4$ phase. Broad diffraction pattern is observed for $NiFe_2O_4$ sample prepared with addition of excess acid to the citric acid urea assisted combustion process. This is also reflected in the calculated crystallite size, which is explained in the reaction mechanism. Hence, X ray diffraction results confirm the formation of nanocrystalline $NiFe_2O_4$. Lowest crystallite size is observed for $NiFe_2O_4$ sample prepared with addition of excess acid.

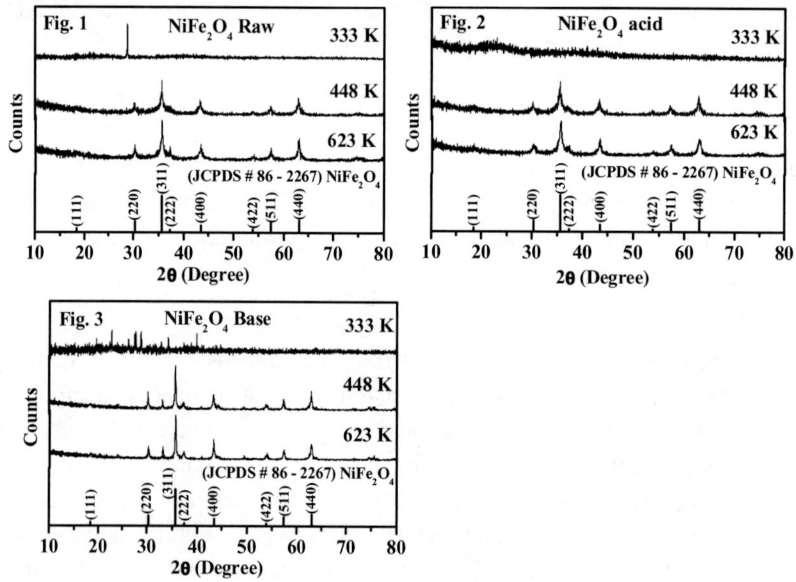

Fig. 1-3: XRD Patterns of Nanocrystalline $NiFe_2O_4$ Sample Prepared with M: CA: U Ratio 1:1:0.5 Without and With Addition of Excess Acid and Base Obtained at Different Temperatures

Fourier Transform Infrared Spectroscopy

Fig 4-6 show the FTIR spectra of nanocrystalline $NiFe_2O_4$ samples prepared with M: CA: U ratio 1:1:0.5 with and without additions of excess acid and base obtained at 333 K and 623 K and $NiFe_2O_4$ sample prepared with M: CA: U ratio 1:1:0.5 with addition of excess acid obtained at various temperatures. From Fig 4, the FTIR spectra of $NiFe_2O_4$ sample obtained at 333 K shows the vibration bands at 3400, 3173, 1725, 1616, 1594, 1400, 1387, 1315, 1071, 905 and 842 cm^{-1}. Characteristic bands at 3400 and 1616 cm^{-1} correspond to the stretching and bending modes of water molecule. The band at 3173 cm^{-1} corresponds to NH_3 vibrations from the ammonia group. The bands at 1594 and 1400 cm^{-1} correspond to symmetric and asymmetric stretching of C=O bond which confirm the chelation of metal ions by citric acid.

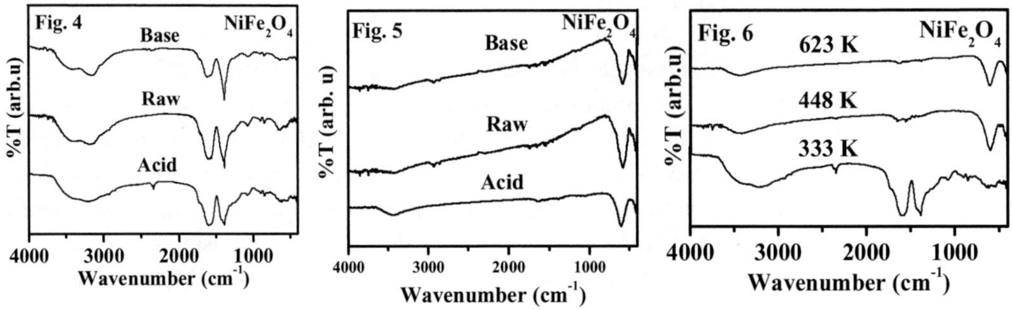

Fig. 4-6: FTIR Spectra of Nanocrystalline $NiFe_2O_4$ Samples Prepared with M: CA: U Ratio 1:1:0.5 without and with Addition of Excess acid and base Obtained at 333 and 623 K, 2 c $NiFe_2O_4$ Sample Prepared with M: CA: U ratio 1:1:0.5 with Addition of Excess Acid Obtained at Different Temperatures

The band at 1725 cm^{-1} corresponds to the asymmetric vibrations of bridging COO^- group and the respective symmetric vibration observed at 1315 cm^{-1}. A band at 1387 cm^{-1} shows the presence of some NO_3 group [8–10]. The observed IR bands confirm the citric acid polymerization, chelating of metal ions and presence of ammonium nitrate. The band due to nitrates is strong in the excess acid added sample and the band due to ammonia is strong in excess base added sample. During calcinations the organic impurities gets decomposed from the sample which is reflected in the absence of corresponding bands in Fig. 5. Figure 6 shows the FTIR spectra of $NiFe_2O_4$ sample prepared with M:CA:U ratio 1:1:0.5 with addition of excess acid obtained at various temperatures. The observed bands at 590 and 414 cm^{-1} corresponds to the octahedral and tetrahedral vibration of ferrites. By comparing the observed vibration bands of $NiFe_2O_4$ sample prepared with and without addition of excess acid and base in citric acid urea assisted combustion process, the sample prepared with the addition of excess acid shows organic impurity free vibration bands corresponds to $NiFe_2O_4$. Hence, FTIR results confirm the purity of $NiFe_2O_4$ sample prepared using citric acid urea assisted combustion process without adding excess acid and base.

Differential Scanning Calorimetry

Figure 7, shows differential scanning calorimetry thermograms for $NiFe_2O_4$ polymeric intermediate samples prepared with M: CA: U ratio 1:1:0.5 additions of excess acid and base.

From Fig. 7, $NiFe_2O_4$ sample prepared using citric acid urea assisted combustion process without addition of excess acid and base shows the ignition temperature at 460 K. Similarly, for $NiFe_2O_4$ sample prepared using citric acid urea assisted combustion process with excess acid shows the ignition at a lower temperature at 463 K. In the case of $NiFe_2O_4$ sample prepared using citric acid urea assisted combustion process with excess base shows the ignition at 480 K. The change in ignition temperature as well as the broadness of the exothermic peak reflects the combustion reaction which is modified based on the addition of excess acid and base to the citric acid urea assisted combustion process. The $NiFe_2O_4$ sample prepared using citric acid urea assisted combustion process without adding acid and base shows a small exothermic peak, which corresponds to the decomposition of citric acid whose decomposition temperature around 460 K. When excess addition of acid neutralize the effect of urea and hinder the formation of ammonium nitrate, which is also reflected in the X ray diffraction results. The addition of excess acid increases the oxidant and modifies the fuel to oxidant ratio and it reflects in combustion reaction. In the case of $NiFe_2O_4$ sample prepared using citric acid urea assisted combustion process with addition of excess base shows ignition at 480 K corresponds to the decomposition of ammonium nitrate whose decomposition temperature is around 475 K. Addition of excess ammonia solution forms more ammonium nitrate compared to the formation of ammonium nitrate in the citric acid urea assisted combustion without addition of acid and base.

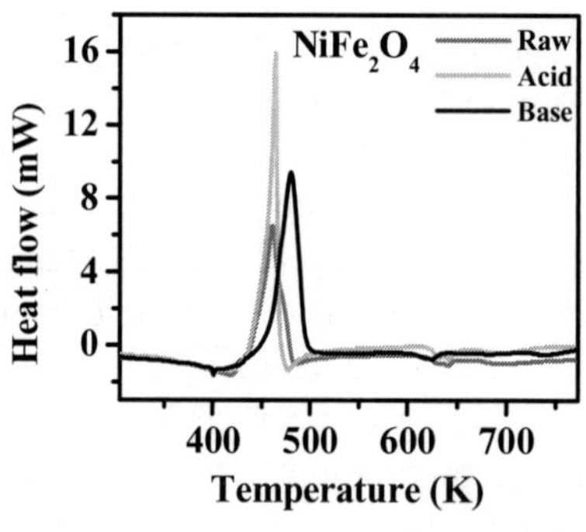

Fig. 7: DSC Thermograms for $NiFe_2O_4$ Polymeric Intermediate Samples Prepared with M:CA:U ratio 1:1:0.5 Additions of Excess Acid and Base

Scanning Electron Microscopy and Energy Dispersive X Ray Spectroscopy

Figure 8 shows the SEM images of $NiFe_2O_4$ sample prepared respectively using citric acid urea assisted combustion process without and with addition of acid and base and 9 shows SEM EDS elemental mappings $NiFe_2O_4$ sample prepared with excess addition of acid obtained at 623 K. From Fig. 8, it is observed that the $NiFe_2O_4$ particles get agglomerated and the particle size was varied form 30 to 100 nm. The elemental spectrum and mapping results confirmed the presence and uniform distribution of Ni, Fe and O elements in the $NiFe_2O_4$ sample. Hence, SEM-EDS results of $NiFe_2O_4$ sample confirmed the formation of $NiFe_2O_4$ nanoparticles.

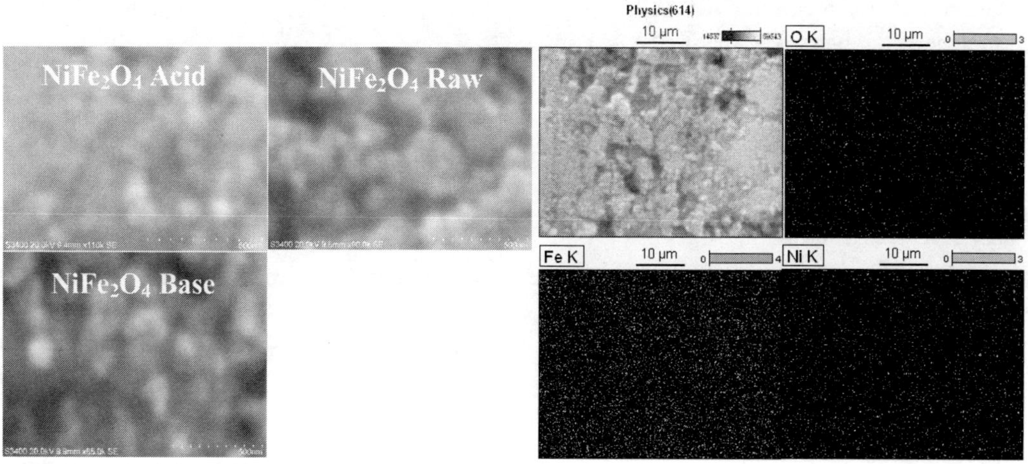

Fig. 8: SEM Images of Nanocrystalline $NiFe_2O_4$ Samples Prepared with M:CA:U Ratio 1:1:0.5 without and with Addition of Excess acid and base Obtained at 623 K and 9. SEM-EDS Elemental Mappings of $NiFe_2O_4$ Sample Prepared with M:CA:U Ratio 1:1:0.5 with Addition of Excess Acid Obtained at 623 K

CONCLUSIONS

Citric acid urea assisted combustion process with different chemical conditions (Addition of excess acid and base) was successfully utilized to prepare nanocrystalline $NiFe_2O_4$. The role of acid and base in the citric acid urea assisted combustion process was explained from the reaction mechanism and the same also confirmed from the characterization results. The X ray diffraction results confirmed the phase formation of $NiFe_2O_4$ in all the three samples. The crystallite size was calculated from the Scherer's formula and confirmed that smallest crystallite size was observed for the $NiFe_2O_4$ sample prepared with addition of excess acid. The results were also correlated with the DSC, FTIR results and the results were according the reaction mechanism. The FTIR results confirmed the formation of $NiFe_2O_4$ and the removal of organic impurities. Complete removal of organic impurities was confirmed for $NiFe_2O_4$ sample prepared with addition of excess acid in the citric acid urea assisted combustion process. The DSC results also confirmed that the addition of acid and base modified the combustion reaction by forming the byproducts like ammonium nitrate, which alters the combustion reaction. The SEM images confirmed the formation of $NiFe_2O_4$ nanoparticles of size less than 100 nm and SEM-EDS mapping results confirmed the presence of Ni, Fe and O elements in the $NiFe_2O_4$ sample.

ACKNOWLEDGMENT

NS is gratefully acknowledged DST, UGC, AICTE, CSIR and DRDO, Govt.of India forreceiving the financial support in the form of major research projects. RI acknowledges UGC-RGNF for receiving JRF fellowship for doing Ph.D. Authors also acknowledges CIF Pondicherry University for using SEM and DSC.

REFERENCES

[1] T. Hyeon, Y. Chung, J. Park, S.S. Lee, Y-W Kim and B.H. Park, *J. Phys. Chem.* 106, 6831, 2002.

[2] J.M. Wesselinowa and I. Apostolova, *J. Phys.: Condens. Matter* 19, 1, 2007.

[3] M. George, Swapna S Nair, K.A Malini, P.A. Joy and M.R. Anantharaman, *J. Phys. D: Appl. Phys*. 40, 1593, 2007.

[4] S-Y Zhao, R. Qiao, X.L. Zhang, and Y. S. Kang, *J. Phys. Chem.* C 111, 7875, 2007.

[5] S Chkoundali, S. Ammar, N. Jouini, F. Fievet, P. Molinie, M. Danot, F. Villain and J-M Greneche, *J. Phys.: Condens. Matter* 16, 4357, 2004.

[6] C.H. Yan, Z.G. Xu, F.X. Cheng, Z.M. Wang, L.D. Sun, C.S. Liao and J.T. Jia, *Solid State Commun*., 111, 287, 1999.

[7] S. Komarneni, M.C.D'Arrigo, C. Leonelli, and G.C. Pellacani, H. Katsuki, *J. Am. Ceram. Soc.*, 81, 3041, 1998.

[8] S. Vivekanandhan, M. Venkateswarlu, N. Satyanarayana, *Journal of Alloys and Compounds* 441, 284, 2007.

[9] Seema Verma and P.A. Joy, *Materials Research Bulletin* 43, 3447, 2008.

[10] G. Socrates, Infrared and Raman Characteristic Group Frequencies, John Wiley and Sons, New York, 2001.

SYNTHESIS AND CHARACTERIZATION OF $LiMn_2O_4$ SPINEL PHASE NANOPARTICLES BY THE PECHINI PROCESS

K. Hari Prasad, D. Narasimulu, Paramananda Jena, M. Venkateswarlu[1] and N. Satyanarayana

Department of Physics, Pondicherry University, Pondicherry
[1]R&D, Amaraja Batteries, Thirupathi
E-mail: nallanis2011@gmail.com

ABSTRACT

Spinel $LiMn_2O_4$ has been receiving a special attention over other materials due to its better electrochemical performance, low-cost, environmentally friendly, and highly abundant cathode material used for Li-ion battery industry. In the present work, nanocrystalline $LiMn_2O_4$ powder has been synthesized successfully by the Pechini process using metal nitrates, citric acid and polyethylene glycol-400. The synthesized samples were characterized using thermogravimetry/differential thermal analysis (TG/DTA), Fourier transform Infrared (FTIR), X-ray diffraction (XRD), scanning electron microscope (SEM) and BET measurements. The observed X-ray diffractogram patterns of the prepared samples confirm the formation of the crystalline $LiMn_2O_4$ spinel phase. Scanning electron microscope images of the synthesized $LiMn_2O_4$ powder showed an agglomeration of spherical shape nanocrystalline $LiMn_2O_4$ particles. Thermal decomposition of polymeric intermediate was investigated through thermogravimetry/differential thermal analysis. FTIR was used to identify the structural coordination of as prepared and calcined polymeric intermediate. BET is used to measure the surface area and porosity of the prepared $LiMn_2O_4$ samples.

V. Rajendran, K. Saminathan, P. Paramasivam and K.E. Geckeler (eds.)
Nano Materials Synthesis and Characterisation, pp. 319-326 (2012).

INTRODUCTION

Lithium-ion batteries are widely used in electronic devices, portable power tools, hybrid electric vehicles, many power supplies, etc., due to their high energy and power density. Spinel $LiMn_2O_4$ is one of the most attractive material in the field of lithium-ion batteries because of its high natural abundance, lower cost, environmental harmlessness and good safety compared with the other materials used in current batteries [1]. For the battery applications, better homogeneity, single-phase, uniform particle morphology with submicron size distribution, and large surface area are considered desired characters for achieving better electrode properties [2]. However, spinel $LiMn_2O_4$ synthesized by the conventional method has several disadvantages, such as inhomogeneity, large particle size, irregular morphology, wide particle size distribution, high synthesis temperature and repeated grinding [3-5]. The small size of nanoparticles can boost reactivity and shorten the diffusion paths, which can increase the ability of electrode to withstand high discharge rates over adequately long cycle life [6-9]. So, the development of nano-structured electrodes is attractive. In some cases, portions of the $LiMn_2O_4$ product are mixed with a desired amount of Li salt in order to improve the cycling performances [10]. Hence, it is necessary to optimize the precursors as well as experimental conditions for the synthesis of nanocrystalline $LiMn_2O_4$ particles with desired properties.

EXPERIMENTAL

Mixing of Citric acid (CA), Ethylene glycol and Aqueous Lithium nitrate/Manganese Acetate solutions (M/CA = 1:1)(M/EG) = 1:2)

Lithium nitrate ($LiNO_3$) and manganese acetate [Mn $(CH_3COO)_2 . 4H_2O$] was used as sources of metal ions. Citric acid ($C_6H_8O_7$) and ethylene glycol ($C_3H_8O_3$) were used as fuels as well as polymerizing agents along with nitric acid. The molar ratio of total metal ions to citric acid ratio was kept as 1:1 and the molar ratio of total metal ions to ethylene glycol was set as 1:2. Standard (16N Concentrated) nitric acid (NA) was added to starting solution (SS) by keeping the volume ratio of SS: NA = 5:1. The resulting transparent yellow color solution was evaporated at 70°C to 80°C under constant stirring condition. Continuous evaporation of the starting solution with nitric acid caused the formation of dark yellowish resin. In order to obtain the dried intermediates, the resulting product (resin and paste) was heated at 150°C for 24 hours in air. On heating, the resin expanded 40 times of its original volume and resulted high porous polymeric intermediate. Flow chart of the nitric acid assisted Pechini process for the synthesis of nanocrystalline $LiMn_2O_4$ powder is shown in Fig. 1.

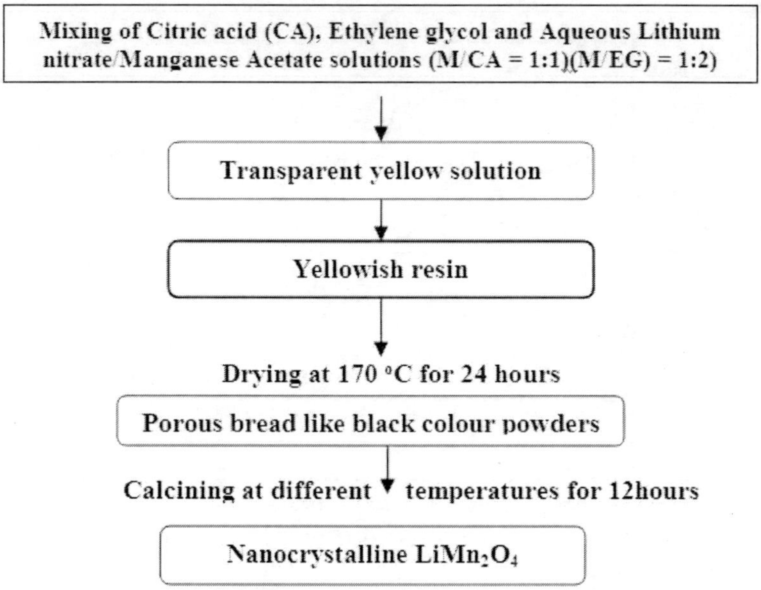

Fig. 1: Flow Chart of the Nitric Acid Assisted Pechini Process for the Synthesis of Nanocrystalline $LiMn_2O_4$ Powder

RESULTS AND DISCUSSION

Thermal Analysis of Polymeric Intermediates by TG/DTA

TG/DTA thermograms of polymeric intermediate of the $LiMn_2O_4$ sample is shown in Fig. 2. The observed two endothermic curves at 58 and 206°C, the corresponding weight loss are attributed to the removal of adsorbed moisture, which indicates the hygroscopic nature of polymeric intermediate. The exothermic curves observed around at 300, 360 and 440°C that correspond to the combustion process accompanied by major weight loss due to decomposition of nitrates and organic residues. After, 500°C no more weight loss was observed in the TG curves for the intermediates, which indicates the complete decomposition of organic derivatives and the formation of $LiMn_2O_4$ structure. DTA curve of polymeric intermediate shows that addition of nitric acid to the starting solution increases the ignition temperature as well as the exothermicity of combustion reaction. Polyesterification between citric acid and ethylene glycol in nitric acid assisted process results the highly polymerized intermediate, which increased the ignition temperature. Similarly, the increased exothermicity of combustion reaction for the intermediate prepared with nitric acid addition is due to the formation of high porous foam like polymeric intermediates. The same result was later confirmed by FTIR, XRD and SEM-EDX [11, 12].

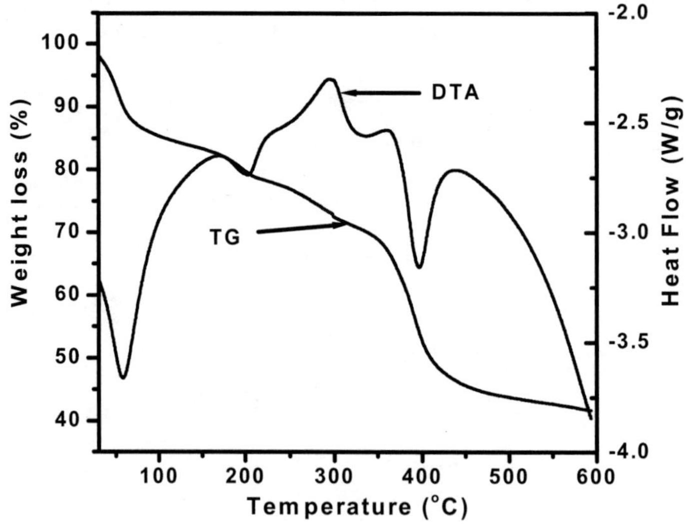

Fig. 2*:* TG/DTA Thremograms of Polymeric Intermediates Prepared Nitric Acid Assisted Pechini Process

XRD

XRD patterns of the polymeric intermediate and calcined polymeric intermediate at 500°C alongwith standard JCPDS data (PDF # 088648) are shown in Fig. 3. From Fig. 3, the observed peaks free XRD pattern of the polymeric intermediate confirms the amorphous nature, which is further confirmed by SEM result [13]. From Fig. 3, the observed XRD pattern of the polymeric intermediate calcined at 500°C is compared with the standard JCPDS data and confirmed the formation of crystalline spinel $LiMn_2O_4$ phase. The crystallite size of $LiMn_2O_4$ powder is calculated using the broadening of the (111) peak data (recorded at the scanning rate of ½° per minute) through Lorentz fitting method and the Scherrer's formula.

$$D = \frac{0.9\,\lambda}{\beta \text{Cos}\theta}$$

Where, λ is the X-ray wavelength (1.54 A^0) and β is the full-width at half maximum (FWHM) of the diffraction line. The crystallite size and lattice parameter of $LiMn_2O_4$ powder are respectively found to be 16 nm and a = 8.1852 Å. The obtained lattice parameter for the $LiMn_2O_4$ powder prepared with nitric acid addition is very much comparable with the reported value.

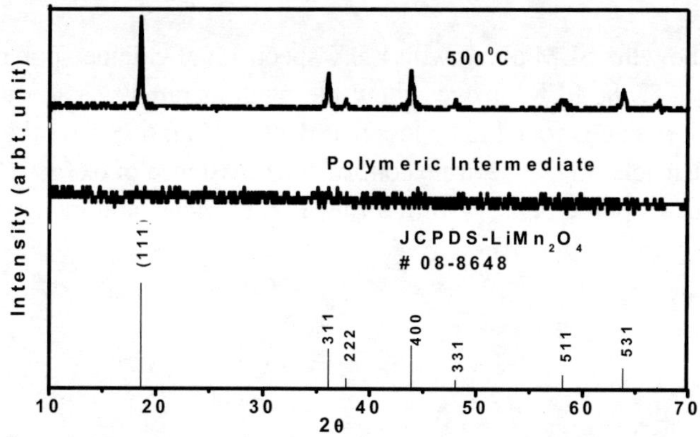

Fig. 3: Show the XRD Analysis of Polymeric Intermediates Prepared by Pechini Process with Nitric Acid

FTIR

Figure. 4 shows the FTIR spectra of the polymeric intermediate and the calcined polymeric intermediate at 500°C. From Fig. 4, the vibration bands at 3423, 1598, 1385, 1318 and 1082 cm^{-1}. From Fig.4, the peaks observed at 3423 cm^{-1} and 1598 cm^{-1} for the polymeric intermediates are attributed to OH stretching due to the adsorbed moisture as well as OH group present in the citric acid derivatives. A band at 1385 cm^{-1} shows the presence of some NO_3 group. The IR bands observed at 1318 and 1082 cm^{-1} are assigned to symmetric and asymmetric vibrations of the COO^- ions respectively, which confirm the chelation of metal ions by citric acid [14, 15]. The peaks observed at 615 and 514 cm^{-1} for the calcined polymeric intermediates at 500°C is attributed to Li-Mn-O stretching vibration band [14, 15].

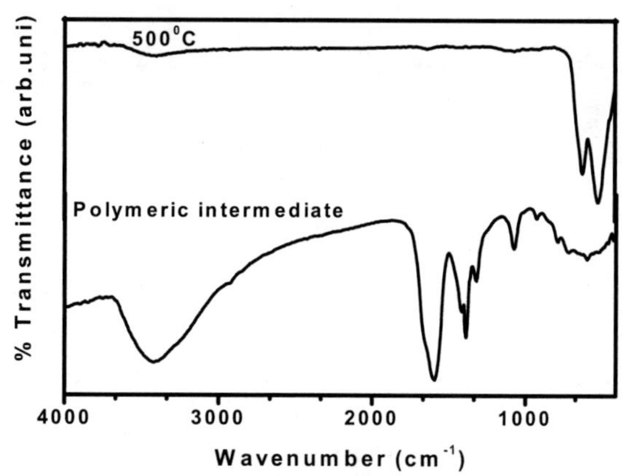

Fig. 4: Show the FTIR Analysis of Polymeric Intermediates Prepared by Pechini Process with Nitric Acid

SEM-EDX

Figures. 5 and 6 show the SEM along with EDX spectrum of calcined polymeric intermediate at 500°C. From Fig.5, the SEM images show the agglomerated of nanocrystalline $LiMn_2O_4$ particles. The particle size is found to be less than 100 nm Fig. 6. Shows the EDX spectrum of the $LiMn_2O_4$ nanoparticle. EDX spectrum confirms the existence of oxygen (O) and manganese (Mn) in the synthesized pure nanocrystilline $LiMn_2O_4$ particles and their weight percentages.

Fig. 5: SEM Images of the Nanocrystalline $LiMn_2O_4$ Prepared by Nitric Acid Assisted Pechini Process

Element Line	*Net Counts*	*Net Counts Error*	*Weight %*	*Atom %*	*Formula*
O K	1485	+/- 114	50.36	77.69	O
Mn K	1656	+/- 78	49.64	22.31	Mn
Mn L	1543	+/- 110	---	---	
Total			100.00	100.00	

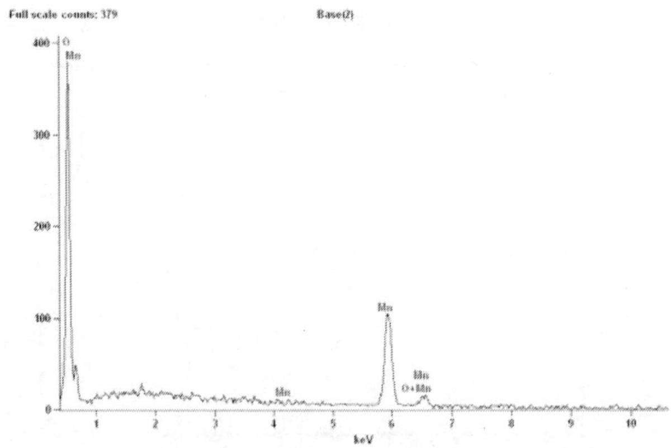

Fig. 6: EDX Pattern of Nanocrystalline $LiMn_2O_4$ Powder Prepared by Nitric Acid Assisted Pechini Process

BET Surface Area Analysis

Surface area of the prepared nanocrystalline $LiMn_2O_4$ particles were measured using Quantachrome instruments Nova 3200e. Measurements were carried out by nitrogen adsorption at liquid nitrogen temperature. The surface area of spinel lithium manganese oxide nanoparticle prepared from pechini process is found to be 11.73 m^2/g [16].

CONCLUSIONS

Nanocrystalline spinel $LiMn_2O_4$ powder was synthesized by Pechini process using citric acid and polyethylene glycol. Calcined polymeric intermediate at 500°C yield phase-pure spinel $LiMn_2O_4$ and it was confirmed by comparing with the standard JCPDS data. FT-IR studies confirm the stretching vibration bands of Li–Mn–O. The SEM images confirmed the formation of $LiMn_2O_4$ nanoparticle of size less than 100 nm and SEM-EDX mapping results also confirmed the presence of Mn and O elements in the $LiMn_2O_4$ sample. The thermal decomposition of the polymeric intermediate behavior was examined by means of thermogravimetry (TG) and differential thermal analysis (DTA).

ACKNOWLEDGEMENTS

Dr. N. Satyanarayana gratefully acknowledges, DST-NANOMISSION, Government of India for providing financial support in the form of major research project DST-NANO sanction letter No: SR/S5/NS-9/2005, dated on 30-07-2010. Authors also acknowledge CIF Pondicherry University for using SEM.

REFERENCES

[1] A. Nagarkatti, *Trans. Soc. Adv. Electrochem. Sci. Technol.* 36 (1) (2002) 1–8.

[2] R. Hahn and J. Mueller, Proceedings of Electronics Goes Green, Fraunhofer IZM, Berlin, Germany, 2000, pp. 727–734.

[3] R.M. Dell, *Solid State Ionics* 134 (2000) 139–158.

[4] B.B. Owens, W.H. Smyrl and J.J. Xu, *J. Power Sources* 81.

[5] Ting-Feng Yi , Yan-Rong Zhu, Xiao-Dong Zhu, J. Shu , Cai-Bo Yue, An-Na Zhou, (2009) *Ionics* 15: 779–784.

[6] Scrosati, B. *Nature,* 1995, 373, 557.

[7] Armstrong, A.R. and Bruce, P.G. *Nature,* 1996, 381, 499.

[8] D.K. Kim, P. Muralidharan, H.W. Lee, Riccardo Ruffo, Y. Yang, Candace K. Chan, H. Peng, Robert A. Huggins and Yi Cui, *Nano Lett.*, 2008, 8 (11), 3948-3952.

[9] Wang, Y. and Cao, G. Z. *Chem. Mater.,* 2006, 18, 2787.

[10] Du, K and Zhang, H., J. *Alloys Compd,* 2003, 352, 250.

[11] W. Lui, G.C. Farrington, F. Chaput and B. Dunn, *J. Electrochem. Soc.*, 143 (1996) 879.

[12] L.W. Tai and P.A. Lessing, J. Mater Res., 7 (2) (1992) 502-510.

[13] H.P. Klug and L.E. Alexander, X-ray Diffraction Procedures for Polycrystalline and Amorphous
Materials, Wiley, New York, 1954.

[14] G. Socrates, Infrared and Raman Characteristic Group Frequencies, John Wiley and Sons, New York, 2001.

[15] G. V. Rama Rao, D.S. Surya Narayana, U.V. Varadaraju, G.V. Rao and S. Venkadesan, *J. Alloys Compd,* 217 (1995) 200-208.

[16] Jena, A.K. and Gupta, K., "Characterization of Pore Structure of Filter Media", *Fluid/ Particle Separation Journal*, vol. 14, pp. 227-241, 2002.

SOLID STATE CHARGE TRANSFER NANOCHELATES OF ISOMERIC PHTHIOCOLATES AND PLUMBAGINATES

M.P. Wadekar, B.L. Khade, Monali Itkar, P.S. Khandagale[1], V.B. Jadhav, V.V. Dhapte and P.K. Khanna[2]

Y.M. College, Bharati Vidyapeeth Deemed University, Pune
[1]Henkel Adhesives Technologies India Pvt. Ltd. Pune
[2]Department of Applied Chemistry, DIAT, Pune
E-mail: m_p_wadekar@yahoo.co.in

ABSTRACT

*Phthiocol (i.e. 3-methyl, 2–hydroxyl, 1,4 naphthoquinone) and plumbagin (i.e. 2-methyl 5-hydroxy, 1,4 naphthoquinone) are the most important isomeric ligands belonging to the Juglone series. Both these ligands possess powerful chelating ability and some unique and characteristic properties which include (a) exhibition of ring isomerism in their metal chelates, caused exclusively due to the difference in ring size (b) formation of nanometric metal chelates whose crystallite size lies between (3-48) nm (c) formation of intence coloured chelates with non-transition and pseudonontransition metal ions. In this communication, the synthesis, characterization and structural investigations of isomeric **nano** phthicolates and plumbaginates with Zn (II), Cd(II), Hg(II), Sn (II) and Pb (II) are reported with special reference to explore the nature and origin of the intense visible colors.*

INTRODUCTION

Hydroxyl 1,4 naphthoquinone, in which the hydroxyl group is placed adjacent to one of the carbonyl group constitute a unique and interesting series of ligands which has been named as

V. Rajendran, K. Saminathan, P. Paramasivam and K.E. Geckeler (eds.)
Nano Materials Synthesis and Characterisation, pp. 327-334 (2012).

Juglone Series. Isomeric phthiocol (II) and plumbagin (III) are the most important ligands of this series.

Some of the characteristics features of juglone series include their natural occurrence in medicinal plants as active constituents [1,2], powerful chelating ability [3] due to which they can interact with all most all types of metals from the periodic table, exhibition of a unique type of isomerism, caused exclusively due to the difference in ring size, recognized as 'ring isomerism'[4], formation of nano-sized metal chalets whose crystallite size lies between 10-90 nm, formation of intense visible coloured metal chelates with non-transition and pseudo non-transition metal ions like Zn (II), Cd(II), Sc (III), significant antimicrobial activity of 1,4 hydroxy naphthoquinone derivatives against different microorganisms [5]. Among these, two of their characteristics are of our present interest. These are (i) formation of nano-chelates and (ii) formation of coloured chelates with non-transition and pseudo non-transition metal ions.In this communication, we would like to report our recent work on the synthesis, chemical characterization, structural investigations of the isomeric phthiocolates and plumbaginates with Zn (II), Cd(II), Hg(II), Sn (II) and Pb (II) with special reference to identification of their nanometric size and exploration of the nature and origin of their intense visible colours.

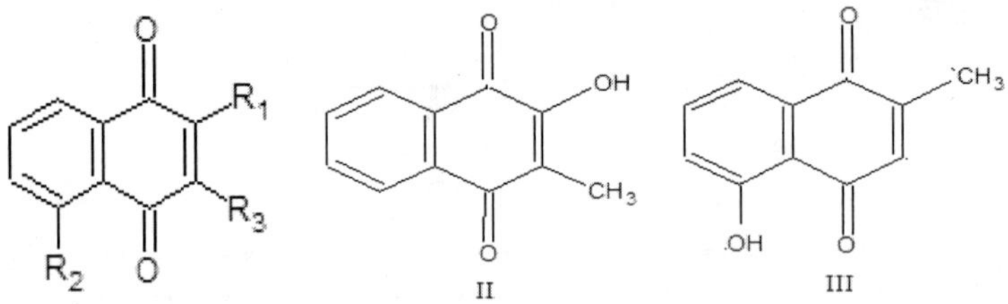

I II III

RESULTS AND DISCUSSION

The chemical composition of the isomeric phthiocolates and plumbaginates with Zn (II), Cd(II), Hg(II), Sn (II) and Pb (II) has been established with the help of elemental analysis which corresponds to $ML_2.2H_2O$ (Where M = Metal ion L = Phthiocol or Plumbagin) for most of the chelates except those of Hg (II) which are anhydrous having their composition ML_2. These solid powders were further analysed by powder XRD measurement to see if there would be possibilities of obtaining any information related to crystallite size/particle size, etc. Although the powder XRD is not so common for transition metal complexes but rare earth based comp[lexes have been analyzed by XRD by some researchers. XRD patterns are shown in Figure 1. The particle size of all these chelates is found to be 3.0 to 48 nm based on Scherrer equation. The sizes obtained from XRD pattern are shown in Table 1.

Table 1: XRD of the Zn (II), Cd (II), Hg (II), Sn (II) and Pb (II) Chelates

Sr. No.	*Compound*	*FWHM*	*2theta*	*Size*
1	Phtiocol	0.33598	13.565	23.821
2	Plumbagin	0.59388	26.409	13.745
3	Zn- Phthiocolate	0.19499	45.491	44.198
4	Zn- Plumbaginate	0.21540	22.961	37.650
5	Cd- Phthiocolate	3.93590	12.841	03.031
6	Cd- Plumbaginate	0.56250	26.458	14.516
7	Hg- Phthiocolate	1.96620	12.771	04.067
8	Hg- Plumbaginate	2.18490	12.780	03.660

The complexes were further analyzed by electronic spectroscopy like UV-Visible and IR. The most interesting and attractive feature of these isomeric nanosized chelates is exhibition of marvelous visible colours. These colours, which are quite stable in solid state for a very long period and such coloured chelated were first observed by Rawal [6] and Kulkarni-Wadekar [7] for alkaline earth chelates. All these coloures not only cover the entire visible spectrum ranging from red to violet, but cover several of their intermediate shade.

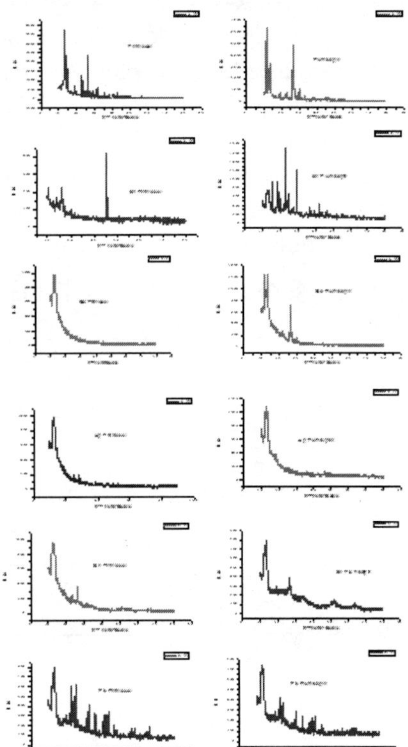

Fig. 1: XRD Patterns

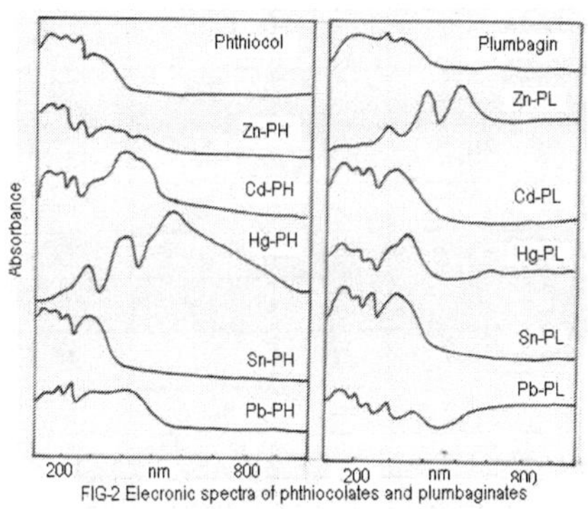

Fig. 2: Electronic Specitra of Phthiocolates and Plumbaginates

Since colour of the chelates are found to be more stable in their solid state than in the solvent, their spectra in transparent spectroscopic KBr pellets are recorded in the region 200–1000 nm against the pure transparent KBr pellets as the reference. These electronic spectra are shown in Figure 2, while the significant absorption peaks and molecular extinction coefficients ε max calculated for λ max in the visible region for some of the complexes are given in Table 3 and Table 4. In general, all electronic spectra can be recognized in terms of three or four principal absorption bands located in the region 200–300 nm, 300–400 nm and 400–800 nm. These are interpretated in terms of benzenoid electron transfer (BET) quinonoid electron transfer (QET), *n*–π* transitions and ligand to metal charge transfer bands [8,9].The first band, which appears in the region 200–270 nm is a BET band involving σ–π* transitions. The second band isattributable to σ-π* transition. The extent of blue shift in metal complexes varies between 20-92 nm with respect to theligands. The third band attributable to π-π* transitions is observed in the region (328–342 nm) for phthiocolates and (411–476 nm) for plumbaginates. This band is metal dependent and accordingly its band position, band shape and band intensity varies from metal to metal.

The fourth band, occurring in the visible region, as referred above, is of special interest and importance for the exploration of the nature and origin of visible colours. Two characteristic features of this band are worth for consideration. The first one is its band position, which depends on the metal and is responsible for the specific visible colour. The large variations in the position of this band from 400 nm to 750 nm explain the corresponding variation in visible colour, covering the entire range of visible spectrum from red to violet. The chelates of Sn and Pb are similar to Zn, Cd and Hg.

Table 3: Electronic Spectra of the Ligands and their Selected Chelates (BET and QET)

Sr. No.	*Compound*	*σ-π* (BET) Transitions*			*π-π* (QET) Transitions*		
		λ max	*O. D.*	*Є Max*	*λ max*	*O. D.*	*Є Max*
1	Phthiocol	247	1.707	163505.7	300	1.627	155842.91
2	Plumbagin	256	2.32	222222.2	307	2.3	220306.51
3	Zn-Phthiocolate	245	0.643	239746.5	295	0.566	211036.5
4	Zn-Plumbaginate	--	--	--	--	--	--
5	Cd-Phthiocolate	247	2.09	703703.7	301	2.017	679124.6
					325	1.9	6397306
6	Cd-Plumbaginate	246	1.883	634006.7	361	1.786	601346.8
					340	1.75	589225.6
7	Hg-Phthiocolate	286	0.491	149603.9	--	--	--
8	Hg-Plumbaginate	239	0.78	237660.0	296	0.646	196831.2

The second and equally important aspect is the molecular extinction coefficient Є, (i.e. absorptivty) whose values are of the order of 10^4 to 10^5. Such high values of å are the indications of charge transfer process. In our case the possible modes for charge transfer include (a) intraligand, (b) intrametal, (c) ligand to metal, and (d) metal to ligand. In the present case, both the ligands are powerful donor of electron density and the divalent metal ions are reluctant to undergo oxidation under the circumstances in which the chelates are synthesized. So it may be concluded that the origin of the visible colours is most probably attributable to **ligand to metal charge transfer**.

Table 4: Electronic Spectra of the Ligands and their Chelates (BET + QET) and CT

Sr. No.	*Compound*	*(BET + QET) Mixed*			*L - M Charge Transfer*		
		λ max	*O. D.*	*Є max*	*λ max*	*O. D.*	*Є max*
1	Phthiocol	3.42	1.582	151532.57	500	0.35	33524.90
		3.92	1.194	114367.82			
2	Plumbagin	389	2.24	214559.39	--	--	--
3	Zn- Phthiocolate	411	0.286	106636.8	446	0.219	81655.5
					755	0.027	10067.1
4	Zn- Plumbaginate	342	0.063	23489.9	506	1.198	446681.6
5	Cd- Phthiocolate	392	1.7	572390.6	530	2.489	838047.1
					575	2.234	752188.6
6	Cd- Plumbaginate	419	1.723	580134.7	--	--	--
7	Hg- Phthiocolate	342	0.406	123705.1	508	1.287	392138.9
		388	0.476	145033.5	648	1.837	559719.7
8	Hg- Plumbaginate	446	1.013	308653.3	755	0.178	54235.2
					755	0.521	147425.0

The solid state infrared spectra in KBr pellets of the ligands and their isomeric chelates are also studied to examine the effect of ring isomerism on the molecular structure of the ligands and their isomeric chelates. These spectra in the region 4000–450 cm^{-1} are shown in Figure 3 while the relevant infrared absorption peaks (in cm^{-1}) which are directly involved in the process of chelation are given in table 5. The spectra of phthiocolates are notably different than those of corresponding plumbaginates. There is a significant peak at 3332 cm^{-1} in the spectra of phthiocol attributable to free O–H stretching but there is no any peak in the spectra of plumbagin in this region indicating strong intramolecular hydrogen bonding and consequently absence of free O–H [10]. All the phthiocolates as well as plumbaginates of Zn(II) and Pb(II) show broad and intense peaks in the region 3600 – 3300 cm^{-1} indicating that these are strongly hydrated [11]. The free C = O stretching frequency of both phthiocol and plumbagin in most of the isomeric phthiocolates as well as plumbaginates remains practically unaltered. But the chelated C = O stretching frequency is shifted to lower frequency region by an 45-60 cm^{-1} in phthiocolates and by 11-13 cm^{-1} in plumbaginates which shows the weakening of this C = O bond due to transfer of electrons density from carbonyl oxygen to central metal ion[4].

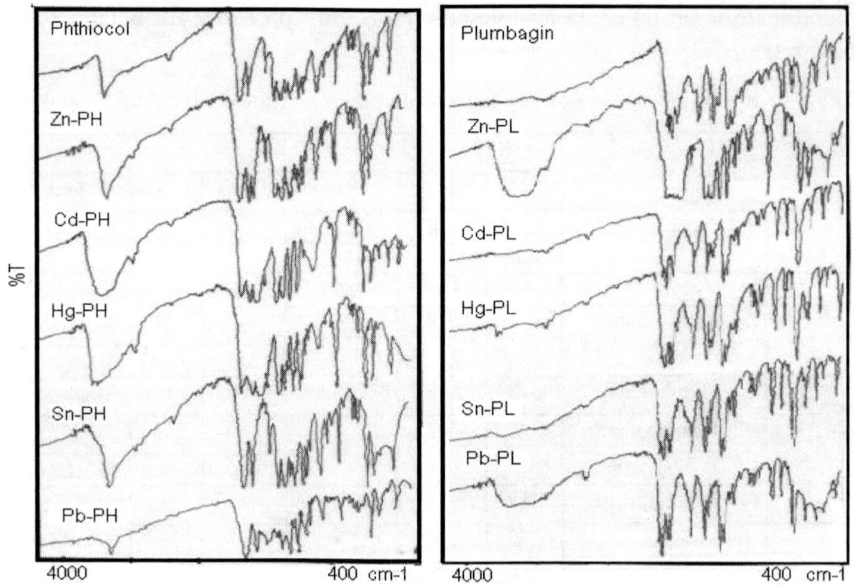

Fig. 3: Infrared spectra of Phthiocolates and Plumbaginates

Table 5: Infrared Absorption Peaks (cm^{-1}) of Metal Phthiocolates and Plumbaginates

Sr.No.	*Compound*	*Type of band stretching (cm^{-1})*				
		O-H	*C=O (Free)*	*C=O (Chelated)*	*C-O*	*CH_3*
1.	Phthiocol	3336	1659	1645	1208	2928
2.	Zn-Phtiocolate	3333	1657	1588	1220	2924
3.	Cd- Phtiocolate	3368	1636	1584	1207	2930
4.	Hg- Phtiocolate	3331	1651	1590	1223	2928
5.	Pb- Phtiocolate	3335	1659	1591	1221	2928
6.	Plumbagin	--	1644	1620	1229	2935
7.	Zn-Plumbaginate	3294	1644	1609	1259	2920
8.	Cd-Plumbaginate	3440	1644	1608	1258	2918
9.	Hg-Plumbaginate	2356	1644	1609	1258	2918
10.	Pb-Plumbaginate	3405	1643	1607	1260	2918

The C-O stretching frequency is shifted to the higher frequency region for phthiocolates (1229-1232) cm-1 as well as plumbaginates (1255-1256) cm-1 indicatig that C-O becomes stronger in the chelates which is consistent with the previous reports[12]. Comparison of these frequencies in isomeric pairs indicate that C-O bonds in plumbaginates are stronger as compare to plumbaginates.The greater strength of chelated C=O and C-O in plumbaginates indicate that the six membered ring in plumbaginates is more stable than five membered ring in phthiocolates.

CONCLUSION

The Crystallite size revels nanometric nature of isomeric phthiocolates. Ring isomerism caused exclusively due to the difference in the size of the chelate ring which is the unique and characteristic property of phthiocol and plumbagin, has also a significant effect on the structural and antimicrobial activities of the isomeric metal chelates The origin of the visible colours is most probably attributable to ligand to metal charge transfer. IR studies indicate six membered chelates of plumbagin under study are thermodynamically more stable than five membered ring chelates of phthiocol.

EXPERIMENTAL

Synthesis of the Isomeric Chelates

Ligand and Metal salts

Phthiocol was synthesized from 2-methyl 1,4 naphthoquinone by using Fiser's method[13]. Plumbagin was a commercial sample purchased from research organics (Madras, Chennai).Both these ligands were recrystallized from methanol before use and their purity was checked through m.p. and TLC. zinc acetate, cadmium acetate and lead (II) acetate were used while for mercury, $Hg(NO_3)_2.4H_2O$ and for tin $SnCl_2.2H_2O$ were used. Aqueous solutions corresponding to 1 X 10^{-3} were prepared in distilled water.

Synthesis of Chelates

Methanolic ligand solution corresponding to (2×10^{-3}) M was taken in a three necked flask provided with an electric heater and magnetic stirrer. The metal ion solution (1×10^{-3}) M was added into the ligand solution with constant stirring and maintaining the temperature of the reaction mixture at 60^0C. After mixing completely the ligand and metal ion the pH was adjusted to 4-5 by 10% ammonia solution and the mixture was kept stirring for about 60 minutes. Then the flask was cooled to room temperature and kept in a refrigerator overnight. Next day the precipitated chelates were filtered under suction. Each filtrate was washed thoroughly first by distilled water, then by methanol and finally by acetone and dried completely under suction and kept in a vacuum desiccator.

Instrumental Techniques

All the chelates were analyzed for carbon and hydrogen using Hosli-Holland carbon, hydrogen microanalyzer while the percentage of metal, except mercury chelates, was determined from the metal oxide residue. The XRD patterns of the ligands and their metal chelates were recorded on the Brukar X-ray diffractometer, Model D_1 in the 2è range from 0 to 90^0.The solid state electronic spectra were recorded in spectroscopic KBr pellets. All the electronic spectra were recorded in the UV region (200 – 1000) nm against pure transparent KBr pellet as a reference on a Shimadzu ultra violet visible spectrophotometer model 160. These were also recoded in KBr pellets in the regin (4000 – 450) cm^{-1} on a Perkin – Elemer model Paragon 500 infrared Spectrophotometer.

ACKNOWLEDGEMENTS

Authors are thankful to University Grant Commission, New Delhi, India for financial assistance through Major research project. They are also thankful to Dr. S.S. Kadam, Vice.Chancellor, Bharati vidyapeeth Deemed University Pune, Prin. K.D. Jadhav, Principal, Y.M. College, Pune and Dr. B. A. Kulkarni for their support.

REFERENCES

[1] Smith GM, Geyer BP, *J. Am. Chem. Soc.*, 63(11), 1941, 3071-3075.

[2] Pendse GS, Iyangar MA, *Planta Med*, 14(3), 1966, 337-351.

[3] Bottei RS, McEachem CP, *J Inorg Nucl Chem*, 32(8), 1970, 2465-2816.

[4] Kanase DG, Kulkarni (Wadekar) M, Kadam SS, Kelkar VD etal, *Asian J. Chem.* 16(3-4), 2004, 1630-1636.

[5] Ambrogi V, Artini D, Carneri DE, Castellino S, Dradi E, Longemann W, Meinardi G, DI Somma M, Tosolini G, *Br. J.Pharmacol.*, 40, 1970, 871-880.

[6] Rawal B.M, Alkaline earth chelates of 3-substituted lawsones, Ph.D Thesis (1986).

[7] Kulkarni-Wadekar M.B., Indicator properties of Juglones and metal complexes of methyl hydroxyl naphthoquinones and nitrosonaphthol derivatives with alkaline earths and lanthanides. Ph.D Thesis (1999).

[8] Singh, Ogata R.T., Moore R.E., Chang CWJ, Scheuer P.J., 24(18), 1968, 6053-6073.

[9] Scott A.I., Interpretation of the Ultraviolet spectra of natural products, McMillan, New York (1968) 8.

[10] Hadzi D, Sheppard, *Trans.Faraday. Soc*, 50, 1954, 911-918.

[11] Sartori G,Furlani C, Damiani A, *J. Inorg. Nucl. Chem. 8*, 1958, 119-125.

[12] Gaikwad SD, Kanase DG, Kelkar VD Kulkarni BA, *Synth. React. Inorg. Met-Org. Chem.* 26(9) 1996, 1487.

[13] Fieser LF, Convineant procedures for the preparation of antihemorrhagic compounds, *J. Biol. Chem.*, 133, 1940, 391-396.

PHOTOLUMINESCENCE IN Fe^{3+} ION DOPED BARIUM TITANATE NANOPARTICLES

P. BARIK AND T.K. KUNDU

Department of Physics, Visva-Bharati University, Santiniketan, West Bengal
E-mail: tapaskumar.kundu@visva-bharati.ac.in

ABSTRACT

Fe^{3+} ion doped barium titanate (BT) nanopowders were synthesized by sol gel route. The average size of t-BT powders lie in the range 16 - 40 nm. The specimens show ultraviolet (UV) emission (peak at 376 nm) along with emission in violet, blue, green and yellow colour. The blue band (peak at 452 nm) is seen to grow significantly with annealing temperature. This band arises due to Ti^{3+} defects which are stabilized by Fe^{3+} ions in BT specimens. The samples show two distinct Electron Paramagnetic Resonance (EPR) bands of g-value around 4.11 and 1.98 originated from Fe^{3+} ions and Ti^{3+} defects respectively. A correlation between the photoluminescence (PL) and EPR band intensity, grown from same species, is found.

INTRODUCTION

Pure Barium Titanate ($BaTiO_3$, BT) and transition metal ion doped BT have important applications in electromechanical devices such as capacitors, transducers, thermistors, actuators, humidity sensors etc [1-4]. It has also interesting optical properties with potential application in high-density optical data storage, laser power combining and optical computing [5]. Of all the known defects into BT, iron is easily soluble and it has significant effect on optical properties of BT [6-8]. Katyar et al observed a structure-less visible emission band in BT at 500 nm at low temperature revealing lower emission energy than its band gap [9]. BT nanocrystals (nc-BT) show relatively enhanced photoluminescence (PL) emission [9, 10]. Various reports were

V. Rajendran, K. Saminathan, P. Paramasivam and K.E. Geckeler (eds.)
Nano Materials Synthesis and Characterisation, pp. 335-342 (2012).

published to describe the origin of the visible emission shown by BT [11-13]. It was reported that the emission is related to the charge transfer between titanium and oxygen atoms. The luminescence from bulk BT was ascribed to the presence of self-trapped excitons (STE) within TiO_6 octahedra, viz Ti^{3+}-O^- STE [10]. Zhang et al established that the PL of BT is due to localized states promoted by the intrinsic defect distribution (oxygen vacancy) [14]. Those states essentially decrease the band gap energy of BT. Recently a group has established that both five-fold and six-fold oxygen-*Ti* coordination exist in amorphous BT systems, which is responsible for PL in the disorder materials [15-17]. Therefore, from the above discussions it seems that more investigations are required in this field to understand the origin of the emission shown by BT.

PL emission from acceptor doped or Fe-doped BT system are particular of interest in this work. Here, we report the emission from Fe^{3+} doped nc-BT powders. The emission appears both in ultraviolet and visible range of wavelength. PL emissions have correlation with electron paramagnetic resonance signals shown by the specimens.

EXPERIMENTAL

Iron ion doped BT ultrafine powder were synthesized by the sol-gel hydroxide method using barium hydroxide [$Ba(OH)_2$, $8H_2O$, Merck], tetraisopropyl orthotitanate [$Ti(C_3H_7O)_4$, Merck] and iron(III) nitrate [$Fe(NO_3)_3$, $9H_2O$, Merck] as starting materials [18]. Here, the powders were doped with 2 mole% of Fe^{3+}. The concentration of the dopant was chosen within the solubility limit of Fe^{3+} in BT [19]. $Ba(OH)_2$, $8H_2O$ was dissolved in ethanol and glacial acetic acid (1:1 volume ratio) by stirring and heating the solution at 50-70°C. After cooling the solution to room temperature a stoichiometric amount of tetraisopropyl orthotitanate was added drop wise. Then, calculated amount of iron nitrate, prepared separately, was added to the above solution. A clear solution, obtained after stirring for a few minutes, was allowed to gel at room temperature. The dried gel-powders were calcined at different temperature in the range 400-700°C for two hour in a tube furnace to obtain doped nc-BT powders. XRD patterns of the specimens were recorded using Phillips powder Diffractometer. The microstructures of the samples were investigated by a TEM of JEOL. A Perkin-Elmer luminescence spectrometer was used to study the light emission from the samples. The EPR spectra were recorded at an X-band frequency (~ 9.43 GHz) with a Varian Associates spectrometer.

Results and Discussions

Figure 1 shows X-ray diffraction patterns of 2 mole % iron ion doped BT powders obtained by heating the gel-powders at different temperature.

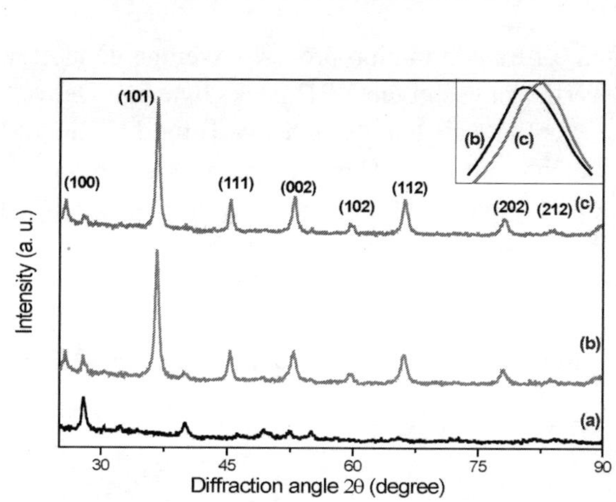

Fig. 1: XRD Patterns for 2% Fe Doped BT Nanopowders after Heating a sol-gel Precursor at (a) 400°C and (b) 600°C (c) 700°C for 2 hr in air. A Close-up in the Inset Compares the Shift in (101) Peak of Sample (b) and (c)

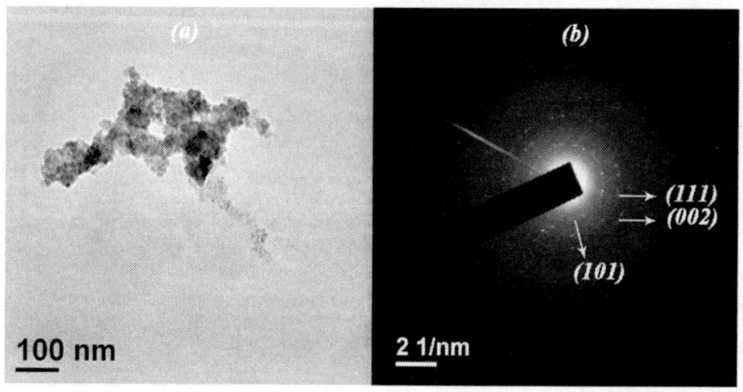

Fig. 2: A Typical (a) TEM Image and (b) Diffraction Pattern Taken from Specimen '*c*' Heat-treated at 700°C for 2hr. The Diffraction Rings in the Pattern (from Center to edge) can be Indexed as the (101), (111), and (002) Peaks of a Tetragonal (P_{4mm}) BT Phase

We have calculated interplanar spacings (d_{hkl}) values corresponding to each of the diffraction peaks. Specimen '*a*', gel-powders annealed at 400 °C, shows a set of d_{hkl} values that matches with an orthorhombic (*o*-BT-II) crystal structure (P_{nma} space group), with average lattice parameters a = 0.6435 nm, b = 0.5306 nm, and c = 0.8854 nm. The observed d_{hkl} values are reproduced with a standard deviation of ± 0.0005 nm. The orthorhombic phase (*o*-BT-II) is structurally different from conventional orthorhombic BT. A thorough analysis by using Rietveld method was given in our previous reports [19, 20]. Specimens '*b*' and '*c*', heating the gel-powders at 600 °C and 700 °C respectively, contain two sets of d_{hkl} values. One set is due to standard tetragonal BT phase and other set of d_{hkl} values arise due to orthorhombic crystal structure. A large fraction of *o*-BT-II is promoted to t-BT at a temperature of 700°C via a surface diffusion limited o!t transformation process. Average diameter (D) of the grains was determined from the FWHM values in the XRD peaks by using Debye-Scherrer relation $d_{XRD} = k\lambda/\beta(\theta)\cos\theta$, where λ is the wavelength of X-ray, β the FWHM of the (200) peak, θ the diffraction angle and the constant $k \approx 1$. The average sizes of 16 nm and 40 nm for the t-BT samples annealed at 600°C and 700°C respectively are obtained. The grain size of the nanocrystalline ceramics was also estimated using transmission electron microscope (TEM). Typical micrograph from specimen '*c*' is shown in Fig. 2a. Corresponding diffraction patterns are shown in Fig. 2b. There is a distribution of particle sizes in the specimens. The results estimated from TEM reasonably agree with those obtained from XRD.

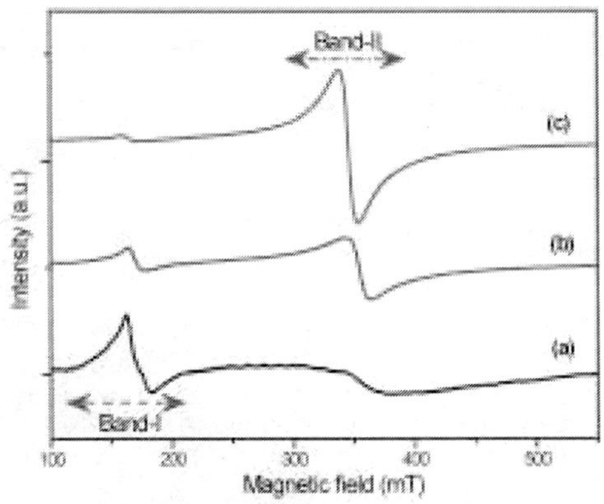

Fig. 3: EPR Spectra for 2% Fe Doped BT Nanopowders after Heating a Sol-gel Precursor at (a) 400°C, (b) 600°C, and (c) 700°C for 2 hr in Air

Figure 3 shows EPR spectra of the specimens after heating the powders at (a) 400°C, (b) 600°C, and (c) 700°C for 2 h in air. The samples show two distinct EPR bands (Band I and Band II) of *g*-value around 4.11 and 1.98. The intensity of Band-II increases by 55% as the sintering temperature rises from 400°C to 700°C. The positions of the signals are not altered significantly. The reverse trend is observed for Band I signal. Its intensity is decreased by 58% with the same change in sintering temperature. Thus, when Band I signal is strong Band II signal is weak and vise versa. The O_i^- is a common EPR sensitive species ($g \sim 2$) in oxides [21, 22]. A peak at $g = 1.97$ was reported in BT due to localized Ti^{3+} states [23]. In the present case, intensity of the signal at $g = 1.98$ have increased with the annealing temperature and the doped specimens show an extra EPR peak ($g = 4.11$). EPR study of Fe^{3+} $(3d^5, S = 5/2)$ in the tetragonal phase was reported earlier [24-26]. In a large number of Fe^{3+}-containing materials, a symmetric and isotropic line at $g \approx 4.0$-4.2 as well as a line at $g = 2.0$ was observed. In our case, the *g*-value at 4.11 indicates that the iron ions are in trivalent state and the site symmetry is slightly distorted octahedral. Here, the Fe^{3+} occupies the Ti^{4+} sites which produce distorted octahedral symmetry. This case produces a low symmetry site structure. As the Fe^{3+} ions occupy the Ti^{4+} sites, EPR response of Fe^{3+} doped BT may be expected to exhibit changes upon crystallization due to removal of distortion around various ions in the BT. This argument is supported by our experimental observation where the signal intensity of Band I decreases with annealing temperature. Band II that grows with annealing temperature can originate either due to species Fe^{3+} or Ti^{3+}. In absence of Fe^{3+} ions, the same signal dies with increasing annealing temperature. The appearance of a few emission bands from the doped BT specimens, when excited properly, corroborates the EPR results.

The emission spectra from the specimens, irradiated at 250 nm by a xenon lamp, are shown in Fig. 4. The excitation energy was chosen from the excitation spectra of the specimens which show an intense peak at 250 nm.

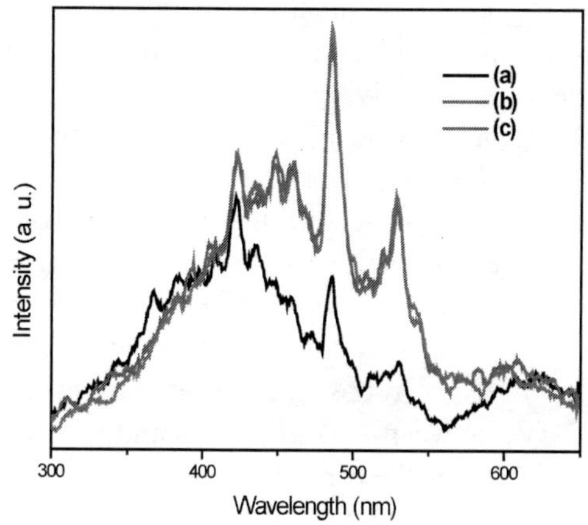

Fig. 4: Photoluminescence Spectra for 2% Fe Doped BT Nanopowders after Heating a Sol-gel Precursor at (a) 400 °C, (b) 600 °C, and (c) 700 °C for 2 hr in Air

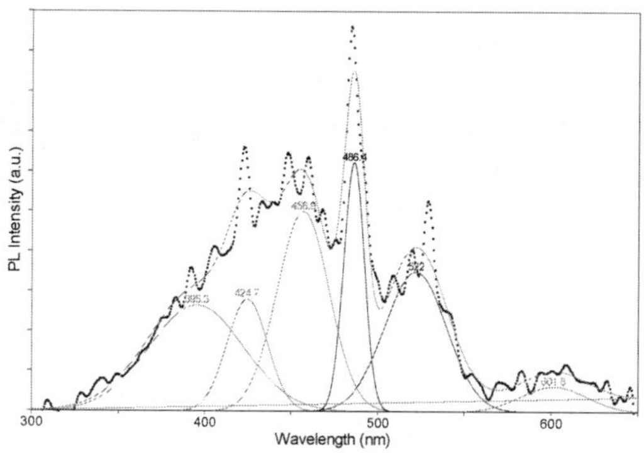

Fig. 5: Photoluminescence Emission Spectra of the Specimen Heated at 700 °C for 2 hr Fitted with Gaussian Curves

The spectra are analyzed by fitting with multiple Gaussian curves. A typical Gaussian fitting of PL spectrum from the specimen '*c*' is shown in Fig. 5. The composite specimens, '*b*' and '*c*', also show same bands with different intensity level and shifted peak positions. The UV emission peak shifts significantly to higher wavelength with increasing annealing temperature. Typically the peak at 376 nm in specimen '*a*' is shifted to 395 nm in specimen '*c*'. It is evident from Gaussian fitting data that the band intensity of 421 nm and 619 nm line decreases and other bands grow on annealing the powder at temperature as high as 600 ºC or above. Among the bands, the growth of 452 nm line is distinct and significant.

The band gap of BT lies between 3 - 3.2 eV as obtained from optical absorption edges [27]. Thus, the energy corresponding to 250 nm line makes a direct transition and UV emission band is assigned to a direct band gap transition. The band gap energy of nc-BT has increased to 4.9 eV according to the excitation maxima. Luminescence at ~2.96 eV (421-424 nm) is caused by the transition from near conduction band edge to valance band. It is argued that the electronic band structure of BT has low-lying narrow conduction bands from defects [28]. In amorphous materials nonbridging oxygen, in which the oxygen is bonded to a single main group atom, participates in ionic interactions to form a typical network like *Ti-O*. Crystallization into a single o- or t-BT polymorph can lead to the elimination of defects or imperfections, particularly of nonbridging oxygen of a TiO_5 group, in which Ti is penta coordinated by oxygen. A decrease of band intensity of 421 nm line, as observed experimentally, supports these statements. The recombination process corresponding to the charge transfer from the central Ti^{3+} to adjacent V_o^+ states in the TiO_5 groups leads to luminescence peak centered at 452 nm. The energy interval (~2.75 eV) between $Ti^{3+}(3d^1)$ states and states supports the emission at a wavelength of 452 nm. Experimentally the band is observed to grow considerably with annealing at higher temperature. If it is considered that EPR Band II is originated from Ti^{3+} species then a correlation between the PL band and EPR band intensity is found. Intensity of the both the signals increase with the rise of annealing temperature. In that case certainly Fe^{3+} ions play an important role in stabilizing Ti^{3+} defects in BT even sintered at higher temperature. In amorphous and composite materials, a large amount of "broken" surfaces or dangling bonds and color centers intrinsically existing either in a distorted structure or a highly strained lattice (as in this example of o-/t-BT-nanocrystals) involve series of E_{em} levels localized within the forbidden gap. They tailor very sensitively the light emission in multiple bands at room temperature. Blue emission at 485 nm (~2.57 eV) is attributed to the intraband transitions from self-trapped excitons [28, 29]. It is also evident Gaussian fitting data that a prominent yellow emission is centered at ~2.38 eV (520-524 nm) and its intensity increases as the sintering temperature increases. The electrons polarons interact with holes, possibly trapped near crystal defects or impurities, and form an intermediate state: self-trapped exciton (STE). These self-trapped excitons were contributed to 485 nm and 520 nm emission [10, 30]. The band at 620-600 nm is explained in terms of the transition from Fe^{3+} sates to valence band of BT. The intensity of this band decreases with annealing temperature and a correlation with EPR band I grown from the same species is found. The existence of Fe^{3+} energy states in BT was earlier established from optical absorption data [8]. PL measurements and optical absorption study on nc-BT powders heat treated at different atmosphere would extract some useful information.

CONCLUSIONS

BT nanopowders doped with 2 mole % Fe^{3+} ions were synthesized by sol gel route. The gel-powders annealed at 400°C shows orthorhombic BT (*o*-BT-II) crystal structure. The phase *o*-BT-II begins to transform to the t-BT on heating the specimens at temperature as high as 600°C. The average size of t-BT powders lie in the range 16 - 40 nm. The samples show two distinct EPR bands (Band I and Band II) of *g*-value around 4.11 and 1.98. Band I signal, g=4.11, is originated due to isolated Fe^{3+} ions in low symmetry site at low temperature. As the Fe^{3+} ions occupy the Ti^{4+} sites, EPR response exhibits changes upon crystallization due to removal of distortion around various ions in the BT. Band II which grows with annealing temperature originates due to Ti^{3+} defects stabilized by Fe^{3+}. The specimens show UV emission (peak at 376 nm) along with the emission in violet, blue, green and yellow color (421, 452, 485, 526, 619 nm). The redshift of UV emission band are primarily caused due to quantum size effect. The blue band (452 nm) is seen to grow with annealing temperature due to Ti^{3+} defects which are stabilized by Fe^{3+} in the doped specimens. A correlation between the PL band and EPR band intensity, grown from same species, is found.

ACKNOWLEDGEMENTS

The work was supported by the grant from the Nano Science and Technology Initiative of the Department of Science and Technology, and University Grant Commission, New Delhi.

REFERENCES

[1] T. Sahoo, S.K. Tripathy, S. Nandy, B. Pandey, H.C. Verma, K.K. Chattopadhyay and S. Anand, *Mater. Sci. Eng. B* 131, 277 (2006).

[2] J. Yu, J. Sun, J. Chu and D. Tang, *Appl. Phys. Lett.* 77, 2807 (2000).

[3] P.S. Pizani, H.C. Basso, F. Lanciotti Jr., T.M. Boschi, F.M. Pontes, E. Longo and E.R. Leite, *Appl. Phys. Lett.* 81, 253 (2002).

[4] P.S. Pizani, E.R. Leite, F.M. Pontes, E.C. Paris, J.H. Rangel, E.J.H. Lee, E. Longo, P. Delega and J.A. Varela, *Appl. Phys. Lett.* 77, 824 (2000).

[5] P. Gunter and J.P. Huignard, Photorefractive materials and Their Application I and II, *Topic in Applied Physics*, Springer-Verlag, Berlin, Vol. 61 and 62 (1988, 1989).

[6] M.B. Klein and R.N. Schwartz, *J. Opt. Soc. Am. B: Opt. Phys.* 3, 293 (1986).

[7] P.G. Schunemann, D.A. Temple, R.S. Hatchcock, H.L. Tuller, H.P. Jenssen, D.R. Gabbe and C. Warde, *J. Opt. Soc. Am. B: Opt. Phys.* 5, 1685 (1988).

[8] A. Mazur, O.F. Schirmer and S. Mendricks, *Appl. Phys. Lett.* 70, 2395 (1997).

[9] R.S. Katiyar, J.F. Meng and B.K. Rai, *J. Vac. Sci. Technol.*, A 15, 2945 (1997).

[10] J. Meng, Y. Huang, W. Zhang, Z. Du, Z. Zhu and G. Zou, *Phys. Lett. A* 205, 72 (1995).

[11] S.K. Dey and R.Z. Leeg, *Ferroelectrics* 112, 309 (1990).

[12] R. Leonelli and J.L. Brebner, *Phys. Rev.* B 33, 8649 (1986).

[13] B. Bouma and G. Blasse, *J. Phys. Chem. Solids* 56, 261 (1995).

[14] M.S. Zhang, Z. Yin, Q. Chen, W.F. Zhang and W.C. Chen, *Solid State Commun*. 119, 659 (2001).

[15] I.A. Souza, M.F.C. Gurgel, L.P.S. Santos, M.S. Goes, S. Cava, M. Cilense, I.L.V. Rosa, C.O. Paiva-Santos and E. Longo, *Chem. Phys*. 322, 343 (2006).

[16] E. Orhan, J.A. Varela, A. Zenatti, M.F.C. Gurgel, F.M. Pontes, E.R. Leite, E. Longo, P.S. Pizani, A. Beltran and J. Andres, *Phys. Rev*. B 71, 085113 (2005).

[17] M.S. Silva, M. Cilense, E. Orhan, M.S. Goes, M.A.C. Machado, L.P.S. Santos, C.O. Paiva-Santos, E. Longo, J.A. Varela, M.A. Zaghete and P.S. Pizani, J. Lumin. 111, 205 (2005).

[18] H.P. Beck, W. Eiser and R. Haberkorn, *J. Eur. Ceram. Soc.* 21, 687 (2001).

[19] A. Jana, T.K. Kundu, S.K. Pradhan and D. Chakravorty, *J. Appl. Phys*. 97, 044311 (2005).

[20] A. Jana, S. Ram and T.K. Kundu, *Philos. Mag. B* 87, 5485 (2007).

[21] K. Hoffmann and D. Hahn, Phys. *Status Solidi A* 24, 637 (1974).

[22] S. Ram and T.K. Kundu, J. Nanosci. *Nanotechnol. 4*, 1076 (2004).

[23] T.R.N. Kutty and P. Murugaraj, *Mater. Lett.* 3, 195 (1985).

[24] A.W. Hornig, R.C. Rempel and H.E. Weaver, *Phys. Rev. Lett.* 1, 284 (1958).

[25] E.S. Kirkpatric, K.A. Muller and R.S. Rubins, *Phys. Rev.* 135, A86 (1964).

[26] T.V. Waldkirk, K.A. Muller and W. Berlinger, *Phys. Rev*. B 6, 4324 (1972).

[27] M. Cordona, *Phys. Rev.* 140, A651 (1965).

[28] S.W. Kim, S. Fujita, S. Fujita, *Appl. Phys. Lett.* 81, 5036 (2002).

[29] L. Brus, *J. Phys. Chem*. 90, 2555 (1986).

[30] W.S. Cho and E. Hamada, *J. Alloys Compd.* 268, 78 (1998).

THE CHARACTERIZATION OF THE CLAY MODIFIED WITH MYRISTYLTRIMETHYLAMMONIUM BROMIDE (MTAB)/ POLYETHYLENEOXIDE (PEO) COMPOSITES AND THEIR MECHANICAL PROPERTIES

AHMET GÜRSES, MEHTAP EJDER-KORUCU[1] AND ÇETIN DOĞAR[2]

Ataturk University, K.K. Education Faculty, Dep. of Chemistry, 25240 Erzurum, Turkey
[1]Kafkas University, Faculty of Science and Literature, Dep. of Chemistry 36100 Kars, Turkey
[2]Erzincan University, Education Faculty, Department of Elementary Education, Erzincan, 24030, Turkey
E-mail: ahmetgu@yahoo.com

ABSTRACT

In this study PEO/clay composite systems were prepared by using solution intercalation methods. Firstly, the clay sample was modified by adsorption of Myristyltrimethylammonium Bromide (MTAB) based on cation exchange capacity (CEC) of clay. After determining the optimum CEC value for organo-clay, the PEO-clay composites were prepared at different clay contents. The samples were characterized by X-ray diffraction analysis (XRD), Fourier Infrared Spectrum (FTIR) and tensile tests. Based on the result of XRD and FTIR studies, it was deduced, the presence of clay in the composites leads to both a distorted helical structured PEO and a significant improvement in the tensile characteristics compared to virgin polymer.

INTRODUCTION

In recent years, polymer/layered silicate (PLS) nanocomposites have attracted great interest, both in industry and in academia, because they often exhibit remarkable improvement in

V. Rajendran, K. Saminathan, P. Paramasivam and K.E. Geckeler (eds.)
Nano Materials Synthesis and Characterisation, pp. 343-352 (2012).

materials properties when compared with virgin polymer or conventional micro and macro-composites [1]. These improvements can include high module [2,3], increased strength and heat resistance [4], decreased gas permeability [5,6] and flammability [7,8], and increased biodegradability of biodegradable polymers [9]. Although the intercalation chemistry of polymers when mixed with appropriately modified layered silicate and synthetic layered silicates has long been known (1, 10, 11), the field of PLS nanocomposites has gained momentum recently [1]. The commonly used layered silicates for the preparation of PLS nanocomposites belong to the same general family of 2:1 layered or phyllosilicates. MMT, hectorite, and saponite are the most commonly used layered silicates [1]. The physical mixture of a polymer and layered silicate may not form a nanocomposite. In immiscible systems, which typically correspond to the more conventionally filled polymers, the poor physical interaction between the organic and the inorganic components leads to poor mechanical and thermal properties [1]. To render layered silicates miscible with other polymer matrices, one must convert the normally hydrophilic silicate surface to an organophilic one, making the intercalation of many engineering polymers possible. Generally, this can be done by ion-exchange mechanism with cationic surfactants including primary, secondary, tertiary, and quaternary alkylammonium or alkylphosphonium cations. Alkylammonium or alkylphosphonium cations in the organosilicates can provide lower the surface energy of the inorganic host and improve the wetting characteristics of the polymer matrix, and result in a larger interlayer spacing. Additionally, the alkylammonium or alkylphosphonium cations can provide functional groups that can react with the polymer matrix, or in some cases initiate the polymerization of monomers to improve the strength of the interface between the inorganic and the polymer matrix [10, 12].

This study aims reporting the synthesis and characterization of the composites based on a widely used polymer, i.e. polyethylene oxide (PEO) and the results for the mechanical properties of the synthesized composites.

EXPERIMENTAL

Materials Used

The materials used in this study are montmorillonite, polyethyleneoxide (PEO), Myristyltrimethylammonium Bromide (MTAB), and dichloromethane (methylene chloride). The montmorillonite, which is a commercial sample of betonies from Çankırı deposit in Turkey, was obtained from Karakaya Mineral Co., Ankara-Turkey (CEC: 107meq/100g, d_{001}: 1.2620 nm). The polyethyleneoxide (PEO) with molecular weight of 900,000 g/mol was purchased from Aldrich. The Myristyltrimethylammonium Bromide (MTAB) and dichloromethane (methylene chloride) were purchased from Merck, and they were used without further purification.

Preparation of Composites

Preparation of Organo-Clay Samples

In order to expand the basal distance between clay layers and to produce organophilic clay, it is necessary the control of interactions between the polymer matrix and clay particles. For this, the montmorillonite (MMT) clay sample was modified adsorption of Myristyltrimethylammonium Bromide (MTAB) at the concentrations under critical micelle concentration (CMC) (0.5, 0.7, and 1.0 g/g clay).

After a sample of 50.00 g MMT was dispersed in a 100 L container with distilled water, the mixture was shaken in order to swell to 200-250 rpm for 10 h, and then MTAB under critical micelle concentrations were added to the clay suspension for 1 h. The mixture was filtered, washed, and dried at 90°C under vacuum and milled.

Preparation of Composites

Polyethylene oxide/organo clay composites were produced by the solution interaction method. The organically modified clay was dispersed and swollen in methylene chloride for 24 h in a three-necked flask at ambient temperature and sonicated for 15 min. The PEO/methylene chloride ratio, 1: 33.6 (g/mL), was maintained by adding an additional quantity of solvent as well as PEO. The flask and its contents, fitted with a condenser, a mechanical stirrer and a thermometer pocket, were then placed in an oil bath heated to 75°C. The flask was kept under stirring condition for 10 h and then the composites were put into Petri dishes and dried in air overnight at ambient temperature. The same procedure was applied to different clay ratios (2.0, 5.0, and 8.0 %). A schematic overview of the nanocomposites synthesis is shown in Fig. 1.

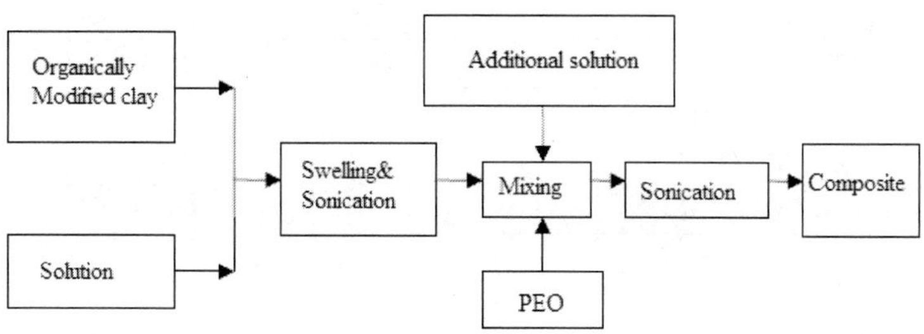

Fig. 1: Schematic Overview of the PEO/Clay Nanocomposite Preparation

Characterization of composites

X-ray Diffraction (XRD)

XRD experiments were conducted on a Rigaku 2200Dmax diffract meter instrument that employed K-α radiation (λ = 1.5404 Å) and performed from 2° to 40°. The scanning rate of the instrument was 2°/min.

Fourier Transform Infrared (FT-IR) Spectroscopy

FT-IR Spectra of samples were collected on a Perkin–Elmer 1600 FT-IR spectrometer from 400 to 4000 cm^{-1} with a nominal resolution of 2 cm^{-1}. For each spectrum 64 runs were collected and averaged. The clay specimen were prepared by adding approximately 1% of the sample powder to dry KBr powder and pressed into disc of 13 mm in diameter and 1–2 mm thickness.

Tensile Testing

Mechanical properties of PEO/organo clay composites, which are such as tensile strength, elasticity module, and strain % were determined. For this, the PEO/organoclay composites were die cut with a rectangular die and tested in a Shimadzu AG-100kNIS test instrument under ASTM 638 M-91a. All tests were carried out at 5 kN at ambient temperature with speed of 2 mm/min.

RESULTS AND DISCUSSIONS

X-ray Diffraction (XRD)

Table 3.1 and Fig.2 show the XRD patterns of MMT and the organoclays. The basal spacing of MMT is 0.69 nm, however, after modification with surfactant, the interlayer height of MMT is obviously increased. With an increase of surfactant concentration in the solution, the basal spacing of the resultant organo clays increase in the following order: 0.69 nm (MMT); 0.82 nm 0.5 CEC OMMT; 0.92 nm 0.7 CEC OMMT, and 1.34 nm 1.0 CEC OMMT.

Table 1: d_{00} Basal Spacing of Montmorillonite and the Resultant Organoclays

Sample	d_{001} *(nm)*	d_{002}*(nm)*
Raw MMT	0.69	0.36
0.5 CEC OMMT	0.82	0.40
0.7 CEC OMMT	0.92	0.41
1.0 CEC OMMT	1.34	0.42

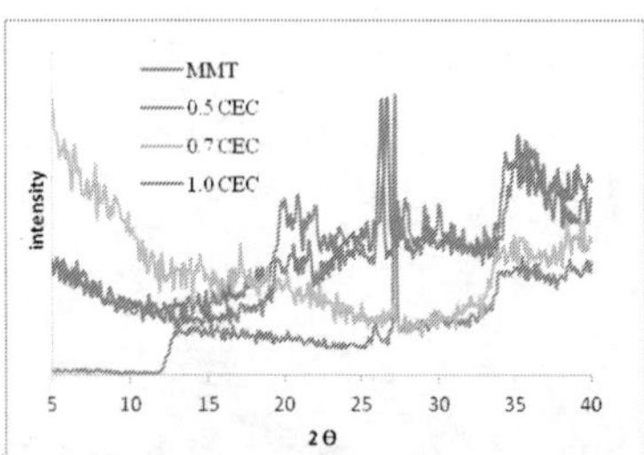

Fig. 2: XRD Patterns of Montmorillonite and the Resultant Organoclays

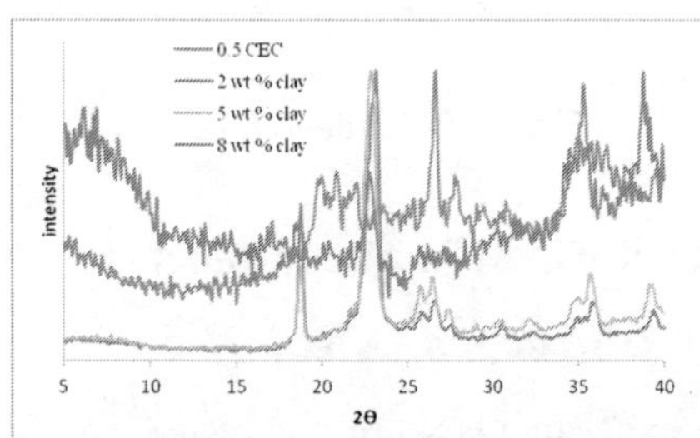

Fig. 3: XRD Patterns of Produced PEO/Clay Nanocomposites at Various Clay Compositions (0.5 CEC)

Figures 3-5 show the XRD patterns of MMT and its PEO/OMMT hybrid products. The modification of clay with organic ions not only makes the clay surface hydrophobic but results in a tremendous increase in d-spacing, which facilitates the penetration of PEO into the interlayer galleries. From Fig.3, it can be seen that at the clay ratio of 8.0% the dispersion degree of clay into polymer matrix decreased. Figure 4 shows that at all of clay ratios, the difractograms of the samples resemble each other, indicating the decreased charge intensity of organo clay particles may lead to the orientation of polymer chains as helical into the interlayer region.

It can be concluded that at the higher clay ratios, the dispersion degree of clay particles into polymer matrix increased, probably due to the positive charge acquired by clay particles at 1.0 CEC and so the intercalation with a different configuration of the polymer chains (see Fig. 5).

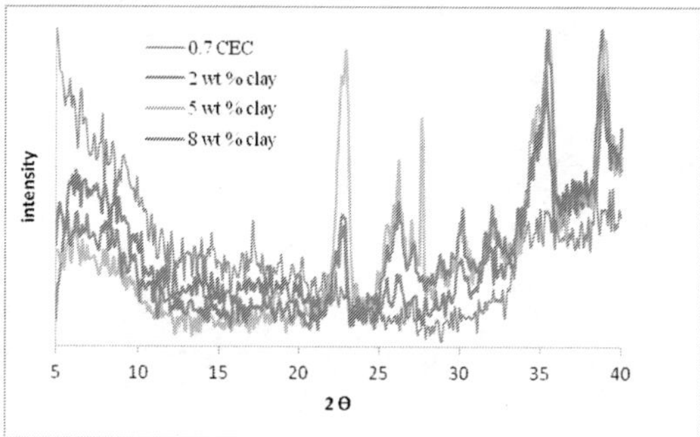

Fig. 4: XRD Patterns of Produced PEO/Clay Nanocomposites at Various Clay Compositions (0.7 CEC)

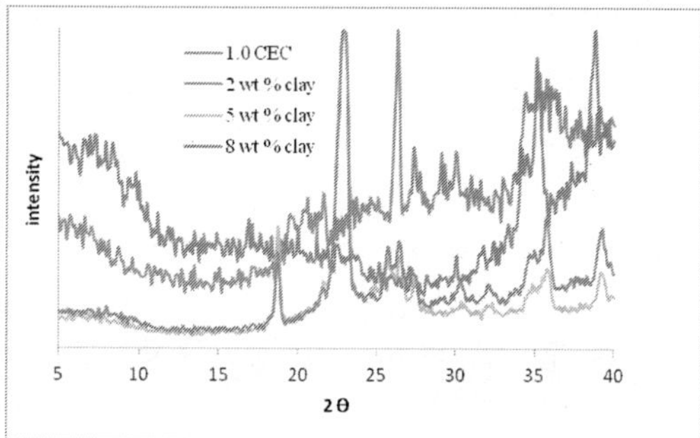

Fig. 5: XRD Patterns of Produced PEO/ Clay Nanocomposites at Various Clay Compositions (1.0 CEC)

FT-IR Analysis

The polymer/clay composite samples were also characterized by FT-IR spectroscopy. FT-IR spectroscopy provides important information regarding the difference between polymer and polymer clay hybrid. The FT-IR spectra of raw MMT and OMMT sample are shown in Fig.6. The spectrum of MMT shows a band near 3634 cm^{-1} owing to the stretching vibrations of structural OH groups and a broad band near 3444 cm^{-1} due to the OH stretching vibrations of H_2O present in the smectite. Displacement of hydrated MMT with MTA^+ ions reduces the intensity of H_2O band. Intensity decrease was even more pronounced in Fig. 6 and the lowest amount of water is observed in same Figure.

The C–H stretching bands of alkylammoniumcations occur in the 3020–2800 cm^{-1} region. Two strong bands near 2918 cm^{-1} (νas(CH_2)) and 2850 cm^{-1} (νs(CH_2)) are present in spectra MTAB-MMT samples, however, no clearly resolved CH_3 stretching band. This vibration is relatively weak even in the spectrum of neat MTA^+ salt; consequently, it is not visible in the spectrum of MTA^+ salt adsorbed on a smectite.

A characteristic νas bending vibration of the $(CH_3)_4N^+$ cation at 1487 cm^{-1} is seen in the spectrum of MTA-MMT (Fig.2). Replacement of one methyl group in $(CH_3)_4N^+$ with a longer alkyl chain $(CH_3)_3N^+(CH_2)_nCH_3$ gives two strong bands near 1473, 1477, and 1474 cm^{-1} attributed to νas (C–H) bending of both CH_3 and CH_2 groups. A gradual decrease of the H–O–H bending band (1642 cm^{-1}) from MMT through MTA-MMT confirms diminishing amount of water in the samples resulting from organic cations adsorption. There are at least two reasons why organo-smectites contain less water than inorganic parent minerals-the larger size of the organic cations physically taking up room that water would occupy in an inorganic system, and much lower hydration energies of the organic cations. As a result, the amount of water in organo-smectites is considerably reduced. An important consequence of replacing inorganic with organic cations is that the clay surface takes on hydrophobic character, which greatly enhances their affinity for non-polar organic.

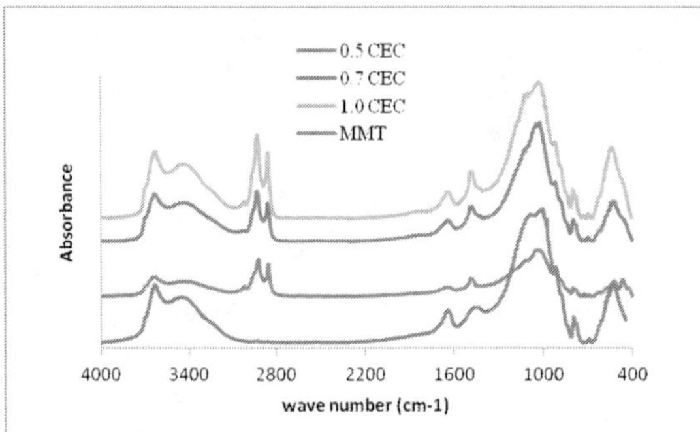

Fig. 6: FTIR Spectra of Produced Organoclays at Various CTAB-Clay Ratios

The FT-IR spectra of all organoclays and PEO/organoclays sample are shown Figs. 7- 9. Pure PEO shows a large broad band of asymmetric $–CH_2$ stretching between 3000 and 2750 cm^{-1} with the two narrow bands at 2739 and 2693 cm^{-1}. In the all composite the band broadens between 3000 and 2800 cm^{-1}, whereas the narrow peaks disappear. In this region, the characteristic peaks of PEO which appear at 2793 and 2952 cm^{-1} are shifted to 2800 and 2936 cm^{-1}. The peaks in PEO at 1342 cm^{-1} shifted to 1347 cm^{-1} in intercalated hybrids. This observation can be attributed to the ion dipole interaction of the interlayer cation of clay and the oxygen atoms of oxyethylene units. The peaks near 862 and 842 cm^{-1} are assigned to the CH_2 rocking vibrations of methylene groups in the gauche conformation required for the helical conformation, so it is generally believed that virgin PEO is in a helical conformation. The absence of the characteristic PEO peak at 1320 cm^{-1} assigned to CH_2 vibrations of the ethylene group in the trans conformation as referred to by Ratna et al. (2006) also confirms the helical structure of PEO[13]. In Figs. 7-9, it can be seen that the PEO peak at 842 cm^{-1} is shifted to 849 cm^{-1} in the intercalated hybrid which is assigned to the CH_2 asymmetric rocking vibration. The peaks at 1279 and 1241 cm^{-1} in PEO merge to give a single peak near 1276 cm^{-1} in the case of the nanocomposite. In the region of 1300–900 cm^{-1}, we note peaks getting broader with the addition of organic clay, particularly the 1116 cm^{-1} peak in PEO which is for asymmetric stretching of the C–O–C groups broadens while the 962 cm^{-1} peak shifts to a lower frequency of 953 cm^{-1}. It is also found that peaks apparent at 945 and 842 cm^{-1} which have been related to helical structure of PEO are present in PEO only and not in the nanocomposite sample. This indicates that in the nanocomposite the helical structure of PEO in distorted/ perturbed as a result of intercalation [13, 14].

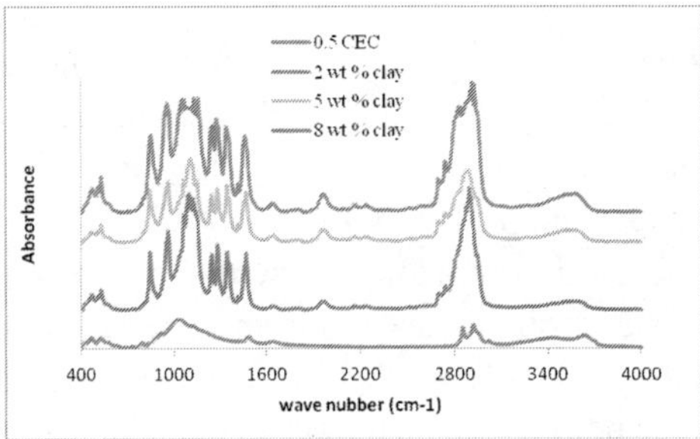

Fig. 7: FTIR Spectra of Produced PEO/Clay Nanocomposites at Various Clay Compositions (0.5 CEC)

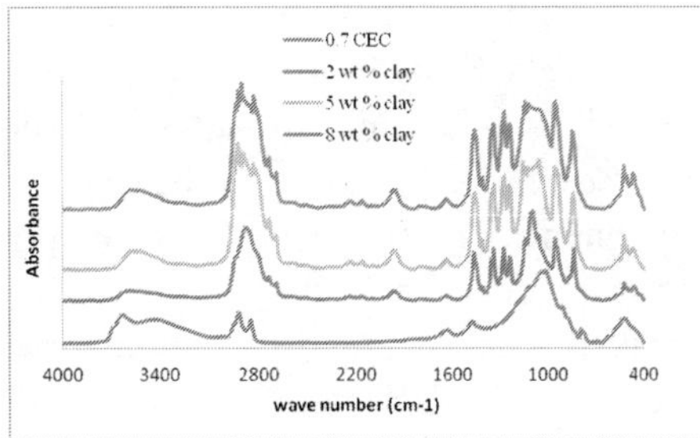

Fig. 8. FTIR Spectra of Produced PEO/Clay Nanocomposites at Various Clay Compositions (0.7 CEC)

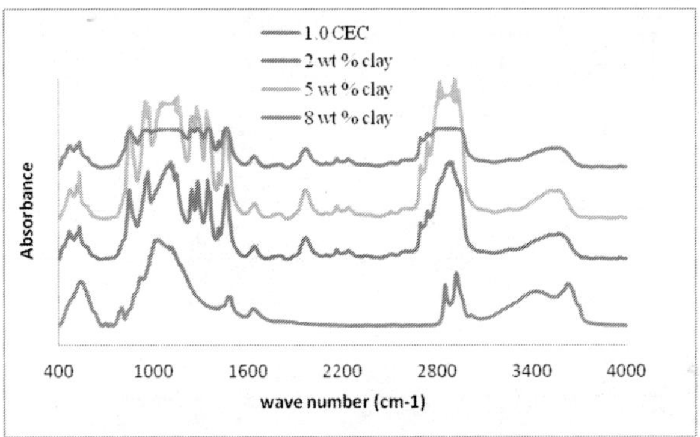

Fig. 9: FTIR Spectra of Produced PEO/Clay Nanocomposites at Various Clay Compositions (1.0 CEC)

Mechanical Analysis

The tensile test was conducted in order to obtain an idea of the effect of intercalation on the mechanical behaviour of composites with different clay loading. The tensile strength, yield strength and strain percentage diagrams of PEO and composite samples are shown Table 2. One can see from Table 2 that tensile strength, yield strength and strain percentage increases with the incorporation of unmodified PEO, attains a maximum value of 2 wt% of clay, followed by a sharp decrease up to 5 wt% of clay. This indicates that incorporation of clay up to an optimum level resulted in a considerable reinforcing effect. It is interesting to note that both tensile strength, yield strength and strain percentage increases simultaneously, i.e. improvement in mechanical property is achieved in both respects. However, at high loading of PEO the organo clay behaves as ordinary filler with little improvement in strength. It can be said that at higher clay content the response to the extension under load is mainly governed by the hard and brittle clay and the intercalated PEO chains, which are not as extended as freely the PEO chains that are not intercalated. The tensile strengths firstly increases and then decreases with the increased CEC values, indicating the possible positive charge of clay particles in the case of 1.0 CEC only become effective in the change of the configuration of polymer chains at the higher clay ratio. In addition, the yield strengths are sensitive the CEC values and so the charge intensity of clay particles. The optimum CEC value and clay ratio are 0.5 and 2.0 wt %, respectively. The strain percentage of 0.5 CEC increases with the increased clay ratio, whereas at higher CEC's the strain percentages pass from a maximum and then decrease. This clearly introduces a relation between the charge intensity of clay particles and the both interaction effectiveness and configuration of polymer chains. It was concluded that in the case of 8.0 wt % clay ratio the interaction effectiveness becomes weaker and clay platelet or tactoids occurs in the polymer matrix

Table 2: Tensile Properties of PEO and PEO/Clay Composite Samles

Sample	*Tensile Strength (MPa)*	*Yield Strength (MPa)*	*Strain Percentage*
PEO	0.781	0.547	54.64
0.5 CEC OMMT			
2 wt % clay	2.64	1.15	113
5 wt % clay	1.88	1.63	212
8 wt % clay	2.36	1.87	247
0.7 CEC OMMT			
2 wt % clay	4.41	3.85	471
5 wt % clay	2.47	1.84	398
8 wt % clay	1.98	1.50	135
1.0 CEC OMMT			
2 wt % clay	2.01	1.63	358
5 wt % clay	1.60	1.35	411
8 wt % clay	3.96	2.26	181

It can be seen that at 0.5 CEC the strain percentage increases with the increased clay ratio, whereas at higher CEC's the strain percentages pass from a maximum and then decrease. This clearly introduces a relation between the charge intensity of clay particles and the both interaction effectiveness and configuration of polymer chains. It was concluded that in the case of 8.0 wt % clay ratio the interaction effectiveness becomes weaker and clay platelet or tactoids occurs in the polymer matrix

CONCLUSIONS

PEO/clay nanocomposites were synthesized using polar organo clay by a solution intercalation method. Intercalation is achieved by simple mechanical mixing and the extent of Intercalation increases due to sonication.

Strong reinforcing effect was observed for clay content up to 2 wt% and after which the behaviour is like ordinary filler. Hence, the concentration of clay plays an important role. Based on the results of FT-IR studies the presence of a distorted helical structure of PEO in the nanocomposite is confirmed.

REFERENCES

[1] Ray, S., Okamoto, M., Polymer/layered silicate nanocomposites: A review from preparation to processing. *Progress in Polymer Science*. 28, 1539-1641. 2003

[2] Okada, A., Kawasumi, M., Usuki, A., Kojima, Y., Kurauchi, T., Kamigaito, O. Synthesis and properties of nylon-6/clay hybrids. In: Schaefer DW, Mark J.E., editors. Polymer based molecular composites. *MRS Symposium Proceedings, Pittsburgh*, 171; pp. 45–50, 1990.

[3] Biswas, M., Ray, S. Recent progress in synthesis and evaluation of polymer–montmorillonite nanocomposites, *Adv Polym Sci*., 155:167–221, 2001.

[4] Giannelis, E.P. Polymer-layered silicate nanocomposites: Synthesis, properties and applications, *Appl Organomet Chem*., 12,675–80, 1998.

[5] Xu, R., Manias, E., Snyder, A.J., Runt J. New biomedical polyurethane uera-layered silicate nanocomposites. *Macromolecules*. 34, 337–339, 2001.

[6] Kojima, Y., Usuki, A., Kawasumi, M., Fukushima, Y., Okada, A., Kurauchi, T., Kamigaito, O. Mechanical properties of nylon 6–clay hybrid. *J Mater Res* 1993; 8 : 1179–84.

[7] Gilman, J.W., Kashiwagi, T., Lichtenhan, J.D. Flammability studies of polymer-layered silicate nanocomposites. SAMPE J., 33, 40–5, 1997.

[8] Gilman, J.W., Jackson, C.L., Morgan, A.B., Harris, Jr. R., Manias, E., Giannelis, E.P., Wuthenow, M., Hilton, D., Phillips, S.H.. Flammability properties of polymer-layered silicate nanocomposites. Propylene and polystyrene nanocomposites. *Chem Mater*; 12, 1866–73, 2000.

[9] Ray, S., Yamada, K., Okamoto, M., Ueda, K. New polylactide/layered silicate nanocomposite: A novel biodegradable material. *Nano Lett*., 2, 1093–6, 2002.

[10] Blumstein, A. Polymerization of adsorbed monolayers: II. Thermal degradation of the inserted polymers. *J Polym Sci A*.,3, 2665–73, 1965.

[11] Theng, B.K.G.. Formation and properties of clay–polymer complexes. Amsterdam: Elsevier; 1979.

[12] Krishnamoorti, Vaia, R.A., Giannelis, EP. Structure and dynamics of polymer-layered silicate nanocomposites. *Chem Mater*., 8, 1728–34, 1996.

[13] Ratna, D., Divekar, S., Samui, A.B., Chakraborty, B.C., Banthia, A.K., Polyethylene oxide/ clay nanocomposite: thermomechanical properties and morphology, *Polymer*, 47, 4068-4074, 2006.

[14] Aranda, P. and Ruiz-Hitzky, E., Polyethylene oxide-silicate intercalation materials. *Chemical Material*, 4, 1395-1403, 1992.

SYNTHESIS AND CHARACTERIZATION OF CuO NANO POWDERS BY CO-PRECIPITATION METHOD

S.M. Sathiya, P. Deepa[1], K. Ravichandran[2], A. Moses Ezhil Raj[3], C. Sanjeeviraja[4] and C. Ravi Dhas[1]

Department of Physics, Holy Cross College (Autonomous), Trichirappalli
[1]Department of Physics, Bishop Heber College (Autonomous), Trichirappalli
[2]Department of Physics, AVVM Sri Pushpam College (Autonomous), Thanjavur
[3]Department of Physics, Scott Christian College (Autonomous), Nagercoil
[4]Department of Physics, Alagappa University, Karaikudi
E-mail: cravidhas@gmail.com

ABSTRACT

CuO nano powders for various concentrations were synthesized by Co-precipitation method. The micro structure and morphological arrangements of CuO nano powders were characterized by X-ray diffraction (XRD) and Scanning Electron Microscopy (SEM). The particle size of CuO nano powders were calculated as 27 to 46 nm and the lattice parameters were calculated. The elemental quantification and stoichiometry ratio of CuO nano particles were confirmed by X-ray Dispersive Analysis (EDAX) which shows the presence of uniform distribution of Copper and Oxygen. Electrical measurements depict the thermally activated conduction within the nano powders. The band gaps of CuO were studied using UV-Vis spectrometric measurements. Phase analyses were carried out using Micro-Raman Scattering (MRS).

INTRODUCTION

Nano particles or nano crystals made of metals, semiconductors or oxides are of particular interest for their mechanical, electrical, magnetic, optical, chemical and other properties [1].

V. Rajendran, K. Saminathan, P. Paramasivam and K.E. Geckeler (eds.)
Nano Materials Synthesis and Characterisation, pp. 353-362 (2012).

Nano particles are of great scientific interest as they are effectively a bridge between bulk materials and atomic or molecular structures. Nano particles exhibit a number of special properties relative to bulk material. Most current work looks at making nano particles (particles with nanometer size) that have special properties, such as the way they scatter light, absorb X-rays, transport electrical currents or heat, etc. [2]. The first observations and size measurements of nano-particles were made during the first decade of the 20th century [3]. Nano particles may or may not exhibit size-related properties that differ significantly from those observed in fine particles or bulk materials [4, 5].

In past few years, the different morphologies of CuO nano particle have attracted more interest for practical application. Copper oxide (CuO) is a well known p-type semiconductor with a narrow band gap of 1.2 eV [6] has gained increasing attention not only for its unique properties but also for its applications in many fields such as gas sensors [7], photo catalyst and electrochemistry sensors, etc. [8]. CuO comes under the category of transition-metal oxides [9]. Nano particle characterization is necessary to establish the understanding and control of nano particle synthesis and applications. Characterization is done by using a variety of different techniques, such as electron microscopy (TEM, SEM), atomic force microscopy (AFM), powder X-ray diffraction (XRD) [10] and Micro Raman Scattering (MRS) [11].

In this work we report a simple route for the wet synthesis of CuO nano powders by continuous stirring of Copper (II) Chloride dehydrate and Sodium Hydroxide. We have investigated the powder preparation and the optimization conditions to synthesis CuO nano powders.

EXPERIMENTAL

CuO nano powders for various concentrations (0.1M - 0.5M) were prepared by Co-precipitation method [12]. The chemicals we used were Copper (II) Chloride dehydrate ($CuCl_2.2H_2O$) and NaOH platelets. In a typical synthesis process, Copper (II) Chloride dehydrate (0.1M-0.5M) were mixed with 100 ml double distilled water to obtain Copper Chloride solution. Then the solutions were stirred in a magnetic stirrer at room temperature for getting a well dissolved solution. After stirring the pH value of the solutions were increased to 12 by adding NaOH platelets one by one (4 gm – 20 gm). In order to get CuO nano powder the solutions were stirred for another 1hr. Then the solutions were kept into the microwave oven at 120°C for 1hr. After microwave processing the mixtures were cooled to room temperature and the resulted black coloured CuO precipitates were separated by centrifugation, washed with distilled water and Ethanol several times to remove the impurities and the residual materials. Then the CuO nano powders were dried in microwave oven at 200°C for 1½ hrs. The obtained CuO nano powders were annealed at 700°C for 2hrs in a Muffle Furnace for characterization.

Experimental Flow Chart for CuO Nano Powder

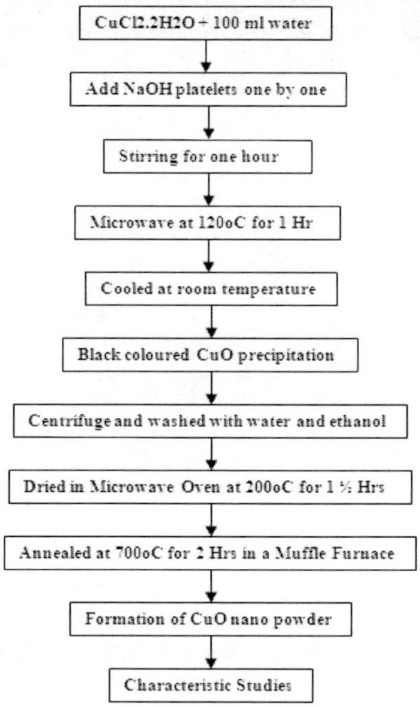

RESULTS AND DISCUSSION

Figure (1) shows the XRD pattern of the CuO nano powders prepared by Co-precipitation method. XRD analysis was employed to study the crystal structures of the CuO nano powders which show the polycrystalline nature of the CuO crystal. All the peaks of the samples are corresponding to the monoclinic structure of standard CuO (JCPDS–80–0076). This also provides information on growth orientation of crystal. The Miller-indexed $(\bar{1}111)$ and (111) reflections are the strongest at $2\theta = 35.6^{\circ}$, which indicate that the preferential crystal planes of nano balls. In addition small peaks corresponding to the CuO $(\bar{2}20)$, $(\bar{1}13)$ plane was also observed. All diffraction peaks can be clearly indexed to the monoclinic CuO phase. The increase in concentration of the solution causes an increase in the intensities of the peaks $(\bar{1}11)$ and (111). The different peaks for concentration 0.3–0.5 M are identical. Since the peaks under comparison are chosen from the same XRD spectrum, the peak height ratio must reflect the amount of ratio of each phase in the composite nano particles. As mentioned above, the relative peak height can be taken as a semi quantitative measure of the corresponding phase in the composite nano particles [13]. The size of the nano particles are also calculated as 27 – 46 nm Table (1) from the Scherrer's formula. The lattice strains (e) were calculated using the relation $\varepsilon = (\beta \cos \theta)/4$. The XRD calculations show that, when the concentration increases the lattice strain also increases. The graph Fig. (2) concentration vs. dislocation density shows that the concentration increases the dislocation density ($\delta = 1/D^2$) also increases. As the molar values increases, the grain size decreases and increase the density of grain boundaries. This in turn decreases the density of the nano powder. The prepared higher molar powders are therefore less dense.

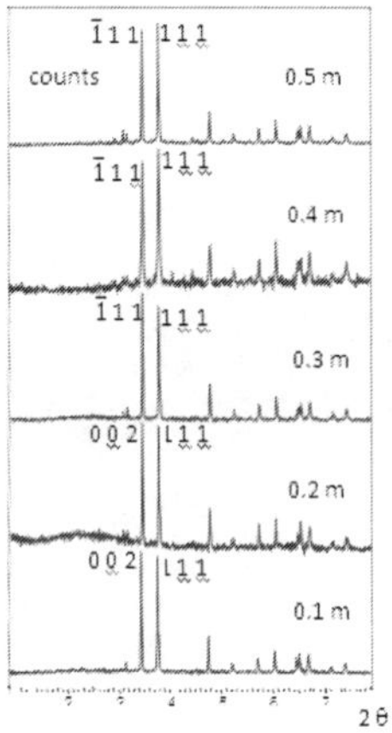

Fig. 1: XRD Patterns of CuO Nano Powder for 0.1M – 0.5M

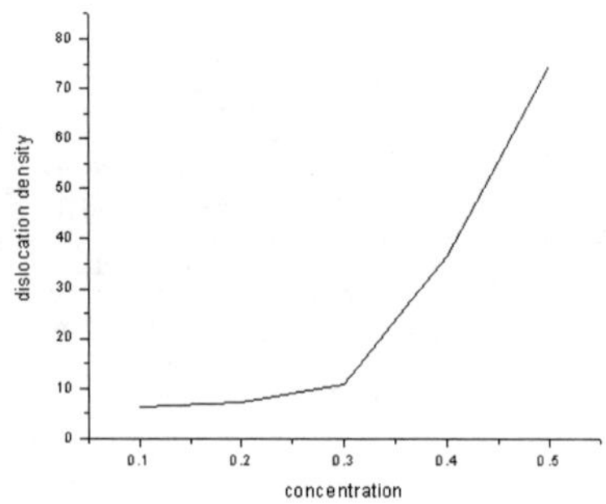

Fig. 2: Plot for Concentration vs. Dislocation Density

Table 1: The XRD Calculation of CuO Nano Powders for 0.1 M to 0.5 M

Sl. No.	*Mole (M)*	*2θ*	*Intensity*	*d Values*	*hkl*	*Grain Size (D) nm*	*Lattice Strain (ε) mm*	*Dislocation Density (δ) x 10^{15} Lines / m*
1.	0.1	35.443 38.648	100 95.42	2.53066 2.32782	0 0 2 1 1 1	43.4339	0.85176	0.644163
2.	0.2	35.443 38.651	100 89.23	2.53062 2.32765	0 0 2 1 1 1	46.3764	0.87782	0.720992
3.	0.3	38.702 35.496	100 98.92	2.32467 2.52696	1 1 1 $\bar{1}$ 1 1	39.7816	1.04809	1.09674
4.	0.4	39.034 35.815	100 90.21	2.3057 2.50516	1 1 1 $\bar{1}$ 1 1	27.322	1.69538	3.66131
5.	0.5	38.721 35.516	100 87.61	2.3236 2.52562	1 1 1 $\bar{1}$ 1 1	30.6811	0.91309	7.47976

Figure (3) shows the EDAX spectrum of CuO nano particles. The EDAX spectrum further proves that synthesized sample is composed of Cu and O elements. This shows that the copper and Oxygen peaks are obtained and atomic ratio (%) of Cu to O is 66.84/33.16 for 0.3 M. and for 0.5 M it is in the ratio of 65.76/34.24 (Table 2). The EDAX measurements give further evidence for the purity of the product that the Cu and O elements are present almost in stoichiometric ratio of (2:1), which shows the surface of the sample is rich in Copper for both 0.3 M and 0.5 M. but in 0.4 M we observed a non-stoichiometric nature (1:6) in the powder formation. This may be due to the reason of deficiency of Cu atom during the synthesis process.

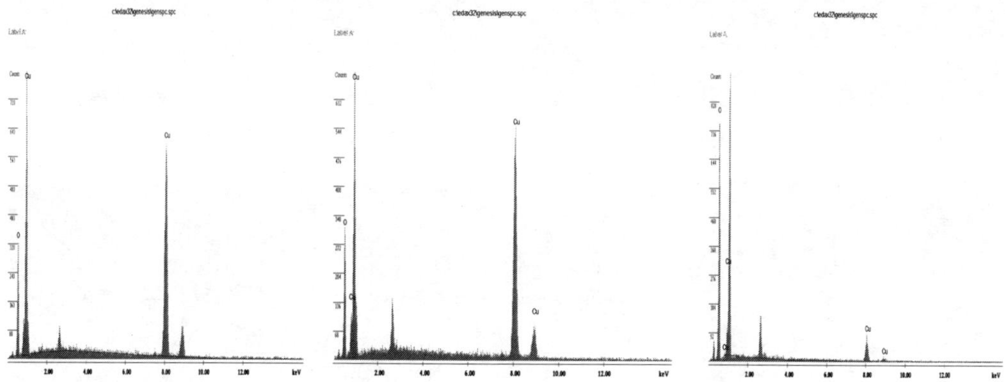

Fig. 3: EDAX Spectra of the CuO Nano Powder (a) 0.3 M (b) 0.4 M (c) 0.5 M

Table 2: EDAX Results of CuO Nano Powders for 0.3 M to 0.5 M

Sl. No	*Mole M*	*Element K*	*Wt %*	*At %*
1.	0.3	O	11.10	33.16
		Cu	88.90	66.84
2.	0.4	O	58.93	85.07
		Cu	41.07	14.93
3.	0.5	O	11.59	34.24
		Cu	88.41	65.76

Figure (4) shows the SEM image of the fine CuO nano powders for 0.3 M–0.5 M concentrations prepared at 700°C for 2hrs. The appearance of the grains is regular and microstructure consists of neatly arranged many round shaped particles (nano balls). As the molar concentration of the CuO nano powder increases the width of the particles also decreases. The average width of the nano particles for 0.3 M, 0.4 M and 0.5 M is 6.8 mm. As the length of the particle reduces from mm to nm the width of the particle also decreases [14]. The salient features of these powders are spherical and their diameters are uniformly distributed. It can also be seen that the given CuO nano powders having relatively large spherical particles are of uniform size, however, in addition there were many tiny particles adhering to the spherical particles. The size of the particles decreases ($\cong$ 40 nm) with increase in molar concentration which is confirmed in the XRD calculations. When the concentration was increased to 0.3–0.5 M the small particles disappear and the relatively large particles became dominant. This shows that the higher concentration would result in particle growth.

Fig. 4: (a) SEM Image for 0.3 M, 0.4 M and 0.5 M (mag 60,000)

Fig. 4: (b) SEM Image for 0.3 M, 0.4 M and 0.5 M (mag 30,000)

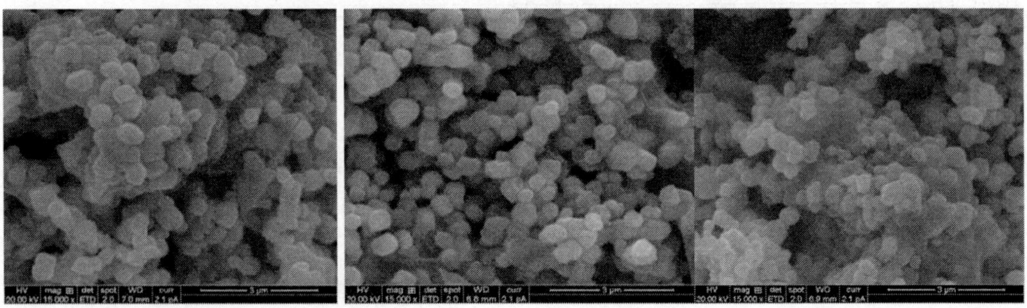

Fig. 4: (c) SEM Image for 0.3 M, 0.4 M and 0.5 M (mag 15,000)

Figure (5) shows the Absorption spectra of CuO nano powder for different molar Concentrations. The Absorption Spectra in the Wave length range 200 nm – 1200 nm were recorded for CuO nano powder prepared at 700°C. Fig (5) shows a broad Absorption peak whose centre is at about 380 nm. It also shows that the absorption decreases exponentially as the wavelength increases and the absorption spectrum is in the visible region. In this spectrum the lower cut-off region is near to 400 nm for 0.2 and 0.5 M. When the wavelength of the absorption spectra increases or above the cut-off region the absorption is constant, which shows that the transparent nature of the CuO nano powders.

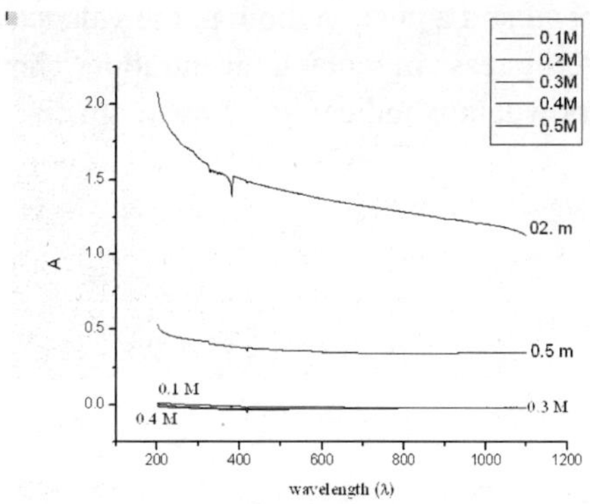

Fig. 5: Absorption Spectra of CuO Nano Powder for 0.1 M to 0.5 M

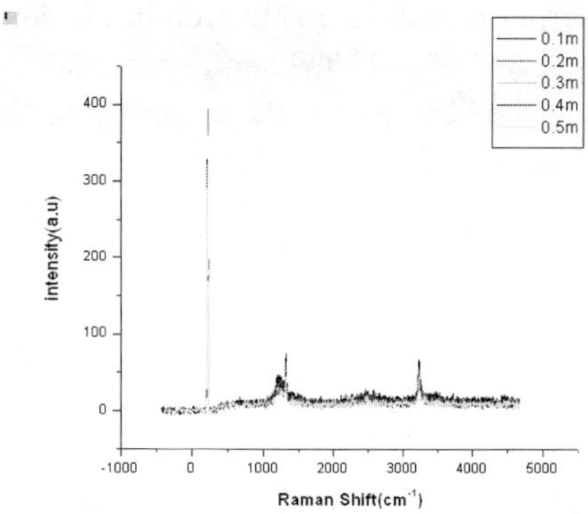

Fig. 6: Micro Raman Scattering Spectra of CuO Nano Powders Prepared by Co-precipitation Method (0.1 M to 0.5 M)

Raman Spectroscopy has been widely used as sensitive probe to investigate the microstructure, namely the local atomic arrangement and vibration of the nano sized materials [15]. CuO belongs to C^6_{2h} group with two molecules per primitive cell. One can find the zone centre Raman Active normal modes Γ_{RA} = 4Au + 5Bu + Ag + 2Bg. There are three acoustic modes (Au + 2Bu), six IR active modes (3Au + 3 Bg), and three Raman active modes (Ag + 2Bg). Three Raman active optical phonons have been identified and comparable to that reported in the literature by other methods [16]. Figure (6) shows the Raman Spectra of the prepared CuO nano powder, at 700° C. The Raman peaks are broadened; the peak at 219 cm^{-1} and 224 cm^{-1} can be assigned to be Ag mode, while the peaks at 1317cm^{-1} and 3227 cm^{-1} can be assigned to be Bg modes. It also shows that the intensity of the spectrum increases with decrease in Raman Shift, indicating the increase in crystalline sizes of CuO.

The fundamental absorption, which corresponds to electron excitation from the valance band to conduction band, is used to determine the nature and value of the optical band gap. The relation between the absorption coefficients (a) and the incident photon energy (hν) are related by the relation $(\alpha h\nu)^{1/n} = A[h\nu - E_g]$. Where A is a constant and E_g is the band gap of the material and the exponent 'n' depends on the type of transition [17]. For direct allowed transitions n = 1/2, for indirectly allowed transitions n = 2, and for directly forbidden transitions n = 3/2. To determine the possible transitions, $(\alpha h\nu)^{1/n}$ versus hν were plotted and the corresponding band gaps were obtained from extrapolating the straight portion of the graph on 'hν' axis at $\alpha = 0$.

Figure (7) gives the information of the Variation of $(\alpha h\nu)^2$ vs hu for CuO nano powders. The values of direct band gap for CuO nano powders synthesized with different molar concentrations were shown in Table (3). The band gap energy was calculated from the absorption spectra. It was found that in direct band gap of CuO nano powder samples deviate between 1.5eV to 3.6 eV, which shows the higher direct band gap as compared to the bulk value. The light absorption leads to an electron in the conduction band and a positive hole in the valance band. The Band gap of CuO nano powders increases with increase in molar concentration. The increase in the band gap of the as-prepared CuO nano powders is indicative of quantum size effects [18].

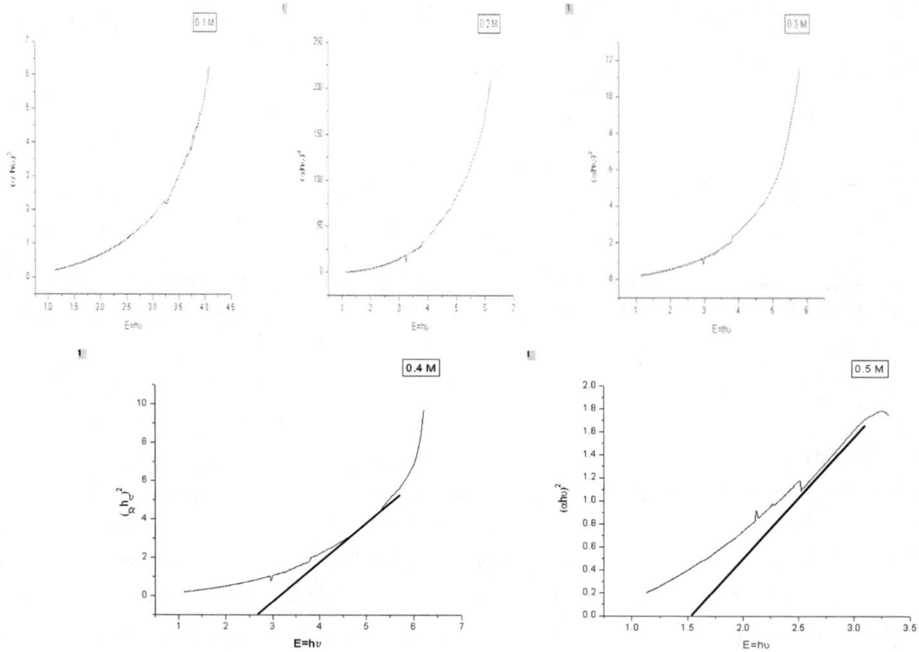

Fig. 7: Variation of hv vs $(\alpha hv)^2$ for CuO Nano Powder Prepared by Co-Precipitation Method for 0.1 M to 0.5 M Concentrations

Table 3: The Values of Direct band for CuO Nano Powders Synthesized with Different Molar Concentrations

Mole (M)	*0.1 M*	*0.2 M*	*0.3 M*	*0.4 M*	*0.5 M*
Band Gap for hυ vs $(\alpha h\upsilon)^2$ (Direct region)	2.5 eV	3.6 eV	3.7 eV	2.6 eV	1.5 eV

The temperature dependence of the semi conductor materials conductivity is expressed by the equation: $\sigma = \sigma_o \exp[-E_a / kT]$. Where σ_o is the pree exponential factor E_a is the activation energy for this thermally activated process and k is the Boltzmann constant [19]. Clearly a plot of log (σ) vs. 1000/T Fig. (8) will be nearly a straight line up to the moderate temperature. The straight line nature of the Arrhenius plot indicates the ohmic conduction, as often found in semiconductors [20, 21]. From the plots it is clear that the conductivity of the nano powders increases with increase in temperature indicating the semiconducting behavior of the samples. The increase in conductivity attributed to the decrease of grain size leading to an increase of the density of grain boundaries resulting in the increment of grain barrier. The conductivity mechanism in the temperature range studied (100 K–300 K) can be explained using Seto's classical grain boundary effect (Seto. 1975), which presumes that the conduction process is governed by the thermionic emission over the potential barriers across the grain boundaries.

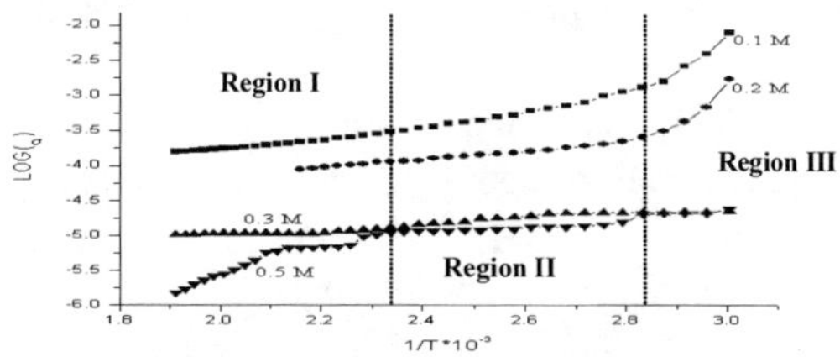

Fig. 8: Variation of Electrical Conductivity σ as a Function of Temperature T of CuO Nano Powders

In our study three regions are very distinctly found. The small increase in conductivity in (region I) low temperature region (190–230K) is related to the vigorous oxygen adsorption on the surface, causing decrease in the carrier concentration, which is consistent with the adsorption of oxygen on the surface of the CuO nano powder [22]. In (region II) moderate temperature region (230–280 K), the conductivity is not much affected by the temperature change probably due to the equilibrium achieved between the two competing process of thermal excitation of electrons and oxygen adsorption. Finally, the conductivity in (region III) high temperature region (280–300 K) increases with increase in temperature this is because of the thermal excitation is the main cause of increase in electrical conductivity.

CONCLUSION

CuO nano powders for various concentrations were synthesized by Co-precipitation method. The micro structure and morphological arrangements of CuO nano powders were characterized by X-ray diffraction (XRD) and Scanning Electron Microscopy (SEM). The sizes of the grains are regular and microstructure consists of many spherical shaped particles. The particle size of CuO nano powders were calculated as 27 to 46 nm and the lattice parameters were calculated using XRD. EDAX results confirm the elemental presentation of the CuO nano powders. Electrical measurements depict there is an increase conductivity with increase in temperature. The band gaps of CuO were studied using UV-Vis spectrometric measurements. Using Micro-Raman Scattering the Phase analyses of CuO nano powders were carried out.

REFERENCES

[1] Chang M.H., Liu H.S., Tai C.T. Preparation of copper oxide nano particles and its application in nano fluid. *Powder Tech 2011*; 207: 378 and 86.

[2] L.B. Kiss, J. Söderlund, G.A. Niklasson, C.G. Granqvist (1999). "New approach to the origin of lognormal size distributions of nano particles". *Nanotechnology* 10: 25–28. Bibcode 1999 Nanot. 10.25K. doi : 10.1088/0957-4484/10/1/006.

[3] Cristina Buzea, Ivan Pacheco and Kevin Robbie (2007). "Nano materials and Nano particles: Sources and Toxicity". *Biointerphases* 2 (4): MR17–MR71.

[4] Henrich, Victor; Cox P.A. (1996). *The Surface Science of Metal Oxides*. Cambridge University Pre. ISBN 0521566878.

[5] S.H. Kim, B.Y.H. Liu and M.R. Zachariah, Synthesis of nanoporous metal oxide particles by a new inorganic matrix spray pyrolysis method, *Chem.Mater*. 14 (2002) 2889–2899.

[6] Hsieh C.T., Chen J.M., Lin H.H., Shih H.C. Field emission from various CuO nanostructures. *Appl Phys Lett* 2003; 83 : 3383-5.

[7] D.W. Kim, B. Park, J.H. Chung and K.S. Hong, *Jpn. J. Appl. Phys*. 39 (2000) 2696–2700.

[8] Jinhuai Liu, Xingjiu Huang, GangYe, Wei Liu, Zheng Jiao, Wanglian Chao, Zhongbai Zhou and Zengliang Yu, *H2S detection sensing characteristic of CuO/SnO$_2$ sensor*, Sensors 2, (2003) 110.

[9] Murphy, C.J. (2002). "Materials science. Nano cubes and nano boxes". *Science* 298 (5601): 2139–41. doi:10.1126/science.1080007. PMID 12481122.

[10] Rodríguez, J.A. and Fernández-García, M. (Eds.) *Synthesis, Properties and Applications of Oxide Nano Particles*. Whiley: New Jersey, 2007.

[11] C. Lingk, M.E. Gross, W.L. Brown, X-ray diffraction pole figure evidence for (111) sidewall texture of electroplated Cu in submicron damascene trenches, *Appl. Phys. Lett*. 74 (1999) 682.

[12] Patnaik, P. *Dean's Analytical Chemistry Handbook*, 2nd ed. McGraw-Hill, 2004.

[13] H.X. Ji and P.M. Marquis, *J. Mater. Sci*. 28 (1993) 1941.

[14] C. Duval, Inorganic thermogrvimetric analysis (Elsevier, Amsterdam, 1963).

[15] Yoshikawa M., Mori Y., Obata H., Maegawa M., Katagiri G., Ishida H. and Ishitani A. 1995 *Appl. Phys. Lett.* 67, 694.

[16] H. Yanagi, S. Inoue, K. Ueda, H. Kawazoe and H. Hosono, *Journal of Applied Physics* 88 (2000) 4159.

[17] J.I. Pankove, Optical Processes in Semiconductors, Englewood Cliffs Prentice – Hall, NJ, 1971.

[18] A.N. Banerjee, R. Maity, P.K. Ghosh, K.K. Chattopadhyay, *Thin Solid Films*, 474 (2005) 261.

[19] S.N. Mott, Conduction in Non-Crystaline Materilas, Oxford University Press, New York, 1987.

[20] K. Tonooka, K. Shimokawa. O. Nishimura, *Thin Solid Films*, 411 (2002) 129.

[21] C. Bouzidi, H. Bouzouita, A. Timoumi, B. Rezig, *Mter. Sci. Eng*, B118 (2005) 259.

[22] Yu T., Zhao X., Shen Z.X., Wu Y.H. and Su W.H. 2004, *J. Cryst. Growth* 268-590.

AUTHOR INDEX